LOW AND INTERMEDIATE ENERGY KAON-NUCLEON PHYSICS

LOW AND INTERMEDIATE ENERGY KAON-NUCLEON PHYSICS

Proceedings of the Workshop held at the
Institute of Physics of the University of Rome,
March 24-28, 1980

Edited by

E. FERRARI and G. VIOLINI
Istituto di Fisica "G. Marconi"
Università degli Studi di Roma

D. REIDEL PUBLISHING COMPANY
DORDRECHT : HOLLAND / BOSTON : U.S.A.
LONDON : ENGLAND

Library of Congress Cataloging in Publication Data

CIP

Workshop on Low and Intermediate Energy Kaon-nucleon Physics, Rome, 1980.
Low and intermediate energy kaon-nucleon physics.

Includes indexes.
1. Particles (Nuclear physics)–Congresses. 2. Kaons–Congresses. I. Ferrari, Ezio. II. Violini, G. III. Title.
QC793.W67 1980 539.7'2162 80–39982
ISBN-13: 978-94-009-8383-0 e-ISBN-13: 978-94-009-8381-6
DOI: 10.1007/978-94-009-8381-6

Published by D. Reidel Publishing Company,
P.O. Box 17, 3300 AA Dordrecht, Holland.

Sold and distributed in the U.S.A. and Canada
by Kluwer Boston Inc.,
190 Old Derby Street, Hingham, MA 02043, U.S.A.

In all other countries, sold and distributed
by Kluwer Academic Publishers Group,
P.O. Box 322, 3300 AH Dordrecht, Holland.

D. Reidel Publishing Company is a member of the Kluwer Group.

TABLE OF CONTENTS

P R E F A C E

The papers collected here were presented in Rome, last March, at the Workshop on low and intermediate energy kaon-nucleon physics.

According to our original plans, the workshop was mostly intended to review the status of a number of theoretical aspects of the field, like the application of analyticity to kaon-nucleon and kaon-nucleus scattering and phase-shift analyses. Moreover we thought that it was important to review also a number of related topics, like the current experimental work in the field and its perspectives, the physics of exotic atoms and πN phase shift analyses.

The workshop was announced at the end of November 1979, and despite the very short notice, it met a considerable favour which led to a considerable expansion of the original program. In particular there was a special interest among experimental groups which considered it as very appropriate to present their latest results.

Thus the present collection of papers constitues a rather complete and updated view of the current activity and problems in the field; to this completeness a substantial contribution comes from the record of the discussion, whose merit is due to our scientific secretaries Drs. G. K. Atkin and B. Di Claudio.

An indication of the favour met by the workshop is provided by the proposal made by several participants after its conclusion, that a similar one may take place in 1982.

The workshop would not have been possible without the support of a number of institutions. We want to express our gratitude for their generous financial support to

- The University of Rome, its Rector Professor A. Ruberti and the Dean of the Faculty of Sciences, Professor G. Tecce.
- The Istituto Nazionale di Fisica Nucleare, its President Professor A. Zichichi and the Director of its Sezione di Roma, Professor L. Paoluzi.
- The International Centre for Theoretical Physics and its Director Professor A. Salam.

The workshop took place at the Institute of Physics of the University of Rome and it is a pleasure to thank for the hospitality its Director Professor S. Cunsolo, as well as Professor G. Careri who made available his office to host the Secretariat of the Workshop.

Also on behalf of all the participants who certainly appreciated this type of effective collaboration, we want to thank the secretaries of the workshop, Mrs. E. Di Silvesto and G. Fascetti.

Finally a special thanks goes to Miss A. Centamore for her tremendous work of typing all the contributions and to Dr. G. K. Atkin for his collaboration in the preparation of the final text.

Rome, August 31, 1980

E. Ferrari
G. Violini

E. Ferrari and G. Violini (eds.), Low and Intermediate Energy Kaon-Nucleon Physics, ix.

LIST OF PARTICIPANTS

J. ANTOLIN - University of Zaragoza (Spain)
G. AURIEMMA - University of Rome (Italy)
G. ATKIN - University of Rome (Italy)
G. BARONI - University of Rome (Italy)
C.J. BATTY - Rutherford Laboratory, Chilton, Didcot (U.K.)
J.E. BOWCOCK - University of Birmingham (U.K.)
H. BUCKA - University of Berlin (Germany)
A. CONTI - University of Florence (Italy)
R.E. CUTKOSKY - University of Pittsburg (U.S.A.)
R.H. DALITZ - University of Oxford (U.K.)
A. de LESQUEN - CEN-Saclay (France)
J. DERRE' - CERN (Switzerland)
B. DE TOLLIS - University of Rome (Italy)
B. DI CLAUDIO - University of Rome (Italy)
R. DOMINGUEZ - University of Barquisimeto (Venezuela)
C.B. DOVER - Brookhaven National Laboratory (U.S.A.)
O. DUMBRAJS - University Karlsruhe (Germany)
E. FERRARI - University of Rome (Italy)
M. GASPERO - University of Rome (Italy)
P. GENSINI - University of Lecce (Italy)
M. HASINOFF - University of British Columbia (Canada)
M. JURIC - University of Beograd (Yugoslavia)
R. KOCH - University of Karlsruhe (Germany)
P. KROLL - University of Wuppertal (Germany)
K.S. KUMAR - Mc Master University, Hamilton (Canada)
M. LUSIGNOLI - University of Rome (Italy)
G. MARINI - University of Rome (Italy)
G. MARTELLOTTI - University of Rome (Italy)
A.D. MARTIN - University of Durham (U.K.)
F. MASSA - University of Rome (Italy)
B. MINETTI - University of Torino (Italy)
D.J. MILLER - University College London (U.K.)
M. MUSETTE - University of Brussel (Belgium)
A. NIGRO - University of Rome (Italy)
R. NOWAK - University of Warsaw (Poland)
G.C. OADES - Aarhus University (Denmark)
N.M. QUEEN - University of Birmingham (U.K.)
M. RESTIGNOLI - University of Rome (Italy)
A.M. RODRIGUEZ-VARGAS - Universidad de los Andes, Bogota (Colombia)
A. SCIUBBA - University of Rome (Italy)
C.P. SINGH - University of Varanasi (India)
M. STASZEL - University of Warsaw (Poland)
M. SVEC - McGill University, Montreal (Canada)
F. TAKASAKI - Nat. Lab. High Energy Physics (KEK), Ibaraki-ken (Japan)
T. TYMIENIECKA - University of Warsaw (Poland)
L. van ROSSUM - CEN-Saclay (France)
C. VERZEGNASSI - University of Trieste (Italy)
G. VIOLINI - University of Rome (Italy)
S.J. WATTS - Queen Mary College, London (U.K.)
N. ZOVKO - Ruder Boskovic Institute, Zagreb (Yugoslavia).

THE PRESENT STATUS OF PION-NUCLEON PHASE SHIFT ANALYSIS*

R. Koch
Institut für Theoretische Kernphysik - Universität Karlsruhe

1. Introduction

Since the CERN (1) and the Saclay (2) phase shift analyses (1967-73), which discovered the most important nucleon resonances up to 2 GeV mass, progress has been made in the methods for determining πN amplitudes and partial waves. Especially the problem of ambiguous solutions is understood (3, 14) and can be solved nowadays by using analyticity constraints together with isospin invariance (4). Also the experimental information on $\pi^{\pm}p \rightarrow \pi^{\pm}p$ and $\pi^{-}p \rightarrow \pi^{o}n$ differential cross sections and polarizations and on the $\pi^{\pm}p$ total cross sections has considerably increased since 1973. At present two phase shift solutions are available which, apart from data, are also based on the information derived from the analytic properties of scattering amplitudes. The CMU-LBL group (5) worked in the lab. momentum range .4-2 GeV/c and is extending the investigations up to 3 GeV/c. The collaboration between the Karlsruhe group and E. Pietarinen from Helsinki (KH) resulted in a set of partial waves (6, 7) which runs from threshold to 10 GeV/c. Also fixed-t amplitudes in terms of an expansion developed by Pietarinen (8) are available in a t-range $-1\,\mathrm{GeV}^2 \leq t \leq o$. A new amplitude analysis (9) in the t-range $|t| \leq .5\,\mathrm{GeV}^2$ and up to 200 GeV/c is in progress at Karlsruhe.

Three other analyses must be mentioned. They make less use of model independent analyticity input but rely more on model representations of partial waves or amplitudes. A. Hendry's phase shift analysis (10) in the range 1.6-10 GeV/c is based on an impact parameter representation. Zidell, Arndt and Roper (11) have performed a low energy analysis in the elastic domain and use an empirical parametrization with the right threshold behaviour for every phase shift. D. Chew (12) has analysed π^+p data between .6 and 2.3 GeV/c with an ansatz including zeroes of transversity amplitudes and an empirical phase parametrization.

In section 3 the methods and some remarkable results of the above mentioned authors will be described and compared. Section 4 contains new results (13) of the Karlsruhe group obtained in the range 2-10 GeV/c. In section 5 these results will be partly discussed in the impact parameter

E. Ferrari and G. Violini (eds.), Low and Intermediate Energy Kaon-Nucleon Physics, 1–29.

picture. In section 2 we will start reviewing some general aspects of amplitude determination from data.

2. The determination of πN amplitudes

Instead of the usual invariant amplitudes, one often uses the transversity amplitudes in connection with data. We define

$$F_I(\pm) = G_I \pm iH_I \tag{1}$$

(I = isospin index to be specified)
where

$$|G|^2 + |H|^2 = \frac{d\sigma}{d\Omega} \quad \text{and} \quad \frac{2\cdot \mathrm{Im}\,(G\cdot H^*)}{d\sigma/d\Omega} = P \tag{2}$$

with G and H the spin no-flip resp. flip amplitudes. From (1) and (2) we get

$$|F(\pm)|^2 = \frac{d\sigma}{d\Omega}(1 \pm P). \tag{3}$$

Here $d\sigma/d\Omega$ and P belong to one of the 3 reactions $\pi^\pm p \to \pi^\pm p$ and $\pi^- p \to \pi^0 n$. From isospin invariance we have relations between the amplitudes $F_\pm(\pm)$, $F_0(\pm)$ (the lower indices denote the 3 reactions just mentioned).

$$F_+(\pm) = F_-(\pm) + \sqrt{2}\; F_0(\pm). \tag{4}$$

If we represent the transversity amplitudes by arrows in a complex diagram and use eq. (4) we have the isospin triangles

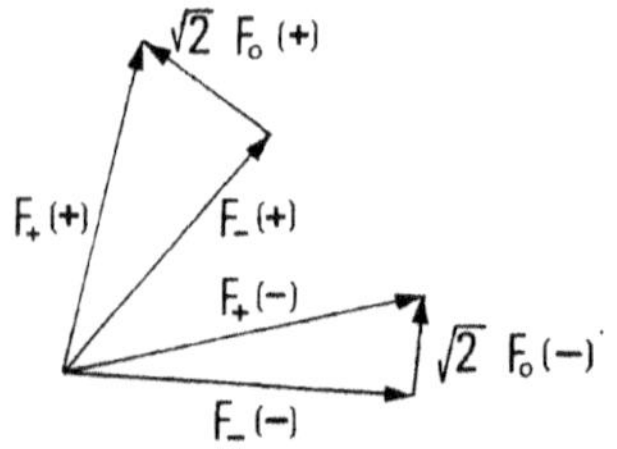

The above mentioned 6 measurable quantities allow us to determine the lengths of the arrows. The overall phase and the relative phase between the 2 triangles remain still unknown. The latter however can be fixed by measuring the spin rotation parameter for one of the reaction $\pi^\pm p \to \pi^\pm p$. Reflections of a triangle around the directions of $F_\pm(\pm)$ are still possible. The measurement of a second spin rotation parameter reduces the number of reflections (36). And we are always left with the overall phase ambiguity. Coulomb-interference measurements can only remove this ambiguity in the forward direction.

It is interesting to study how the analytic properties of the scattering amplitudes can be exploited to resolve continuous and discrete ambiguities.

First let us start with the moduli of the amplitudes known from data (eq. 3) and use analyticity in t or cosθ at fixed energy. Many authors have considered this problem (14). At energies where elastic unitarity is valid and can be used as an additional constraint on the partial waves we are left with a restricted number of discrete ambiguities. At higher energies, i.e. lab. momentum values $k \geq .4$ GeV/c, where unitarity looses its constraining power, there is a continuum of ambiguities in addition to discrete ones.

An alternative way of including theoretical constraints is to use explicitly the analytic properties of the invariant amplitudes in the cut s-plane at fixed t for instance, while making a data analysis. Thus forward dispersion relations together with the optical theorem provide a possibility to find the overall phase at $t = 0$. Here one has clearly used unitarity.

The work of E. Pietarinen consisted in applying the fixed-t analiticity constraints to $t < 0$. If the moduli are again given by data in some energy interval and if unitarity cannot be fully exploited, the phases of the amplitudes are still not uniquely determined by fixed-t analyticity constraints alone, because we do not know the zeroes of the amplitudes.

I.S. Stefanescu however claimed recently, that together with fixed-s analyticity, isospin invariance and 2 assumptions on the high energy behaviour of the amplitude C^+ and the amplitude A^+, the problem of extracting unique πN amplitudes from $\frac{d\sigma\pm}{d\Omega o}$ and $P^{\pm}_{o}$ data can be solved (for details see ref. 4).

Two problems with the inclusion of fixed-t analyticity for instance via fixed-t dispersion relations must be mentioned :

i) the high energy behaviour of the considered amplitudes has to be known.
ii) If t becomes more negative the threshold point runs further away from the beginning of the physical region. Therefore some continuation of the moduli or at least the imaginary part of the amplitudes is needed before fixed-t dispersion relations can be used.

Concerning the first point, the rise of the modulus of the isospin even amplitude $|C^+|$ is given up to 200 GeV/c by differential cross section measurements (15) and $|\mathrm{Re}\, C^+(s, t)| / \mathrm{Im}\, C'(s, t)$ goes to zero as $s \to \infty$. The latter is clearly seen in the forward direction and is also true for some negative t-interval, $|t| \leq .5$ GeV^2. Furthermore the polarization and spin rotation measurements (16) at 6, 16 and 40 GeV/c for $\pi^{\pm}p \to \pi^{\pm}p$ suggest $A^+ \to 0$ at high energies.

At low energies the imaginary parts of the amplitudes can be continued outside the physical region with the help of the partial wave expansion. Low energy partial waves are assumed to be well known. The extrapolation is in principle limited by the condition for the convergence of the Legendre expansion. The t-limit is $t \approx -.5\ GeV^2$ but it can be extended without much increasing the uncertainties to $t \simeq -1\ GeV^2$.

3. The present amplitude and phase shift analyses

3.1 The CMU-LBL analysis (5)

As part of this work will be presented at the conference, I will only briefly mention the main aspects for a comparison with the below described Karlsruhe-Helsinki analysis.

The CMU-LBL analysis consists of 3 steps :

i) "amalgamation" of data,
ii) partial wave analysis using analyticity constraints along curves inside the physical region,
iii) resonance parameter determination.

The result of the amalgamation procedure, which consists of interpolation and combination of data inside bins, is a highly correlated data set. This data set is used for a large number of single energy χ^2 fits. The fits are performed with an ansatz composed of a Born-term part and a polynomial in the Cutkosky-Deo variable. The result of the fits are clusters of χ^2 minima. Their number and size reflect the ambiguities which still remain. A selection of clusters is then made by asking consistency with dispersion relations along 6 curves inside the physical region including the forward and backward directions. The CMU-LBL analysis provides a full error matrix.

3.2 The Karlsruhe-Helsinki analysis (6, 7, 8, 13) (KH 78 and KH 80)

The idea is to combine the fixed-s, fixed-t and fixed c.m. angle (see below) analyses in an iterative procedure.

a) Fixed-s

We use the standard partial wave expansion for the spin no-flip and flip amplitudes

$$G^I = \frac{1}{q} \sum_\ell \left[(\ell+1)\ T^I_{\ell+} + \ell\ T^I_{\ell-}\right]\ P_\ell\ (\cos\theta)$$

$$H^I = \frac{1}{q} \sum_\ell \left[T^I_{\ell+} - T^I_{\ell-}\right]\ P_\ell'\ (\cos\theta) \cdot \sin\theta \qquad (5)$$

$$T^I_{\ell\pm} = (\eta^I_{\ell\pm} \exp\ (2i\ \delta^I_{\ell\pm})\ -1)/2i \qquad (6)$$

(q = c.m. momentum, θ = c.m. scattering angle, ℓ = orbital angular momentum; η = inelasticity parameter, δ= phase shift).

The data are shifted to fixed-s bins and electromagnetic corrections are applied before fitting. For this purpose we use a current solution and interpolate the partial waves $T^I_{\ell\pm}(s)$. We make the e.m. corrections using the results of the Nordita group (23).

The consistency of neighboring data and normalizations are checked in the course of the analysis.

It is important to truncate the expansions (5) smoothly. Therefore a convergence test function was added to the normal χ^2. Also the low energy partial waves were constrained to be elastic i.e.$\eta^I_{\ell\pm} \equiv 1$.

b) Fixed-t

The invariant amplitudes at fixed-t are represented as follows

$$C^+(\nu,t) = \frac{g^2 m\,(t-2\mu^2)}{4m^2-t}\left(\frac{1}{m^2-s}+\frac{1}{m^2-\mu}\right) + C_L^+(\nu,t) + \frac{R(z,t)}{1+z}\sum_{n=0}^{\infty} c_n^+ z^n$$

$$C^-(\nu,t) = \frac{g^2 m(s-u)}{4m^2-t}\left(\frac{1}{m^2-s}+\frac{1}{m^2-u}\right) + C_L^-(\nu,t) + (s-u)\,(1+z)^{1-\alpha_\rho(t)}\,.\sum c_n^- z^n$$

$$B^+(\nu,t) = \frac{g^2}{m^2-s} - \frac{g^2}{m^2-u} + B_L^+(\nu,t) + (s-u)\,(1+z)\,R(z,t)\sum_{n=0}^{\infty} b_n^+ z^n \qquad .(7)$$

$$B^-(\nu,t) = \frac{g^2}{m^2-s} + \frac{g^2}{m^2-u} + B_L^-(\nu,t) + (1+z)^{1-\alpha_\rho(t)}\sum_{n=0}^{\infty} b_n^- z^n$$

$$\text{where } z = \frac{.72 - \sqrt{\nu_{th}^2 - \nu^2}}{.72 + \sqrt{\nu_{th}^2 - \nu^2}}, \qquad \nu = \frac{s-u}{4m} \qquad (8)$$

maps the ν^2-plane onto the inside of a unit circle.
(m = proton mass, μ = pion mass, s, t, u are the Mandelstam variables, g^2 = πNN coupling constant = $14.28 \cdot 4\pi$)

The $C_L^\pm(\nu,t)$, $B_L^\pm(\nu,t)$ take into account the overlapping parts of the s- and u-channel cuts.

Before fitting, the data are shifted to fixed-t bins with the help of spline fits. The χ^2 to be minimized in the subsequent analysis has the following form

$$\chi^2 = \chi^2_{data} + \lambda\cdot\chi^2_{\substack{partial\\ waves}} + \sum_{n=0}^{N}(n+1)^3\left[\beta_1 c_n^{+2} + \beta_2 c_n^{-2} + \beta_3 b_n^{+2} + \beta_4 b_n^{-2}\right] \qquad (9)$$

The last term is the convergence test function. The β_i determine the smoothness of the solution. The second term constrains the fit towards the results of the previous partial wave solution. With the parameter λ a more or less strong coupling can be chosen.

c) Fixed c.m. angle θ

In the KH 78 analysis a set of transversity amplitudes has been defined by Pietarinen (see refs. 6, 8). For these amplitudes a parametrization at fixed centre of mass angle θ of the following type has been used

$$F(q^2, \cos\theta) = F_L(q^2, \cos\theta) + \frac{1}{\pi}\int_0^{q^2_{max}} \frac{\text{Im } F(q'^2, \cos\theta)}{q'^2 - q^2}\, dq'^2 + F_R\,(q^2, \cos\theta) \quad (10)$$

The integral term removes the cut singularities at low energies. F_L and F_R are parametrized using conformal mapping. F_L contains the left hand cut contributions and F_R contains the remaining right hand cut contributions. F_L was partly calculated in the backward direction by inserting $\pi\pi N\bar{N}$ partial waves. F_L also contains the nucleon pole terms (for details see refs. 6 and 8). In ref. 24 an attempt has been made to study the contributions of the left hand singularities in more detail. The χ^2 function is similar to the fixed-t case.

The fixed-θ results are of importance at lab. momenta higher than $\simeq$ 3. GeV/c where the fixed-t curves tend more and more to forward scattering angles. Thus in the central region of the physical scattering domain fixed-t amplitudes can no more be used as constraint input in the fixed-s analysis. However the high energy behaviour of the fixed-θ amplitudes represents a new uncertainty.

The iteration consists in using finally the fixed-t and fixed-θ amplitude results in the fixed-s partial wave analysis. Figure 1 shows a flow diagram of the KH program.

With the results of the different types of analyses various applications have been made :

a) resonance masses, widths and elasticities have been determined from the partial waves (6),

b) low energy parameters (scattering lengths, σ-term, πNN coupling constant ...) were calculated (6, 17),

c) $\pi\pi$-$N\bar{N}$ partial waves have been studied (6, 17, 18) using extrapolations of the fixed-t invariant amplitudes to the unphysical region and $\pi\pi$ phase shifts for $t > 4\mu^2$,

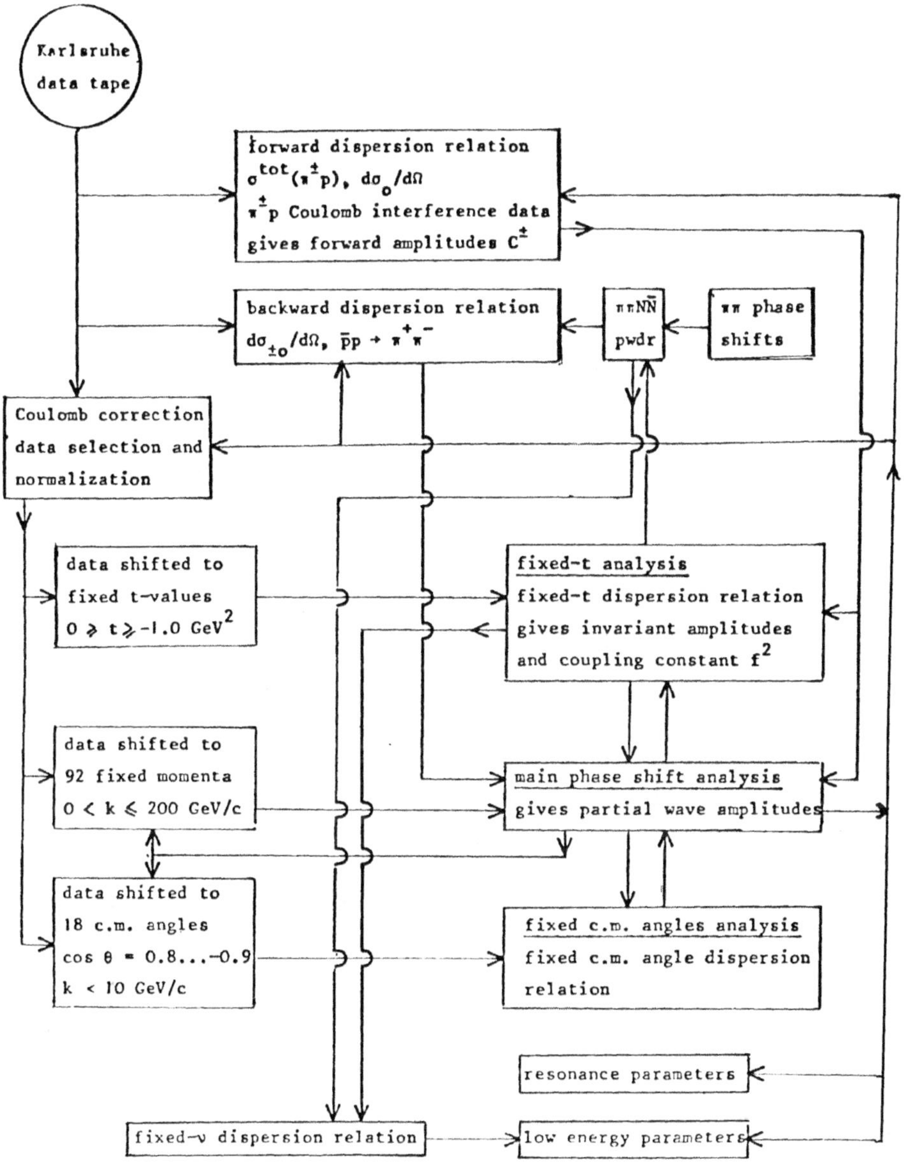

Figure 1 - Flow diagram of the Karlsruhe-Helsinki πN analysis.

d) the zero trajectories for invariant amplitudes have been studied (19). They give interesting information for pion-nucleon scattering models. It is found that the zero patterns for the $\pi^{\pm}p$ and isospin even and odd combinations of the amplitudes A and B are similar to each other and are very simple. However discrepancies with the zero pattern from Odorico's Linear Zero Model are observed.
The trajectories remain near the physical s, t-plane and run mostly along constant t or constant u curves.
The results found from the CMU-LBL 79 and KH 78 phase shift analyses are in good agreement (for details see ref. 19),

e) at present we are testing our partial waves with partial wave dispersion relations (20) and partial wave relations from fixed-t dispersion relations (21).

3.3 Comparisons between the CMU-LBL and KH analyses

The two analyses agree on the existence of all 2, 3 and 4 star resonances in the mass range 1500 to 2200 MeV. However the agreement on the location and widths is not always good. In some cases the masses differ by almost 100 MeV. In the worst cases the widths differ by 100%. Such a case is demonstrated in figure 2, where we show the S11 partial wave. It is seen that the elasticity is rather different.

In figure 3 we show as another example the D35 where CMU-LBL finds stronger evidence for a resonance than KH.

By means of the F17 partial wave in figure 4 we want to illustrate the difficulties with the interpretation of broad structures as resonances (see also the discussion in section 5).

Finally in a number of partial waves there are structures quoted as resonances (1 star) by CMU-LBL, which have not been taken into account by KH and vice versa (third resonance in S11, see figure 2).

In general we think that the CMU-LBL and KH analyses have resolved the ambiguities which occur in principle in phase shift analysis in much the same way and that remaining differences can only be a consequence of treating the data differently.

3.4 The phase shift analysis of D. Chew (12)

In this work all $\pi^{+}p$ data below 2.2 GeV/c have been analysed using a method proposed by E. Barrelet (22). A parametrization in terms of zeroes of transversity amplitudes is used at fixed energy

$$F(w) = \sqrt{\frac{d\sigma}{d\Omega}}\,\Bigg|_{t=0} \frac{e^{i\phi}}{w^{\mathrm{int}\,(N/2)}} \prod_{i=1}^{N} \frac{w - w_i}{1 - w_i} \tag{11}$$

$$w = e^{i\theta}\,, \quad N \le 6$$

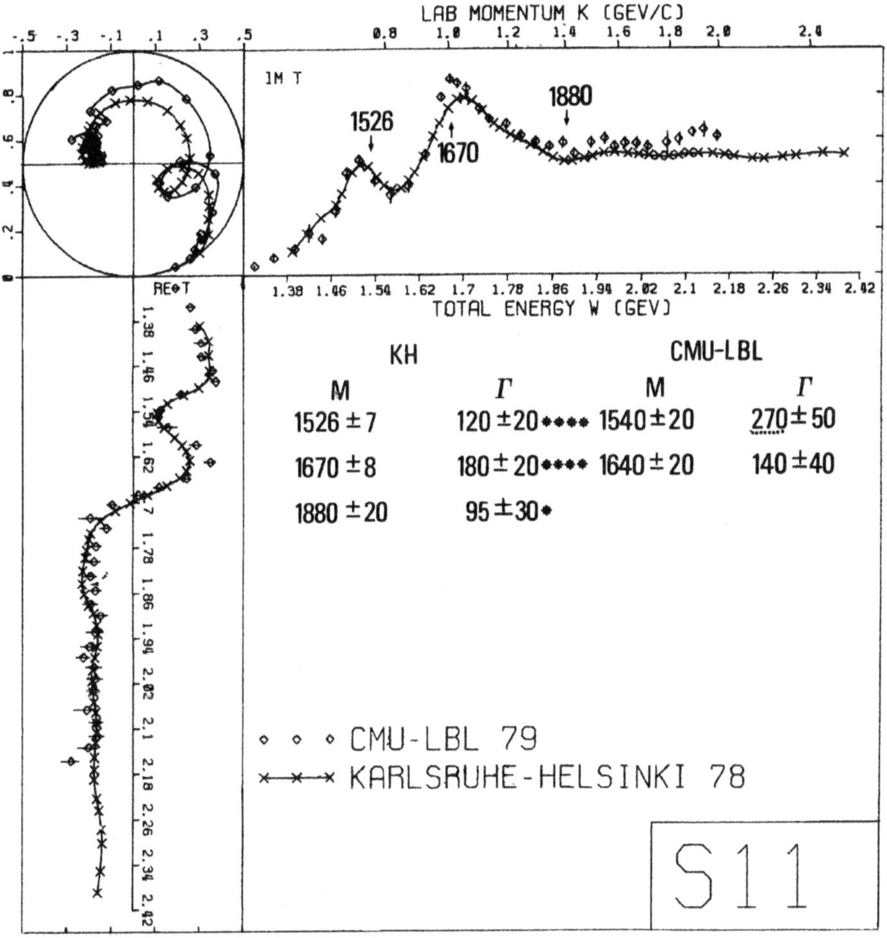

Figure 2 - Comparison between the KH 78 —x— and CMU-LBL 79 ϕ phase shift analysis: S11 partial wave

The w_i are obtained from fits to the data and $d\sigma/d\Omega|_{t=0}$ from forward dispersion relations. For the angular dependence of the phase the following ansatz is taken

$$\phi = \phi_o \left[1 + A\,(\cos\theta - 1) + B \sin\theta\right] \tag{12}$$

A and B are constants to be determined.

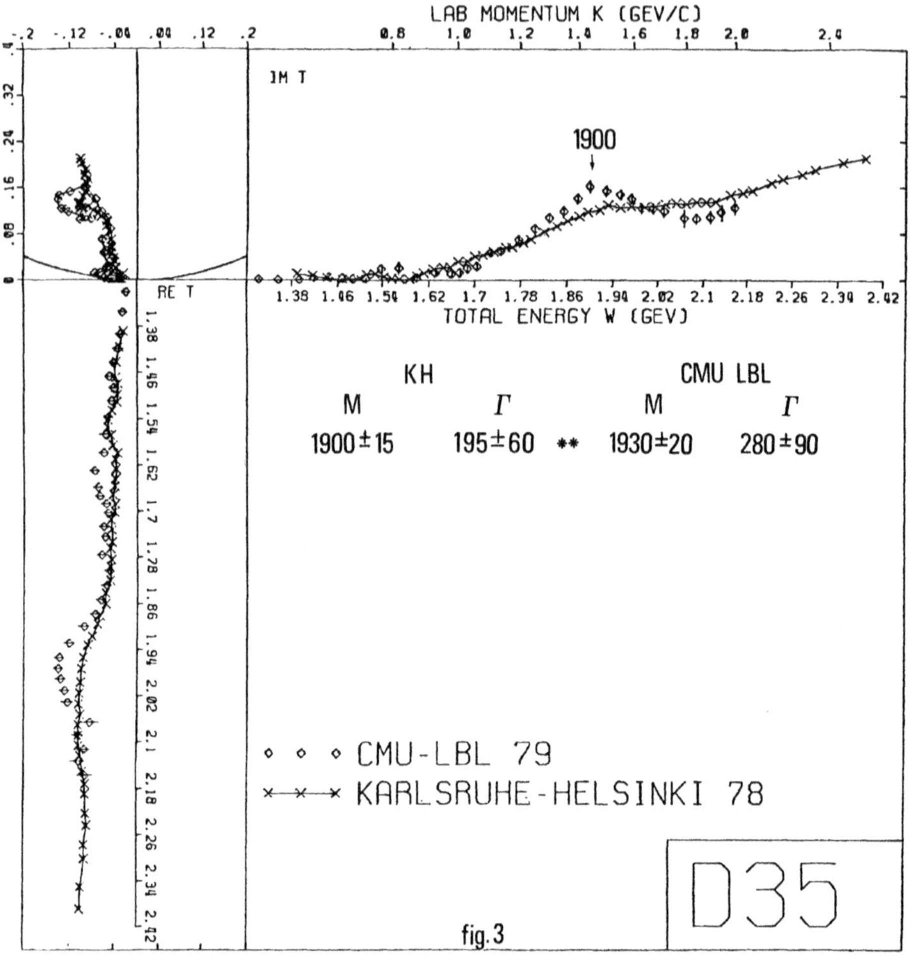

Figure 3 - Comparison between KH 78 and CMU-LBL 79 : D35 partial wave

A comparison between the results of this analysis and the CMU-LBL and KH analyses shows remarkable deviations concerning nucleon resonances. Two 4 star objects, S31(1610) and P31(1888) are not resolved in the D. Chew analysis. Furthermore the resolution of other 4 and 3 star states is weak. Up to now it has not been checked if the reconstructed invariant amplitudes are compatible with fixed-t dispersion relation constraints. The absolute phase is considerably different from KH and CMU-LBL (37).

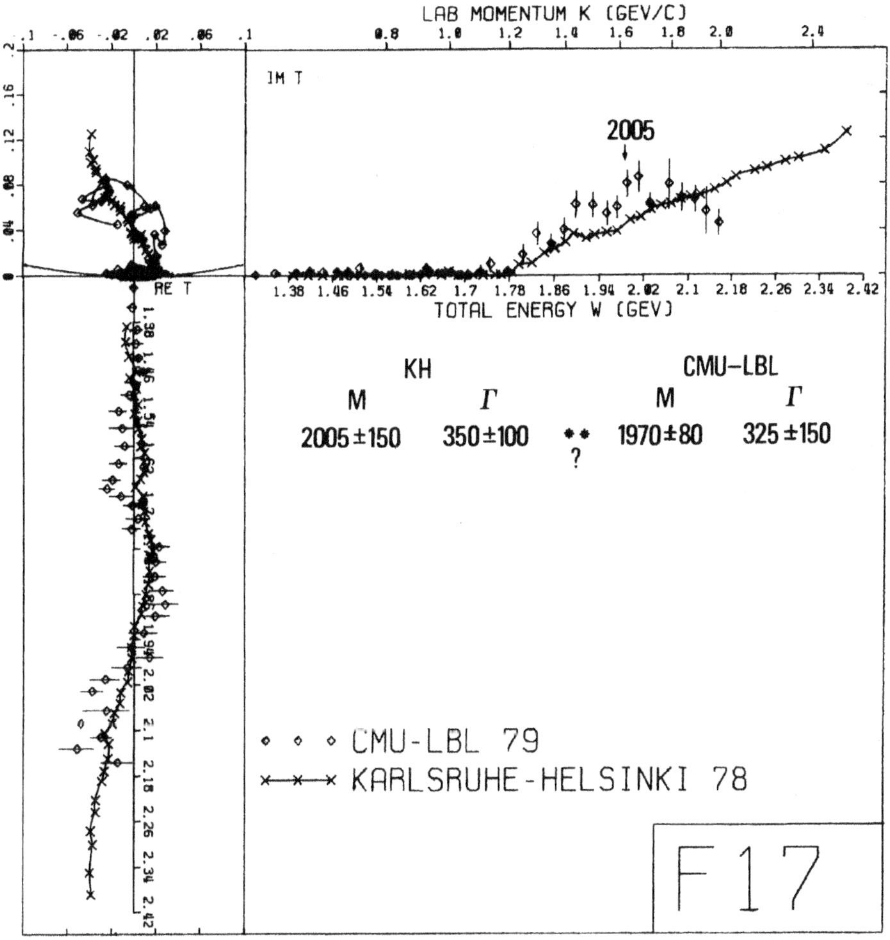

Figure 4 - Comparison between KH 78 and CMU-LBL 79: F17 partial wave

3.5 The low energy phase shift analysis of Zidell, Arndt and Roper (11)

The authors parametrize each phase shift by means of an empirical ansatz. It has the correct threshold behaviour

$$\text{S31:} \quad \tan \delta_{0+} = \rho_0 \Big/ \sum_{n=1}^{3} P_n \, T_\pi^{\,n-1} \qquad (13)$$

$$\text{all others except P33} \qquad \tan \delta_{\ell\pm} = \rho_\ell \cdot \sum_{n=1}^{N} P_n \, T_\pi^{\,n-1} \quad ; \quad N \le 3 \tag{14}$$

$$\text{where} \quad \rho_\ell = (2m\, T_\pi)^{\ell+\frac{1}{2}} \qquad \text{and} \qquad T_\pi = \sqrt{K_{lab}^2 + \mu^2} - \mu$$

The P_n are the parameters to be determined. P33 is parametrized by means of a K-matrix resonance formula. About 30 parameters have been fitted to a selected data set below ≈.5 GeV/c. No additional theoretical input except unitarity is used.A first check with dispersion relations for the B-amplitudes reconstructed from the resulting phase shifts shows that the analytic properties at fixed-t are not quite fulfilled. Figure 5 shows the right hand side of the following equation which is used for the determination of the πNN coupling constant g^2. In this plot all points should be on a straight line. A rather strong deviation is observed at very low energies.

$$\frac{g^2}{m} + \tilde{B}^-(o,t)\,(\nu_B \pm \nu) = (\nu_B \pm \nu) \cdot \tag{15}$$

$$\cdot \left\{ \mp \operatorname{Re} B_\pm(\nu,t) \pm \frac{\nu}{\pi} \int_{\nu_{th}}^{\infty} \left[\frac{\operatorname{Im} B_+}{\nu' \mp \nu} + \frac{\operatorname{Im} B_-}{\nu' \pm \nu} \right] \frac{d\nu'}{\nu'} \right\}$$

$\nu = (s-u)/4m$, $\nu_B = (t-2\mu^2)/4m$, $\tilde{B}^- = B^-$ minus nucleon pole term.

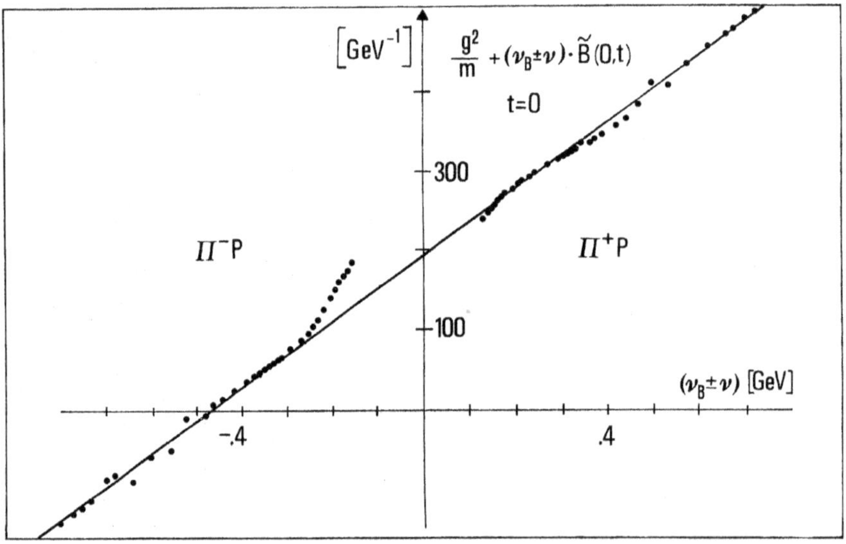

Figure 5 - Consistency check with fixed-t disp.rel. for the Zidell et al. low energy phase shift analysis (see eq. 15)

Checks using partial wave dispersion relations should also be made before the results of the described analysis can be accepted.

4. New phase shift analysis (13) at lab. momentum values 2-10 GeV/c

The data gaps which existed at the time of the KH 78 analysis for lab. momenta $k \geq 3$ GeV/c were partly filled by the new high precision data of the Argonne group (Jenkins et al., ref. 25). They measured differential cross sections in the angular range $|\cos\theta| \leq .35$ and the momentum range $2 \leq k \leq 6.3$ GeV/c for $\pi^+p \rightarrow \pi^+p$ and $2 \leq k \leq 9.7$ GeV/c for $\pi^-p \rightarrow \pi^-p$.

Moreover charge exchange differential cross sections and polarizations of the Kyoto group (refs. 26 resp. 27), measured at KEK in the momentum range 2-3.5 GeV/c became recently available. Also π^-p polarizations have been measured at KEK by the Nagoya group (28).

The momentum dependence of the Argonne data shows remarkable structures at fixed-t and fixed- c.m. angle for energies larger than the masses of the well established resonances (see figure 6).

We have reanalysed all data between 2 and 10 GeV/c and found resonance-like behaviour of partial waves with total spin 13/2 to 21/2. Possible resonances in I and K waves (spin 11/2-15/2) had already been reported by KH 78.

In general a great similarity can be observed between the Argand-behaviour of the G, H and I waves containing established resonances and the K, L, M, N waves, where new effects appear.

In order to represent the relative significance of the structures found in the Argand diagrams of all partial waves we choose a classification according to Regge trajectories as well as of the trajectories themselves. The states on the N_α and N_γ trajectories cause the largest loops in the Argand plots. For higher spins the loops decrease but these particular trajectories can be pursued to spin 21/2. Smaller effects are created by the N_β and Δ_δ trajectories if we exclude the low mass states for the latter. Δ_β and Δ_α states cannot be clearly observed in partial waves higher than F35 resp. G39. On the N_δ and Δ_γ trajectories only the P13 resp. D33 states can be well recognized.

Table 1 gives an overwiew of the situation. We have underlined the states according to the importance of structures produced in Argand plots: full line = 3 or 4 star resonance, dashed line = good candidate, dotted line = candidate but weakly resolved, no line = no structure.

In figure 7 we illustrate the similar behaviour of series of partial waves containing states from the N_α and Δ_δ trajectories. The partial waves have been scaled by adequate factors.

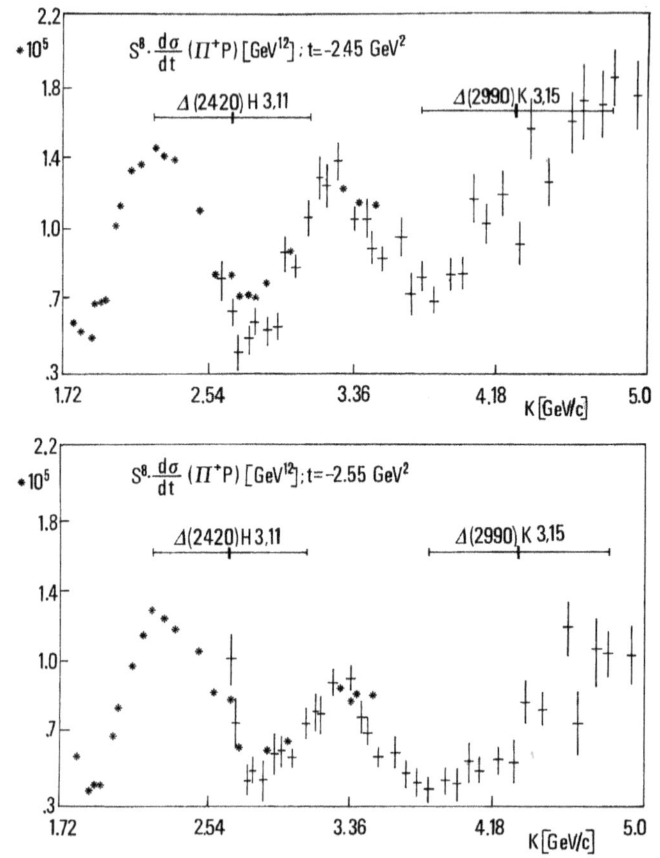

Figure 6 - Energy dependence of π^+p elastic cross sections at fixed t = -2.45 and -2.55 GeV^2 (ref. 25). The cross sections have been multiplied by s^8 in order to compensate the rapid decrease. ✱ : values reconstructed from our phase shift analysis

Figure 8 shows the new resonance candidates not yet taken into account in KH 78: L1,15; M1,17; N1,19; O1,21 and L3,17; M3,19. The masses of possible resonances in these waves range from ∿ 3 to ∿ 4 GeV.

It is also interesting to notice that the impact parameter for the K, L, M, N candidates is approximately 1.1 fm, a value which is also obtained for resonances in the F, G, H, I waves.

Table 1

States on leading Regge trajectories. Depending on the signature the perity is given by $(-1)^{J-\frac{1}{2}}$ called natural parity or $(-1)^{J-\frac{1}{2}-1}$ called unnatural parity

PARITY							
+, n	N_α	P11(938)	F15(1687)	H19(2205)	K1,13(2610)	M1,17	01,21
−, n	N_γ	D13(1520)	G17(2140)	I1,11(2580)	L1,15	N1,19	
+, u	Δ_δ	P33(1233)	F37(1913)	H3,11(2420)	K3,15	M3,19	03,23
−, u	N_β	S11(1526)	D15(1680)	G19(2270)	I1,13	L1,17	N1,21
+, n	Δ_α	P31(1890)	F35(1900)	H39(2220)	K3,13	M3,17	03,21
−, u	Δ_β	S31(1610)	D35(1900)	G39(2470)	I3,13	L3,17	N3,21
+, u	N_δ	P13(1710)	F1,7(2000)	H1,11	K1,15	M1,19	01,23
−, n	Δ_γ	D33(1680)	G37(2220)	I3,11	L3,15	N3,19	

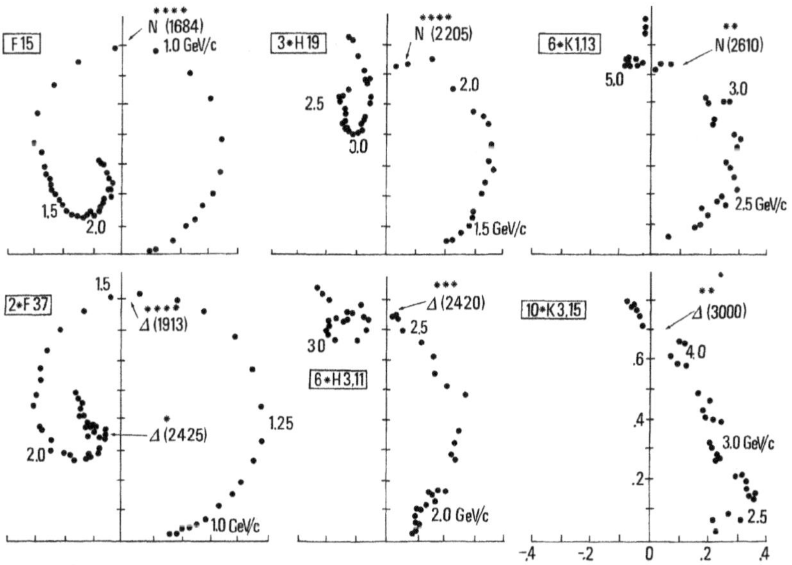

Figure 7 - Argand diagrams of peripheral partial waves F15, H19, K1,13; F37, H3,11; K3,15. The quantities have been scaled to approximately the same size

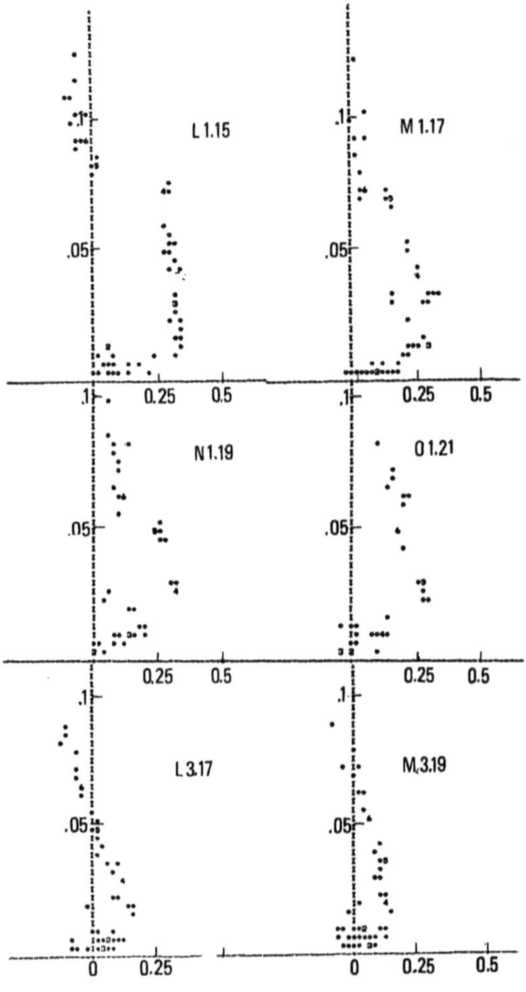

Figure 8 - Argand diagrams for the partial waves L1,15; M1,17; N1,19; O1,21 and L3,17; M3,19. The numbers are lab. momenta

Similar results have been found by A.W. Hendry (10). Therefore we will briefly discuss his analysis.

The dual absorption model (29) is the starting point of Hendry's work. In this model the scattering amplitude has two components: Pomeron-exchange corresponding to non-resonant s-channel background and ordinary t-channel Regge pole exchange, dual to the exchange of peripheral resonances in the s-channel.

The following ansatz for the partial waves written in terms of the impact parameter $b \approx \ell/q$ contains a central part which can be identified with Pomeron-exchange and a peripheral part.

$$\mathrm{Im}\, T_{\ell\pm} = x_1^{\pm} \exp(-x_2^{\pm} b^2) + (x_3^{\pm} + (-1)^{\ell} x_4^{\pm}) \exp\left[-x_5^{\pm}(b - x_6^{\pm})^2\right], \qquad (16)$$

$$\mathrm{Re}\, T_{\ell\pm} = y_1^{\pm} \exp(-y_2^{\pm} b^2) + (y_3^{\pm} + (-1)^{\ell} y_4^{\pm}) \exp\left[-y_5^{\pm}(b - y_6^{\pm})^2\right]. \qquad (17)$$

In a first step the x_i, y_i (48 parameter at each energy, if isospin is taken into account) are determined by fitting the available experimental data and also real parts from forward dispersion relations in the lab. momentum range 1.6-10 GeV/c. The $T_{\ell\pm}$ are then varied again around the parametrization in order to improve the fit.

The Argand diagrams of Hendry's partial waves look in some cases similar to ours. Masses and widths of I, K, L, M, N resonances are also given in ref. 10. According to Hendry the widths range from 500 (I-states) to $\approx$ 2000 MeV (N-states).

Two remarks must be made concerning the analysis :

i) for the real parts of the partial waves Hendry uses the same type of parametrization as for the imaginary part. According to ref. 29 only the ansatz for the imaginary parts is motivated by geometrical ideas and duality;

ii) as the ℓ-distributions of the real and imaginary parts of the partial waves are restricted by the choosen impact parameter form, it should be checked if the single partial waves are after all consistent with partial wave dispersion relations and if the reconstructed invariant amplitudes satisfy fixed-t dispersion relations.

5. Impact parameter results for the diffraction amplitude from KH 80

A large background has been observed in the Argand plots of the high partial waves shown in the previous section. Therefore it is our aim to study the background in detail before making resonance parameter determinations. We start with some results for the impact parameter dependence of the diffraction amplitude.

We consider the s-channel helicity formalism, where partial waves are defined as

$$2F^{J}_{+\pm}(s) = T_{(J-\frac{1}{2})+} \pm T_{(J+\frac{1}{2})-} \qquad (18)$$

The index of T is $\ell\pm$ and $J = \ell \pm 1/2$

For the impact parameter amplitudes the following approximation can be used at high energies

$$f_{+\pm}(s, b) \simeq F^{J}_{+\pm}(s) \; ; \quad b = (J + \tfrac{1}{2})/q \tag{19}$$

whereas the exact Fourier-Bessel transform of the helicity amplitudes in the Cottingham-Peierls (30) version reads

$$f_{+\pm}(s, b) = \frac{1}{4\sqrt{\pi}} \int_{-4q^2}^{0} F_{+\pm}(s, t) \; J_{\Delta\lambda}(b\sqrt{-t}) \, dt \tag{20}$$

($\Delta\lambda = 0$ for ++ and $\Delta\lambda = 1$ for +-) (for a review see ref. 31).

In figure 9 we have plotted the impact parameter dependence of the imaginary part of the isospin even amplitude Im $f^{0}_{++}(s, b)$.

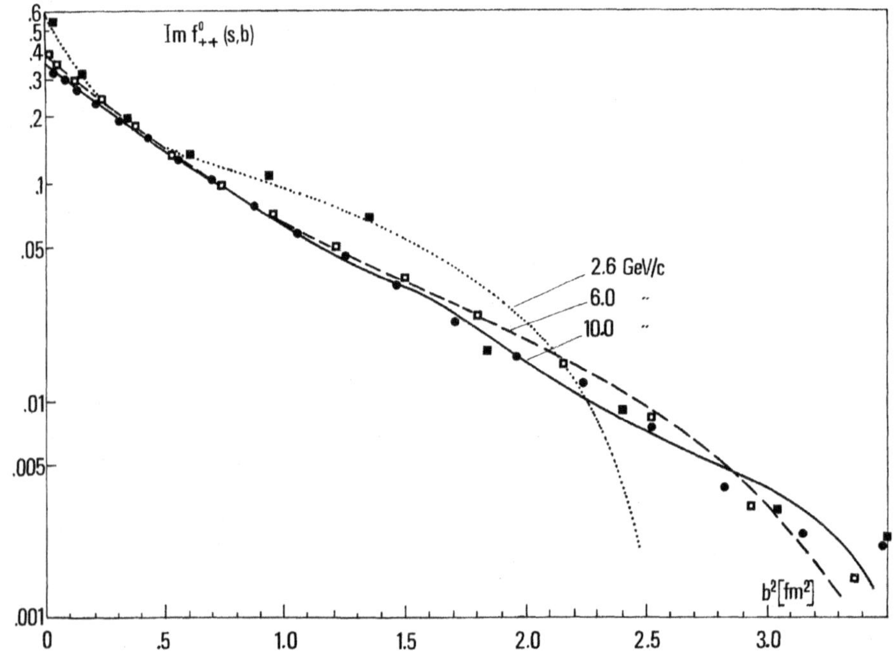

Figure 9 - Im $f^{0}_{++}(s,b)$ as calculated from the isospin even combination of partial waves Im$^{J}_{++}$ at 2.6 (■), 6.0 (□), and 10.0 (•) GeV/c. The curves follow from the evaluation of the impact parameter transform according to eq. (20)

We see that the shape is roughly in agreement with Harari's ansatz.

$$\mathrm{Im}\, f^{o}_{++}(s,\, b) \approx Ae^{-b^2/2B} + \frac{D}{\sqrt{b}}\, e^{-(b-R)^2/2c} \tag{21}$$

Namely besides the large central part there is a peripheral peak at $b \approx 1.1$ fm which decreases with increasing energy.

It is also of interest to consider the real part Re $f^{o}_{++}(s,\, b)$. As figure 10 shows, for $b \lesssim 1.1$ fm the real parts are strongly negative and become positive for $b \gtrsim 1.1$ fm. The behaviour can be expected, because the real parts of the peripheral partial waves are positive below and negative above the resonance energy. But even at 10 GeV/c, where no resonance could be detected, a zero around 1.1 fm is seen.

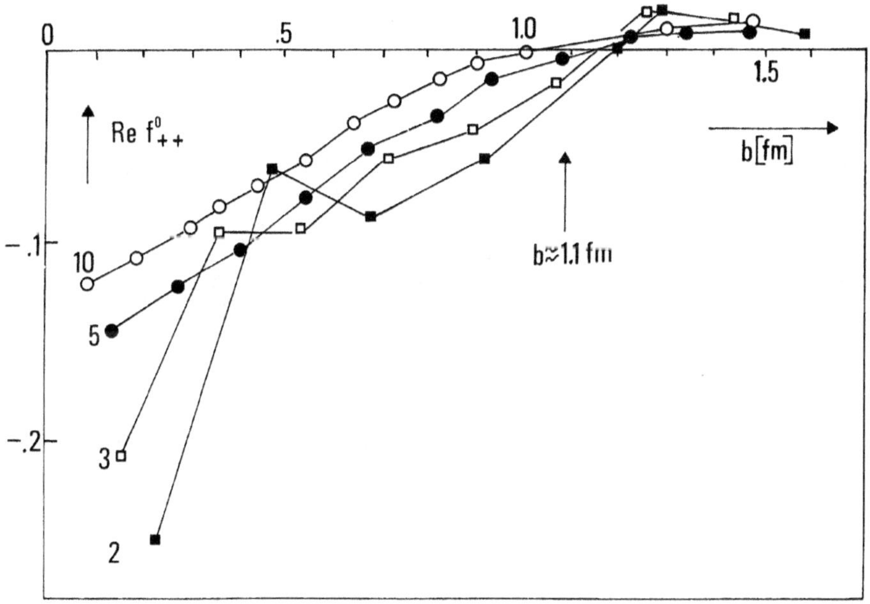

Figure 10 - Re $f^{o}_{++}(s,\, b)$ as calculated from the isospin even combination of partial waves. The numbers refer to the pion lab. momentum in GeV/c

The Argand diagram for $f^{o}_{++}(s, b)$ is shown in figure 11. Here we observe a remarkable simplicity in the non-peripheral region. The points lie approximately on a straight line !

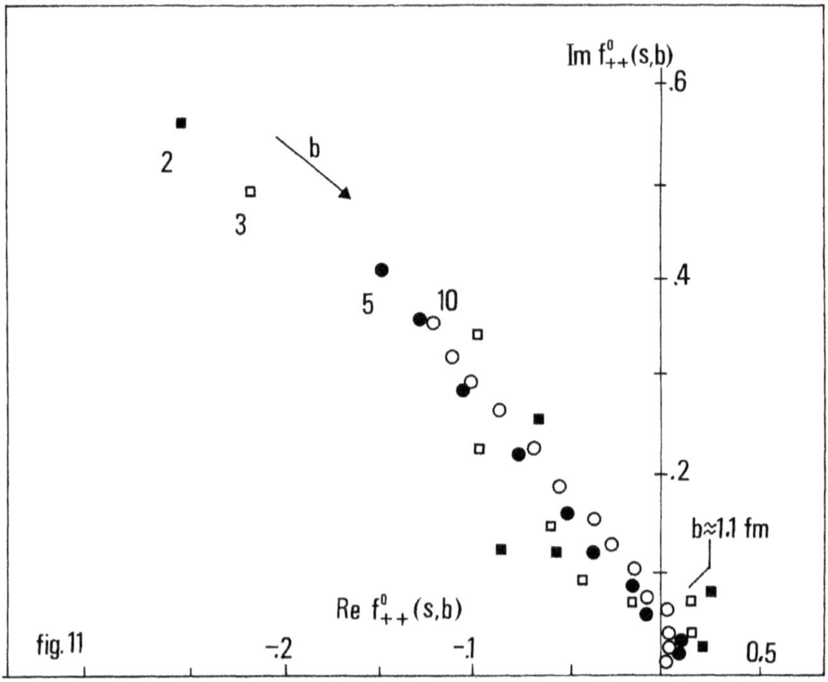

fig. 11

Figure 11 - Argand diagram of $f^{o}_{++}(s, b)$ as calculated from the isospin even combination of T^{J}_{++}. The numbers refer to pion lab. momenta 2(■), 3(□), 5(●) and 10(○) GeV/c. At each momentum J = 1/2 belongs to the point on the upper left side, whereas the intersection with the ordinate occurs at b ≈ 1.1 fm

Another important quantity often considered is the inelastic overlap function for diffraction scattering.

From models (refs. 32-34) it is expected to have a simple behaviour both in s and b.

We define :

$$\rho(s, b) = \frac{1}{4q} \int_{-4q^2}^{0} Jo\,(b\sqrt{-t})\ \text{Im}\ F_{inel}\ (s, t)\ dt \tag{22}$$

where in the spinless case Im F_{inel} is calculated from

$$\text{Im}\ T_\ell\ (s) - |T_\ell|^2 = 1/4\ (1 - \eta_\ell^2) \tag{23}$$

In van Hove's model a Gaussian $\rho(s, b)$ was proposed

$$\rho_{++}\ (s, b) = \rho(o)\ \exp\ (-b^2/2\,B_{inel}) \tag{24}$$

as a consequence of an exponential decrease of the inelastic contribution to the imaginary part of the isospin even helicity no-flip amplitude

$$\text{Im}\ F_{++\,inel}\ (s, b) = 2\sqrt{\pi}\ \rho(o)\ B_{inel}\ \exp\left[\frac{-B_{inel} \cdot t}{2}\right] \tag{25}$$

In figure 12 we have plotted $\rho_{++}(s, b)$ as calculated from our partial waves. We see that agreement with the ansatz (24) is obtained for $k \geq 6$ GeV/c. At lower momenta again a peripheral peak appears, but also a decrease at $b \approx 0$.

$\rho\ (s, b)$ is also of interest for a version of geometrical scaling (34). The overlap function should be a universal function of the variable b/R

$$\rho\ (s, b) = \hat{\rho}\ (b/R) \tag{26}$$

if

$$R^2(s) = \frac{\sigma_{in}^{tot}(s)}{8\pi} \left[\int_0^\infty \hat{\rho}\ (z)\ z\ dz\right]^{-1} \tag{27}$$

We have not reconsidered this property so far.

Finally we will make a remark about the contributions of high partial waves to the slope of the forward diffraction amplitude.

High partial waves represent the effects of nearby singularities in the cosθ plane. Therefore their tail is important, if invariant or helicity amplitudes are calculated from phase shifts for $t \simeq 0$. As a consequence for instance the forward slope of $d\sigma/dt$ is very sensitive to high partial waves. We will now study the forward log slope of the imaginary part of the diffraction amplitude :

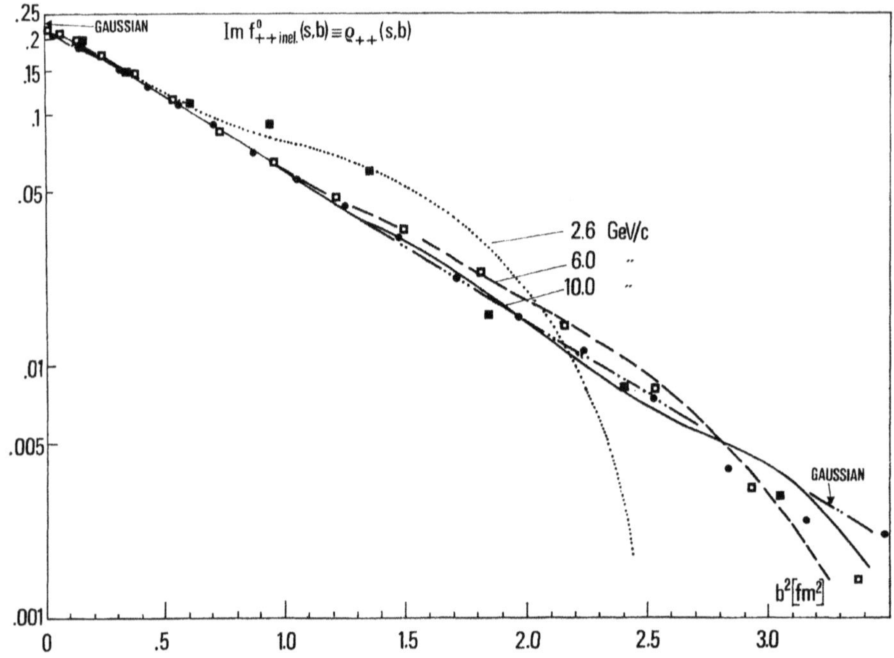

Figure 12 - Inelastic overlap function ρ_{++} (s, b) as calculated from the isospin even combination of the isospin 1/2 and 3/2 amplitudes (eq. 22). The straight line (-·· -) is the Gaussian of eq. (24). Notation as in figure 9.

$$q \left. \frac{\partial/\partial t \quad \text{Im } F^{o}_{++}}{\text{Im } F^{o}_{++}} \right|_{t = 0}$$

(the imaginary part dominates in $d\sigma/dt$).

Figure 13 shows the single J-contributions to this quantity at 2, 5 and 10 GeV/c as calculated from our new results. The abscissa is $b = (J + 1/2)/q$. The distributions are peaked at $b \simeq 1.1$ fm. The peak is narrow at 2 GeV/c. Only a few partial waves contribute significantly.

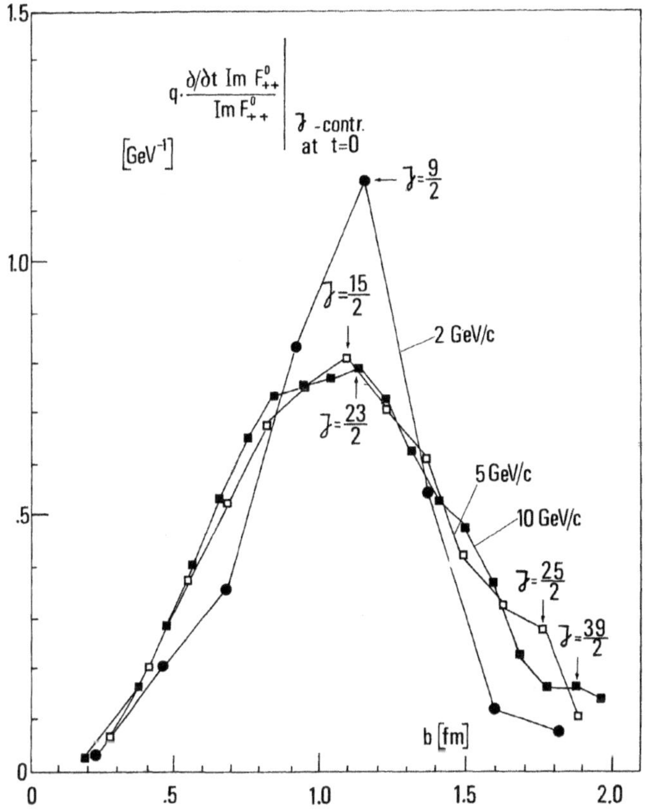

Figure 13 - Contribution of different partial waves to the logarithmic slope of Im F^0_{++} at $t = 0$. Normalization : the slope is approximately given by the area under the curves. The abscissa is $b = (J + 1/2)/q$.

It is broad at 5 and 10 GeV/c, where more than 10 waves become important to build up the correct slope. We notice further that if a partial wave gives the highest contribution to the forward slope, it seems also to be resonating around the same energy.

In ref. 35 Höhler and Jakob found that the partial waves projected out of an exponentially decreasing diffraction amplitude would have a behaviour very similar to our highest ones.

From this it becomes clear that the identification of high mass resonances will be difficult. The resonance effects would consist of a slight enhancement and a small loop as is for instance still seen in H3,11 and K1,13 in figure 7.

Schematically :

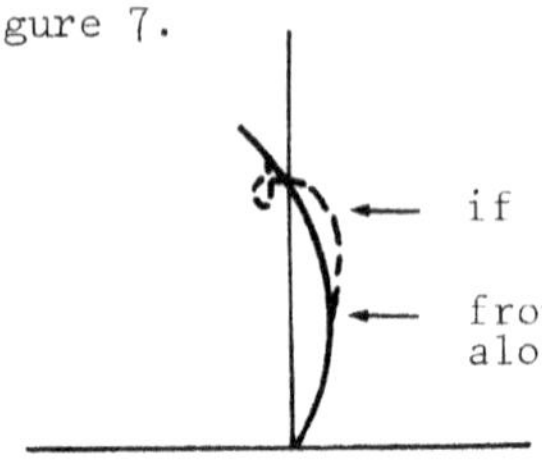

The positive enhancement of the real part follows from a partial wave dispersion relation if the rapidly increasing imaginary part is inserted.

6. Summary and conclusions

- More recent analyses make also use of the analytic properties of scattering amplitudes in order to obtain a unique solution.
- The KH result has approximately the analytic properties of the individual fixed-t, fixed-θ_{cm}, and fixed-s analyses.
- The CMU-LBL group produces an amalgamated data set and keeps track of the error correlations during their analysis. Analyticity is exploited at fixed-s and approximately fixed-θ to select the "right" solution.
- CMU-LBL and KH agree reasonably well.
- Besides resonance determination KH makes a number of applications.
- The results of D. Chew and A. Hendry have to be checked with dispersion relations.
- The result of a low energy analysis based on a threshold expansion is at variance with fixed-t dispersion relations at threshold.
- It is worthwhile studying impact parameter properties of scattering amplitudes at high energies.
- Resonance-like effects in high partial waves at higher energies have to be compared with background contributions.

Preliminary results on the KH 80 analysis have already been reported in ref. 38.

Acknowledgement

I thank Prof. G. Höhler for helpful advices during the preparation of this talk and Dr. I. Sabba Stefanescu for discussions on the uniqueness problems in partial wave analysis.

References and notes

*. Supported by Bundesministerium fûr Forschung und Technologie.

1. Almehed, S., and Lovelace, C.: 1972, Nucl. Phys. 40B, 157.
2. Ayed, R.: 1976, Thesis, Université de Paris-Sud, and CEA-N-1921.
3. Atkinson, D., Kok, L.P., and de Roo, M. : 1978, Phys. Rev. D17, 2492, and references given therein.
4. Sabba Stefanescu, I., Karlsruhe preprint TKP 80/3.
5. Kelly, R.L., and Cutkosky, R.E.: 1979, Phys. Rev. D20, 2782. Cutkosky, R.E., et al.: 1979, Phys. Rev. D20, 2804 and 2839.
6. Höhler, G., Kaiser, F., Koch, R.,and Pietarinen, E.: 1979, Handbook of Pion-Nucleon Scattering, Physics Data 12-1.
7. Koch, R., and Pietarinen, E.: 1980, Nucl. Phys. A336, 331.
8. Pietarinen, E.: 1972, Nucl. Phys. B49, 315; 1972, Nuovo Cimento 12A, 522; 1976, Nucl. Phys. B107, 21.
9. Höhler, G., Hutt, M., Koch, R., and Staudenmaier, H.M., Karlsruhe preprint TKP 80/10.
10. Hendry, A.: 1978, Phys. Rev. Lett. 41, 222.
11. Zidell, V.D., et al.: 1980, Phys. Rev. D21, 1255.
12. Chew, D. : 1979, LBL preprint 10075; 1976, Proceedings of the Topical Conference on Baryon Resonances, Oxford.
13. Höhler, G., and Koch, R.: 1980, Karlsruhe preprint, TKP 80/2 Febr.
14. Atkinson, D., Heemskerk, A.C., Swiestra, S.D.: 1976, Nucl. Phys. B109, 322 and references therein.
15. Schiz, A., et al.: 1979, Fermilab-Pub-79/81 Exp.
Fajardo, L.A., et al., Fermilab. Pub-80/27, submitted to Phys. Rev.
Akerlof, C.W., et al.: 1976, Phys.Rev.Lett. 35,1406; 1976, Phys.Rev. D14, 2864.
Ayres, D.S., et al. : 1977, Phys. Rev. D15, 3105.
16. Borghini, M., et al. : 1970, Phys. Lett. 31B, 405; 1971, 36B, 493.
Gaidot, A., et al. : 1975, Phys. Lett. 57B, 389; 1976, 61B, 103.
de Lesquen, A., et al.: 1972, Phys. Lett. 40B, 277.
Pierrard, J., et al. : 1975, Phys. Lett. 57B, 393; 1976, 61B, 107.
17. Koch, R.: 1978, Thesis, University of Karlsruhe.
18. Pietarinen, E., Helsinki preprint HU-TFT-17-77.
19. Borie, E., Gampp, W., Höhler, G., Koch, R., and Sabba Stefanescu, I.: 1980, Zeitschrift f. Physik C4, 333.
20. Hamilton, J.: 1967, Pion-Nucleon Interaction in "High Energy Physics" ed. by E.H.S. Burhop, Vol. 1, Academic Press.
Hutt, M. : 1979, Diplomarbeit, University of Karlsruhe.

21. Baacke, J., and Steiner, F. : 1970, Fortschritte der Physik 18, 67; and Koch, R., work in progress.
22. Barrelet, E. : 1972, Nuovo Cimento 8A, 331; 1970, Thesis, Paris.
23. Tromborg, B., Waldenström, S., and Overbo, I. : 1977, Phys. Rev. D15, 725.
24. Höhler, G., Koch, R., and Pietarinen, E. : 1977, Karlsruhe preprint TKP 77/12 July.
25. Jenkins, K.A., et al.: 1978, Phys. Rev. Lett. 40, 425; and ANL-HEP-PR-79-39.
26. Suzuki, Y. : 1979, Thesis, Kyoto.
27. Minowa, M. : 1979, Thesis, Kyoto.
28. Fukushima, M., et al. : 1979, Nagoya preprint DPNU-35-79 Nov.
29. Harari, H., and Schwimmer, A. : 1972, Phys. Rev. D5, 2780. Harari, H.: 1971, Ann. Phys. (N.Y.) 63, 1432.
30. Cottingham, W.N., and Peierls, R.F. : 1962, Phys. Rev. 126, 766.
31. Eichmann, G. : 1975, Thesis, University of Karlsruhe.
32. van Hove, L. : 1964, Rev. Mod. Phys. 36, 655.
33. Zachariasen, F. : 1971, Physics Reports 2C, 1.
34. Dias de Deus, J. : 1973, Nucl. Phys. B59, 231. Grein, W., and Kroll, P. : 1975, Phys. Lett. 58B, 79. Dias de Deus, J., and Kroll, P. : 1977, Nuovo Cimento 37A, 67.
35. Höhler, G., and Jakob, H.P.: 1974, Zeitschrift f. Phys. 268, 75.
36. Dean, N.W., and Ping Lee : 1972: Phys. Rev. D5, 2741.
37. Sabba Stefanescu, I., private communication.
38. Höhler, G.: 1979, Invited paper on High Energy πN Phase Shift Analysis. Proceedings of the International Conference on High Energy Physics in Geneva, Vol. 2, 594.

DISCUSSION

- G. Violini :

In your analysis you use isospin invariance. What is the way, if any, in which you include effects which are not isospin invariant, like mass difference effects ? Is this done by some correction method or by using some multichannel formalism?

- R. Koch :

We use the formalism and results of the Nordita group. Their corrections are obtained by the use of partial wave dispersion relations. Mass differences for charged and neutral pions and neutron and proton have been taken into account, also the $n\gamma$ channel in $\pi^- p$ scattering. After applying these corrections still an isospin breaking effect in the P 33 partial wave (Δ^{++}, Δ°) can be observed in total cross sections from SIN. But no clear effect in differential cross-sections is seen.

- R.E. Cutkosky :

You said that peripheral resonances occur at energies such that the momentum is proportional to $J + \frac{1}{2}$. This is quite different from the idea that the leading resonances occur on a linear Regge trajectory, with s proportional to J. How certain is it possible to be, in your analysis, that a linear trajectory is ruled out ? Is it possible that the "peripheral resonance", for large J, is an effect made up of many weak, overlapping states, instead of just one resonant state, in analogy to the "giant dipole resonance" in nuclear physics ?

- R. Koch :

There is still an uncertainty concerning the masses of the highest resonances depending on the background one might choose. But if we make a very rough determination, where we locate the resonance around the position where the real part of the partial wave becomes negative, a deviation from a linear Regge trajectory is seen for states with spins higher than 13/2. The masses tend to be larger than predicted from a linear continuation of the lower part of the Chew-Frautschi plots.

Concerning the second question, we have no indication for overlapping weak states building up the "peripheral resonance".

- J. Bowcock :

Although you mentioned the existence of continuum ambiguities at the beginning of your talk this has not been taken into account in the final presentation of your results. As was reported at the Oxford Conference by both Atkinson and the Birmingham Group the inclusion of an overall energy-momentum dependent phase factor can cause large changes in the partial waves which leads to a big spread in resonance parameters and removes the need for some of the smaller structures. When you have reached your amplitudes at the end of the calculation should you not ask what changes can be made in these by modifying them using an s and t dependent phase factor ?

- R. Koch :

Of course it would be interesting to make such a final test, although I think that our amplitudes are strongly restricted by the use of fixed-t, fixed angle and fixed-s analyticity constraints. I refer to the paper by I.S. Stefanescu, reference 4. To my knowledge Atkinson used the unitarity condition only.

- N. Zovko :

What about including the πd data at high energies and large angles into the πN amplitude analysis ?

- R. Koch :

To my knowledge there are so far no results from an analysis where πd data and πN data together have been used. Further I think that in πN scattering we do not have so large ambiguity problems anymore.

- N. Zovko :

I am certainly not the first one who suggests that the πd data at energies above 1 GeV and momentum transfer not less than, say, $0.3\ (\mathrm{GeV/c})^2$ represents a very sensitive test ground for the quality of different solutions for elementary (πN) amplitudes. It is, in fact, obvious in some sense, that the same sample of πN data, which is used to determine the amplitudes (phase shifts) themselves, is not suitable for discriminating among different solutions. One has, therefore, to get out to πd differential cross section data at large angles. The πN amplitudes enter into the double scattering region with the fourth power and strongly enhance the smallest discrepancies.

- M. Staszel :

The influcence of the deuteron wave function does not allow that.

- N. Zovko :

You are right as long as one is in the so called single scattering region where deuteron form factors dominate the t-dependence. However, in the region of double scattering, which is covered by the interval you talked about, the double scattering term dominates the total amplitude. Deuteron structure enters effectively only through the s-wave form factor while the dominant t-dependence comes from the πN amplitudes.

PION-NUCLEON DATA, DISPERSION RELATIONS, AND RESONANCES

R.E. Cutkosky
Carnegie-Mellon University - Pittsburgh, PA

A description of the methods and of the latest results of the CMU-LBL πN partial wave analysis project has recently appeared in the Physical Review (1-3) so I shall give here only a short summary of the main ideas and of some general conclusions. Many of the ideas were used first in an analysis of K^+p data carried out a number of years ago (4). In addition, after finishing the K^+p analysis we had a better understanding of the difficulties of partial wave analysis and were led to incorporate some improvements into the present work. In describing our work on πN physics, I shall concentrate on those aspects which may be especially relevant to the interests of people who are currently actively working on KN physics.

Our method of preparing an amalgamated data set (1) was a direct outgrowth of our experience with K^+p data. We realized the importance, in an energy-independent fit, of using data on cross sections and polarizations which refer to exactly the same energy. This is especially important when simultaneously analyzing data on different charge states. Thus, we want to interpolate data measured at slightly different lab momenta to a standard reference momentum. When one does this, it turns out to be convenient to unfold the momentum resolution at the same time, and also to interpolate to a set of standard angles (we chose a 3° spacing). We believe that it is important to separate this problem of interpolation, and the resolution of some further problems of consistency among data sets, from the partial wave analysis per se. That is, we propose to carry out the interpolation and make the consistency checks in a way which is as model-independent as possible, unbiased by the results one might expect from the partial wave analysis. Furthermore, it is important to recognize that the interpolation can't be done without introducing some small additional uncertainties into the data, and that these uncertainties need to be taken into account. The basic method we use is to fit the quantities $d\sigma/d\Omega$ and $P(d\sigma/d\Omega)/\sin\theta$ in a limited momentum band by a two-variable polynomial, and then slide the data along the fitted surface. A conformally-transformed angular variable is used, along with a truncation function which terminates the expansion smoothly.

E. Ferrari and G. Violini (eds.), Low and Intermediate Energy Kaon-Nucleon Physics, 31–40.

The data inconsistencies resolved at this stage involve systematic normalization and momentum errors, and also abnormal fluctuations, or scatter, in individual data points. The relative normalizations of different experiments are settled, and, in the case of $d\sigma/d\Omega$, adjusted to the 0° prediction based on dispersion relations. The momenta are also adjusted in a similar way. It is quite standard to take normalization uncertainties into account, but the momentum errors, which are usually just as well documented by the experimenters, and in our experience almost as important in practice, are not usually taken into account in analyses. Data which lie too far from the fitted surface have their errors stretched, so that resulting distribution of confidence-level values will be flat.

In almost all cases, the adjustment of normalization and momentum has been entirely consistent with the quoted uncertainties. One place where a significant momentum adjustment was found necessary was in the charge-exchange polarization data at 1030 MeV/c, where there are two experiments (Shannon et al.: 5, and Brown et al.: 6) which strongly disagree at certain angles. As pointed out by Brown et al. (6), the nature of the discrepancy suggests that one of the momenta was incorrectly measured. Our program indeed shifted one momentum up and one down, each by 20 MeV/c, in preparing the amalgamated data set. When these amalgamated CEX polarization data were used in the energy-independent partial wave fits, the central values of the data turned out to disagree with the other data at 1030 MeV/c. However, the amalgamated data had large and highly correlated errors, associated with the momentum shifts, and this enabled the partial wave fitting program to readjust the amalgamated data in the appropriate way (the resulting values (2) were close to the Shannon et al. measurements (5)). Another instance of significant momentum-shifting occurred in the 180° $d\sigma/d\Omega$ measurements of Rothschild et al. (7), which I shall discuss later.

For energy-independent fitting, we express the real and imaginary parts of spin flip and non-flip amplitudes in the general form

$$F(x) = B(x) + R(x)\, P(z(x)), \tag{1}$$

where $x = \cos\theta$. Here $B(x)$ is a fixed "Born term", P is a fitted polynomial in a variable $z(x)$ obtained by conformal mapping, and $R(x)$ is a modulating factor which builds in the expected asymptotic behaviour $F \sim x^{\alpha(s)}$ at $|x| \to \infty$. For real parts, B is given by the exchange of Reggeized N, Δ, ρ, and f, while for imaginary parts, a non Reggeized Pomeron and a two-pion absorptive term calculated by Alcock and Cottingham (8) was also included.

The use of the variable $z(x)$ and of the exchange term $B(x)$ corresponds to the incorporation of constraints associated with analyticity in the cut t-plane at a fixed value of s. This significantly reduces ambiguities in the energy-independent fitting. The errors on the partial wave amplitudes, and in particular, the error correlations, are also very different than is obtained with equal-quality fits in the conventional scheme.

To include electromagnetic effects correctly and also to impose unitarity constraints, we recast Eq. (1) into a conventional partial wave expansion truncated at $J = 39/2$. The partial wave amplitudes are divided into a "low" set treated explicitly, and a "high" set expressed in terms of the low waves. The fitting involves minimization of the quantity

$$X^2 = \chi^2 + \Phi + D + U \qquad (2)$$

where χ^2 describes the goodness-of-fit to the amalgamated scattering data. The truncation function Φ cuts off the expansion of $P(z)$ smoothly; the effective number of parameters really allowed in the fitting varies from about 25 at 0.4 GeV/c to about 45 at 2.0 GeV/c. Unitarity constraints on the high waves are imposed by the term U, and dispersion-relation constraints by the term D.

We use dispersion relations (analyticity in s) to resolve ambiguities in the single-energy fits and also to reduce errors in the fitted partial wave amplitudes. In the K^+p analysis (4), partial wave dispersion relations (PWDR) were used for this purpose. However, for the present work we decided to abandon PWDR. There were two reasons for this. The main reason is that we wanted to separate completely the problem of finding a well-constrained, unique fit from the problem of identifying resonances. We wanted to be sure that model-dependences and unconscious biases did not influence the results. The second reason is that the partial wave amplitudes (at given energy) are very highly correlated with each other. It would be necessary to take these correlations into account accurately, in order to obtain reliable results. This is possible, but would be very complicated for πN scattering, because so many waves are significant.

The method we adopted was to use dispersion relations along a set of hyperbolic curves lying in the physical region for scattering (HDR). We avoid fixed-t dispersion relations because these necessarily involve an extrapolation outside the physical region. In fact, the near-backward scattering region is generally a place where the amplitudes have a great deal of structure but where the data is often rather erratic. The use of dispersion-relation input is especially valuable for sorting out this region, and we have preferred to use relations directly connecting nearby points in the physical region to each other in doing this.

Another point that I would like to emphasize is that one has to be very careful not to use too much dispersion-relation input. In an iterative scheme, there is a great danger of over-smoothing, i.e. getting locked into a fit which has all the right analyticity properties but is not really the best representation of the data. This can be partially overcome by subtracting certain effects of the input data before using the results as constraints.

For the forward C amplitudes, our dispersion relations have the classical form, in that the imaginary part is given at all energies (by the optical theorem) and one then calculates the real part at each energy by doing an integral. For all the other amplitudes, we have to use input data on both real and imaginary parts, usually with comparable errors, and only over a part of the range. Use of the HDR is therefore really a matter of fitting the energy dependence by functions which have the correct behavior. We fit using an expression of the general form

$$F(E) = b(E) + \Sigma_\alpha \; a_\alpha \; h_\alpha(E), \tag{3}$$

where b(E) is again a Born term, and the function h (E), defined below, are weighted by the expansion coefficients a_α. The term b(E) includes Reggeized s- and u-channel nucleon poles, the second sheet s-channel Δ pole, and also an ηN threshold term.

A method of constructing a good set of functions h_α is described in Appendix B of ref. 2. We have recently developed a technique which is much faster, in terms of computer time, and gives essentially the same results; this new method may be of general interest. We wish to minimize the quantity

$$X^2 = \chi^2 + \lambda||F||^2; \tag{4}$$

more explicitly, we have

$$\begin{aligned} X^2 = \; & \Sigma\left[D_i - \Sigma \, a_\alpha \, h_{\alpha i}\right] W_{ij} \\ & \times \left[D_j - \Sigma \, a_\beta \, h_{\beta j}\right] \\ & + \lambda \, \Sigma \, a_\alpha (h_\alpha, h_\beta) \; a_\beta \end{aligned} \tag{5}$$

where W_{ij} is the weight matrix for the data (which actually includes correlations only between real and imaginary parts at the same energy). The $h_\alpha(E)$ are chosen to be orthogonal, in the sense that

$$\Sigma \, h_{\alpha i} \, W_{ij} \, h_{\beta j} + \lambda(h_\alpha, h_\beta) = \delta_{\alpha\beta}. \tag{6}$$

We start with some selected "seed" funtions h_{-1}, h_{-2}, ..., $h_{-\nu}$, and use the kernel funtion H(E,E'), which satisfies

$$(H(\cdot,E),g) = g(E). \tag{7}$$

We generate a sequence of functions by first defining

$$f_{\alpha+\nu}(E) = \Sigma_{ij} \, H(E,E_i) \; W_{ij} \; h_{\alpha j}. \tag{8}$$

To obtain $h_{\alpha+\nu}$ (E), this function $f_{\alpha+\nu}$ is then orthogonalized to h_0, h_1, ..., $h_{\alpha+\nu-1}$ in the standard way. In the present application, ν=2 gives the best results. An application of this method to spline-smoothing will be described in a forthcoming publication (9). The parameter λ appearing in Eqs. (4-6) is determined by the maximum-likelihood prescription.

As examples of HDR fits, I show in Figure 1 plots of the B amplitudes at $\theta=0°$ and of the C amplitudes at $\theta=180°$. These are reproduced from Ref. 2, along with some added preliminary points between 2.0 and 2.2 GeV/c (not included in the fits shown). The 180° C amplitudes are especially interesting because Im C (180°) in proportional to the difference between the odd-parity and even-parity cross sections, and because $|C(180°)|^2$ is proportional to $d\sigma(180°)/d\Omega$.

The 180° cross section is measureable in principle, and we have, e.g., data from Rothschild et al. (7), although usually a substantial extrapolation from the most backward measured points is required. Our amalgamation program suggests that a substantial momentum shift is needed in the data of reference 7 in order to make them compatible with the extrapolation from several other data sets which cover a wide angular range but do not go all the way back to 180°. We have accepted this calculated momentum shift, and that is the main reason why our 180° cross sections, shown in Fig. 2, are different from those of Pietarinen (10), who has evidently used ref. 7 without modification.

I have wondered whether some of the difficulties with fitting in the 600-900 MeV/c region which are evident in Fig. 1 and Fig. 2 might be associated with a breakdown of isospin invariance, specifically, from π^0-η mixing (11, 12). The strong coupling of the S_{11} (1540 MeV) resonance to the ηN channel enhances such effects in this region. In principle, the effects should be several times quoted experimental errors. However, many of the experimental data in this energy region also seem to be subject to systematic errors which are not understood but are evident in the form of mutual inconsistencies, so the situation is still quite unclear. I think it is important to resolve this question through further work.

The main interest in πN scattering in this energy region lies, of course, in the resonances. Reference 3 describes the parametrization we have used to fit the energy dependence of the partial wave amplitudes and extract information about the resonances. The method, summarized graphically in Fig. 3, involves use of "elementary" resonance terms and separable-potential terms coupled to several meson-baryon channels. Fig. 4 shows the F_{37} resonance as fitted, along with a few preliminary points; this is the cleanest and best determined resonance we have in the energy region studied.

The D_{35} and F_{17} partial waves, shown in Fig. 5 and 6, contain resonances about which there is still some controversy. Here our additional preliminary points seem to continue the trend we have seen at lower energies, except that the D_{35} real part remain somewhat more negative than the extrapolated fit had suggested.

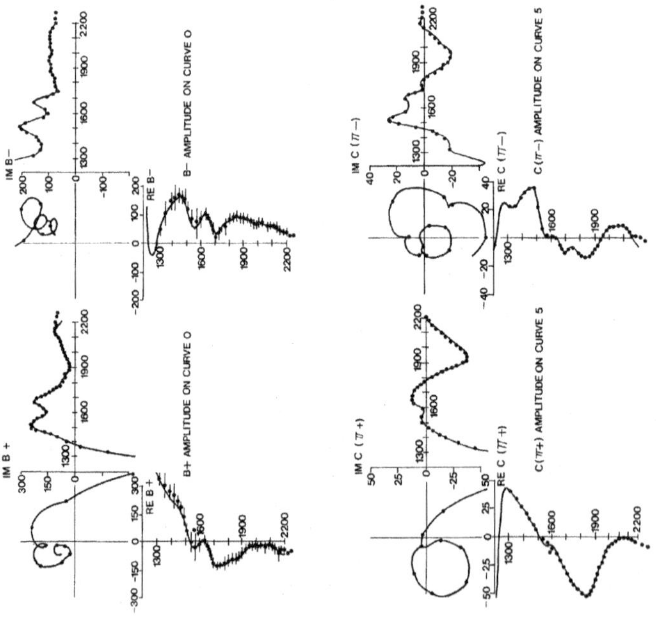

fig.1

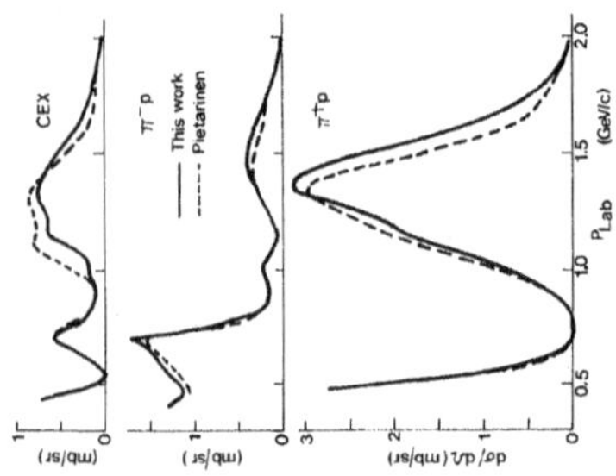

fig.2

fig.3

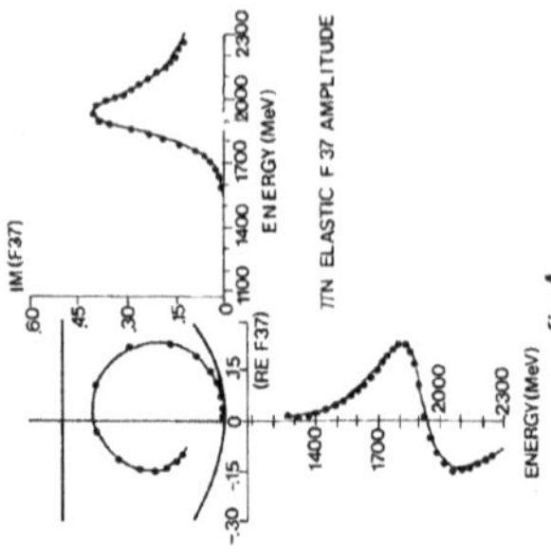

fig.4

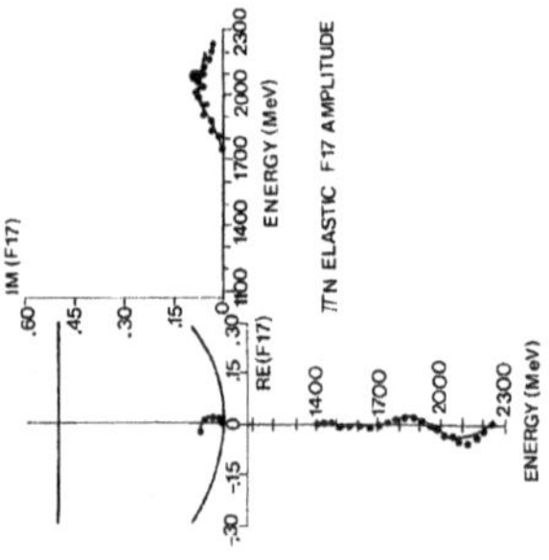

fig.5

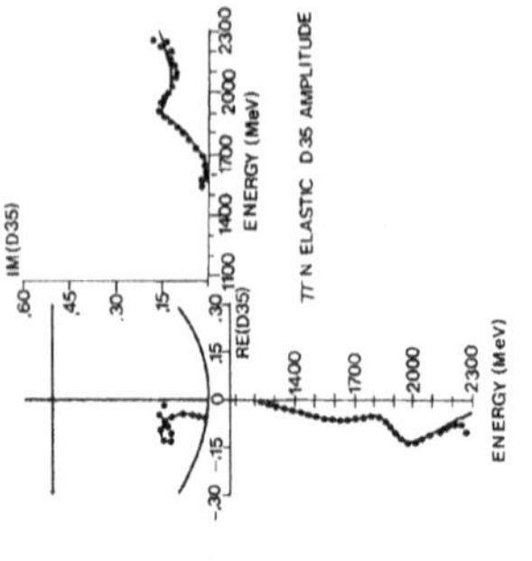

fig.6

Several calculations have recently been made of splittings in the (70,1⁻) multiplet of $SU_6 \times O_3$ using ideas based on QCD (13, 14). Kelly (15) has compared our empirical results with these calculations. As shown in Fig. 7, the results are in quite good agreement except for the splitting between the D_{33} and S_{31} resonances; the calculations agree quite well with the average value but give zero splitting. In these calculations the two-body spin-orbit force arising from one-gluon exchange has been omitted - that is the reason good agreement with the I=1/2 resonances was obtained. Our results on the $(56,2^+)$ multiplet also suggest that there is a negligible net spin-orbit force acting in that multiplet; it is generally assumed that this occurs because the confining potential is scalar and gives a Thomas-precession term of the opposite sign. However, the source of discrepancy in the I=3/2 $(70,1^-)$ state is more mysterious, because it can be shown that the standard spin-orbit term does not contribute in those states.

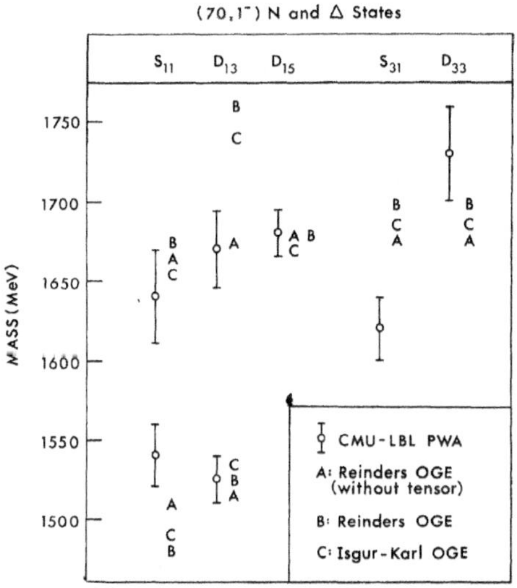

Fig. 7

Kelly (15) has remarked that other partial wave analyses (16) agree with us that there is a substantial D_{33}-S_{31} splitting. The reason the theoretical calculations (13, 14) were able to claim good agreement with empirical results is that they used summaries (16) which included analyses of photoproduction, and this enlarged the spread of quoted values enough to hide the disagreement.

Some general conclusions we have drawn from the results of the CMU-LBL analysis (3) are, first, that "non-minimal" multiplets exist - these are odd-parity 56-plets and even-parity 70-plets. However, these are all weakly coupled to the elastic channel. There are still many missing members of the non-minimal multiplets, especially the $(70,0^+)$ and $(70,2^+)$, and much further work will be needed to completely sort things out. Hadron exchange has a significant effect on the non-resonant background amplitudes in many partial waves, and may also have significant effects on the couplings of the resonances.

The results of partial wave analysis give a number of interesting implications for quark dynamics, although these go somewhat beyond the scope of my talk. Hendrick and I (17) have suggested that light quarks in a linear potential give a better first-order approximation to the spectrum than does the non-relativistic harmonic oscillator model. Forsyth and I have inferred, on the basis of regularities among the couplings, that a diquark model is not acceptable. The effects of gluon-exchange forces are especially interesting, and there is striking evidence for the spin-spin forces given by QCD models. Spin-orbit forces are apparently weak, which supports the idea that the confining potential is a scalar.

What general suggestions for KN partial wave analysis can be made on the basis of experience with πN analyses ? It is clear that determination of the properties of the KN resonances is interesting and important for the study of quark dynamics. The study of KN physics is certainly worth a great deal of effort. In closing, I should also like to emphasize once more the importance, to such a program, of a careful preliminary examination of the experimental data, and also to remind you of the advantages to be obtained from use of non-standard analyticity properties, especially at fixed s, as constraints in the analysis.

References

1. Kelly, R.L., and Cutkosky, R.E.: 1979, Phys. Rev. D20, 2782.
2. Cutkosky, R.E. et al.: 1979: Phys. Rev. D20, 2804.
3. Cutkosky, R.E., et al.: 1979, Phys. Rev. D20, 2839.
4. Cutkosky, R.E., et al.: 1976, Nucl. Phys. B102, 139.
5. Shannon, S.R., et al.: 1974, Phys. Rev. Letters 33, 237.
6. Brown, R.M., et al.: 1978, Nucl. Phys. B144, 287.
7. Rothschild,R.E., et al.: 1972, Phys. Rev. D5, 449.
8. Alcock, J.W., and Cottingham, W.N.: 1972, Nucl. Phys. B41, 141.
9. Cutkosky, R.E., and Pomponiu, C.:in preparation.
10. Pietarinen, E.: 1978, Univ. of Helsinki Report No. HU-TFT 78-13 (unpublished).
11. Cutkosky, R.E.: 1979, Phys. Lett. 89B, 339.
12. Törnqvist, N.A.: 1972, Phys. Lett. 40B, 109.
13. Isgur, N., and Karl, G.: 1979, Phys. Rev. D19, 2653.

14. Reinders, L.J.: 1978, J. Phys. G4, 1241.

15. Kelly, R.L.: Private communication.

16. Particle Data Group : 1978, Phys. Lett. 75B, 1.

17. Cutkosky, R.E., and Hendrick, R.E.: 1977, Phys. Rev. D16, 793.

DISCUSSION

- N. M. Queen :

You pointed out that there are different possible ways of extracting resonance parameters from partial-wave amplitudes. To what extent do the values of these parameters depend on the particular method of extracting them?

- R. E. Cutkosky :

When resonances overlap or when they have energy-dependent factors in their widths, the definitions of parameter values depend on the method of parametrization. An interesting example is given by $S_{11}(1540)$; the presence of the ηN threshold cusp makes the "eyeball" value of the width smaller than the value we obtain whan we allow for the effect of the threshold.

- N. M. Queen :

In an analysis with a sufficiently large number of parameters there is always the danger that one may be able to fit any given set of data, and I am wondering whether it is possible to say what degree of confidence one can have in the analysis. In this connection it seems to me that it may be interesting to apply the method to the analysis of simulated data generated from model functions whose structure is known exactly, to see how accurately one can actually reproduce detailed resonance structures in the amplitudes.

- R. E. Cutkosky :

What you propose would be interesting but it would be a major undertaking. To ensure that the results were ubiased , it should be done by someone else - independent of, but in close communication with experimental groups and analysis groups.

- C. B. Dover :

Are there any missing members of the low-lying $\{56\}^+$ and $\{70\}^-$ multiplets of SU(6) x O(3)?

- R. E. Cutkosky :

All non-strange members of the $(56,0^+)$, $(56,0^+)^*$, $(56,2^+)$ and $(70,1^-)$ multiplets are known.

IS THE POLE APPROXIMATION TO THE UNPHYSICAL PART OF THE LEFT HAND CUT SUFFICIENT ?

P. Kroll
Department of Physics - University of Wuppertal

1. Introduction

In any application of dispersion relations to kaon-nucleon and nucleon-nucleon scattering the existence of an unphysical part of the left hand cut (LHC) is a challenge. If one is, for example, interested in predictions of real parts of scattering amplitudes or more generally if one uses dispersion relations as constraints in phase shift or amplitude analyses in order to resolve ambiguities, one has to assume something about the contributions from the unphysical part of the LHC which clearly weakens the power of the dispersion relations. On the other hand, if an amplitude, that is real as well as imaginary part, for a given t value and for a given energy range is known one may turn around the procedure and use the dispersion relation as a tool to obtain information on the discontinuity of the LHC. The easiest and often used method to exploit this idea is to approximate the discontinuity by a sum of poles and to determine their coupling constants and eventually also their masses. In nucleon nucleon scattering this method has been applied for example by Bugg (1) and by Verwest et al. (2). With the advent of total cross-section data in pure spin states, however, it turned out by a detailed study performed by us (3) that this pole approximation is not sufficient. We found clear evidence for contributions from the 3π continuum. In the following I will report on that investigation. I will start with a description of the available information on the discrepancy functions and will proceed with a qualitative discussion of the physics of the LHC. Thereafter the discontinuity of the LHC will be discussed quantitatively. I will end up with some conclusions.

2. The NN discrepancy functions

Forward nucleon-nucleon scattering is described by three invariant functions F_i. By unitarity their imaginary parts are related to the total cross-sections in pure spin states.

E. Ferrari and G. Violini (eds.), Low and Intermediate Energy Kaon-Nucleon Physics, 41–52.

i) spin averaged cross section

$$\sigma_{tot} = \frac{1}{2}\left\{\sigma\,(\overset{\rightarrow}{\leftarrow}) + \sigma\,(\overset{\rightarrow}{\rightarrow})\right\} = 2\sqrt{\pi}\ \mathrm{Im}\,(\phi_1+\phi_3)\Big|_{t=0} = \frac{4\pi}{p_L}\ \mathrm{Im}\ F_1 \tag{1.a}$$

ii) transversely polarized nucleons

$$\Delta\sigma_T = \sigma\,(\uparrow\downarrow) - \sigma\,(\uparrow\uparrow) = -\,4\sqrt{\pi}\ \mathrm{Im}\ \phi_2\Big|_{t=0} = -\,\frac{4\pi}{p_L}\ \mathrm{Im}\ F_2 \tag{1.b}$$

iii) longitudinally polarized nucleons

$$\Delta\sigma_L = \sigma\,(\overset{\rightarrow}{\leftarrow}) - \sigma\,(\overset{\rightarrow}{\rightarrow}) = 4\sqrt{\pi}\ \mathrm{Im}\,(\phi_1-\phi_3)\Big|_{t=0} = \frac{4\pi}{p_L}\ \mathrm{Im}\ F_3 \tag{1.c}$$

where for comparison the relations to the s-channel helicity amplitudes ϕ_i are also given.

The (singly subtracted) forward dispersion relations for proton-nucleon scattering read

$$\begin{aligned}\mathrm{Re}\ F_{ipN}(\omega) &= \mathrm{Re}\ F_{ipN}(m) + \frac{1}{m}\,\tau_{pN;1}\,R_{i\pi}\,\frac{\omega-m}{\omega+\omega_\pi} \\ &+ f^d_{ipN}\,\frac{\omega-m}{\omega_d-\omega} - \beta_i\,\frac{\omega-m}{\pi}\int_{\omega(2m_\pi)}^{\infty} d\omega'\,\frac{\mathrm{Im}\ F_{i\bar{p}N}(\omega')}{(\omega'+\omega)\,(\omega'+m)} \\ &+ \frac{\omega-m}{\pi}\int_m^{\infty} d\omega'\,\frac{\mathrm{Im}\ F_{ipN}(\omega')}{(\omega'-\omega)\,(\omega'+m)}\end{aligned} \tag{2}$$

Hereby the following notations are used :

$\omega(p_L)$ lab. energy (momentum);

$m_\pi(m)$ pion (nucleon) mass;

$\beta_1=\beta_2 = -\beta_3 = 1$; $\omega(2m_\pi) = -m+2m_\pi^2/m$; $\omega_\pi = -m+m_\pi^2/2m$;

$\mathrm{Re}\ F_{ipN}(m)$ subtraction constants;

f^d_{ipN}, $R_{i\pi}$ are the residues of the deuteron (4) and pion poles respectively (compare table 1).

Table 1 - Meson-nucleon coupling constants

meson	coupling constant	remark
π	$g_\pi^2/4\pi = 14.24$	fixed
η, η'	$g_\eta^2/4\pi = g_{\eta'}^2/4\pi \doteq 0(< 1.0)$	
ω	$g_{V\omega}^2/4\pi = 8.1 \pm 1.5$ $g_{T\omega}/g_{V\omega} = 0.14 \pm 0.20$	
A_1	$g_{A_1}^2/4\pi = 7.3 \pm 3.0$	
f	$g_{1f}^2/4\pi = 2.2 \pm 0.9$ $g_{2f}/g_{1f} = 0.6 \pm 0.9$	taken from ref. 14

τ is the isospin crossing (s-u) matrix defined by

$$\begin{pmatrix} F_{ipp} \\ F_{ipn} \end{pmatrix} = \begin{pmatrix} 1/2 & 1/2 \\ 1 & 0 \end{pmatrix} \begin{pmatrix} F_i^{(I_u=1)} \\ F_i^{(I_u=0)} \end{pmatrix} \tag{3}$$

where I_u denotes the isospin of the $N\bar{N}$ system. The discrepancy functions are defined by

$$\Delta_{ipN} = \text{Re } F^{exp}_{ipN} - \text{Re } F^{DR}_{ipN} = -\beta_i \frac{\omega-m}{\pi} \int_{\omega(2m_\pi)}^{\bar{\omega}_i} d\omega' \frac{\text{Im } F_{i\bar{p}N}(\omega')}{(\omega'+\omega)\ (\omega'+m)} \tag{4}$$

where the Re F^{exp}_{ipN} are the real parts of F_{ipN} which are either reconstructed from phase shifts or taken from Coulomb interference measurements and the Re$^{DR}_{ipN}$ collect all known parts of the dispersion relation (2). $\bar{\omega}_1 = m$, $\bar{\omega}_2 = \bar{\omega}_3 = 3$ GeV as determined by our knowledge on the Im $F_{i\bar{p}N}$ (compare in figure 1).

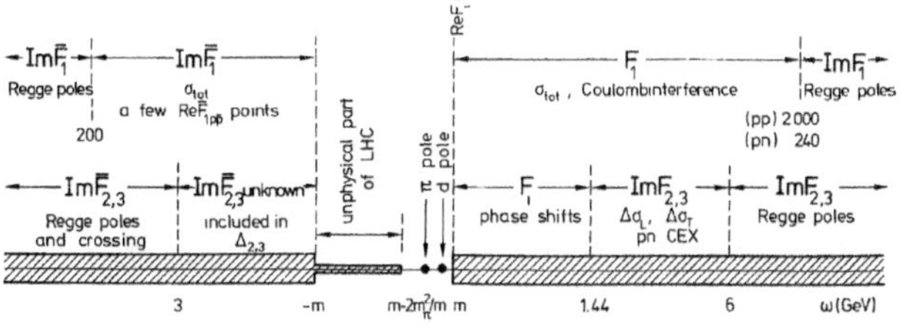

Figure 1 - The singularity structure of the pN forward amplitudes in the ω plane and the input used in our analysis

Since we are aiming at a study of the $\bar{N}N$ cut, it is convenient to form combinations of the discrepancy functions with definite u-channel quantum numbers :

$$\text{i) } \Delta_s^{(I_u)} = \frac{1}{2}\left\{\Delta_{1pN}^{(I_u)} - \Delta_{2pN}^{(I_u)} - \frac{1}{2}\Delta_{3pN}^{(I_u)}\right\} = -\frac{\omega-m}{\pi}\int_{\omega(2m_\pi)}^{\bar{\omega}} d\omega' \frac{\mathrm{Im}\,\bar{F}_3^{(I_u)}(\omega')}{(\omega'+\omega)(\omega'+m)} \quad (5)$$

Only spin singlet partial waves ($\bar{f}_o^{\,J}$) contribute to the $N\bar{N}$ amplitude $\bar{F}_s$, that is only states with unnatural parity (J^P=A).

$$\text{ii) } \Delta_t^{(I_u)} = \frac{1}{2}\left\{\Delta_{1pN}^{(I_u)} + \Delta_{2pN}^{(I_u)} - \frac{1}{2}\Delta_{3pN}^{(I_u)}\right\} = -\frac{\omega-m}{\pi}\int_{\omega(2m_\pi)}^{\bar{\omega}} d\omega' \frac{\mathrm{Im}\,\bar{F}_t^{(I_u)}(\omega')}{(\omega'+\omega)(\omega'+m)} \quad (6)$$

Here, only coupled triplet partial waves ($\bar{f}_{11}^J$) with J^P=N contribute.

$$\text{iii) } \Delta_m^{(I_u)} = \Delta_{1pN}^{(I_u)} + \frac{1}{2}\Delta_{3pN}^{(I_u)} = -\frac{\omega-m}{\pi}\int_{\omega(2m_\pi)}^{\bar{\omega}} d\omega' \frac{\mathrm{Im}\,\bar{F}_m^{(I_u)}(\omega')}{(\omega'+\omega)(\omega'+m)} \quad (7)$$

to which coupled triplet waves $\bar{f}_{22}^J$ (J^P=N) and uncoupled triplet waves $\bar{f}_1^J$ (J^P=A) contribute.

In figure 1 I compile the input we used in the actual calculation of the discrepancy functions. I want to point out that because of lack of $\Delta\sigma_T(pn)$ data, we made use of forward pn charge exchange differential cross-sections. Details of the calculations are given in refs. 3, 5, 6. Our analysis provides us with :

$\Delta_{2,3pp}$, $\Delta_{2,3pn}$ from threshold up to ω = 1.44 GeV

Δ_{1pp} up to 2000 GeV

Δ_{1pn} up to 70 GeV

The high energy parts of the latter two functions are important for any model of the LHC since they keep the high mass contributions under control. The same holds for the $p\bar{p}$ discrepancy function $\Delta_{1p\bar{p}}$ for which we have also some information from the rare $p\bar{p}$ Coulomb interference data.

3. The physics of the LHC

By unitarity the discontinuity of the LHC is built up by the contributions form the allowed intermediate states

$$\mathrm{Im} < N\bar{N}\ |T|\ N\bar{N} > \ = \sum_n < N\bar{N}\ |T^+|\ n > < n\ |T|\ N\bar{N} >$$

$$= < N\bar{N} \,|T^+|\, \pi\pi > < \pi\pi \,|T|\, N\bar{N} > + < N\bar{N} \,|T^+|\, 3\pi > < 3\pi \,|T|\, N\bar{N} > + \ldots \tag{8}$$

The most important contributions to the unitarity sum certainly come from the 2π intermediate states (nearest singularities). We have at our disposal independent information on these contributions which we may check by comparison with the discrepancy functions. In case that they pass this test we can make use of them in the further analysis.

Two groups, Gustafson et al. (7) and Höhler et al. (8) have calculated the lowest $N\bar{N} \to \pi\pi$ partial waves $f_\pm^J$ ($J \leq 1$ and $J \leq 3$ respectively) from the πN amplitudes (9) by analytic continuation. Approximating for larger J the partial waves by the Born term i.e. nucleon exchange, we obtain, from unitarity, the imaginary parts of the $N\bar{N}$ amplitudes for u less than 5o m_π^2 for $I_u = 1$ and 4o m_π^2 for $I_u = 0$. The results from the two groups agree very well. Inserting the imaginary parts into eqs. (5-7), we find the contributions of the 2π cut to the discrepancy functions.

For the other contributions (from the 2π cut for $u > 5o$ (4o) m_π^2 and from the other cuts) no independent informations from other sources are available. The usual procedure is to approximate these contributions by a sum over meson pole terms, each of which produces a term

$$\frac{1}{m} R_{ij} \frac{\omega - m}{\omega + \omega_j} = \frac{1}{m} R_{ij} \frac{1}{1 + m_j^2/2mT} \tag{9}$$

in the discrepancy function $\Delta_i^{(I_u)}$ (ω_j (m_j) is the position (mass) of the pole j and R_{ij} its residue, T is the kinetic energy in the lab.system). Each pole has a characteristic energy dependence determined by its mass. Heavy mesons (mass $\geq$ 1.1 GeV) play only a minor role in the energy range under consideration because the curvature of their pole terms is much weaker than what we observe in the discrepancy functions.

However, as I will demonstrate in a moment, the discrepancy functions show clear evidence for contributions from the 3π continuum. All contributions needed for a sufficiently accurate approximation to the discontinuity of the LHC are graphically represented by

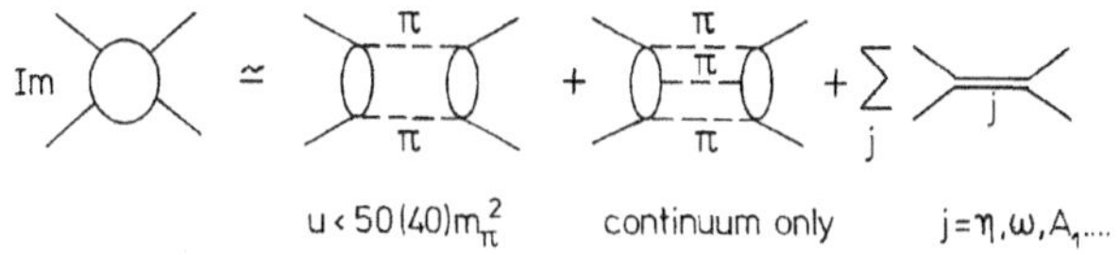

In table 2 the most important contributions to the various discrepancy functions are compiled whereby, for convenience, the lowest angular momentum states of the 3π continuum are classified in terms of $\pi\varepsilon$ and $\pi\rho$ states. ε stands for a $\pi\pi$ S wave ($I = 0$) and ρ for a $\pi\pi$ P wave ($I = 1$).

Comparison of the table with figures 2-4 shows that

Table 2 - Contributions from LHC to the NN discrepancy functions (only masses $\leq$ 1.1 GeV are considered)

discrepancy function	intermediate state	sign of contribution	remark
$\Delta_s^{(1)}$	$\pi\varepsilon$ (S wave) $\pi\rho$ (P wave)	$-$ $-$	π subtracted out no 2π lightest meson : B (1235)
$\Delta_s^{(0)}$	η (549), η'(958) $\pi\rho$ (S wave)	$-$ $+$	no 2π
$\Delta_t^{(1)}$	$f_+^{1,3,..}$ ($N\bar{N}\to\pi\pi$) $\pi\rho$ (D Wave) δ (980)	$-$ $+$ $+$	f_+^1 contains the ρ } probably Zweig forbidden ($u\bar{d}$ $s\bar{s}$) next important pole : A_2 (1310)
$\Delta_t^{(0)}$	$f_+^{0,2,..}$ ($N\bar{N}\to\pi\pi$) $\pi\rho$ (P wave) ω (783) ϕ (1020) S* (980)	$+$ $-$ $-$ $-$ $+$	$f_+^{(0)}$ contains the ε next important pole: f (1271) } probably Zweig forbidden ($s\bar{s}$; $s\bar{s}$ ($u\bar{u}+d\bar{d}$))
$\Delta_m^{(1)}$	$f_-^{1,3,..}$ ($N\bar{N}\to\pi\pi$) $\pi\varepsilon$ (P wave) $\pi\rho$ (S wave) A_1 (1100)	$-$ $+$ $+$ $+$	f_-^1 contains the ρ next important pole : A_2 (1310)
$\Delta_m^{(0)}$	$f_-^{2,4,..}$ ($N\bar{N}\to\pi\pi$) $\pi\rho$ (P wave) ω (783) ϕ (1020)	$+$ $-$ $-$	next important pole : f (1271)

i) the negative sign of $\Delta_s^{(1)}$ and the positive one of $\Delta_s^{(o)}$ require contributions from the 3π continuum. The energy dependences of the discrepancy functions call for contributions with effective masses considerably below looo MeV. Thus heavy mesons cannot be responsible for the signs of $\Delta_s^{(o,\,1)}$.

ii) $\Delta_t^{(1)}$ is completely saturated by the 2π term, there is no need for other contributions. This is in agreement with the 3π continuum idea since its lowest angular momentum state contributing to $\Delta_t^{(1)}$ is a $\pi\rho$ D wave which should be small.

iii) the 2π contribution to $\Delta_m^{(1)}$, essentially the flip P wave f_-^1, dramatically exceeds the data, calling again for the 3π continuum. The lowest angular momentum states have the right sign to fill the gap between the 2π term and the data. Neglect of the 3π contribution would force us to conclude that the 2π term is in severe disagreement with the NN data.

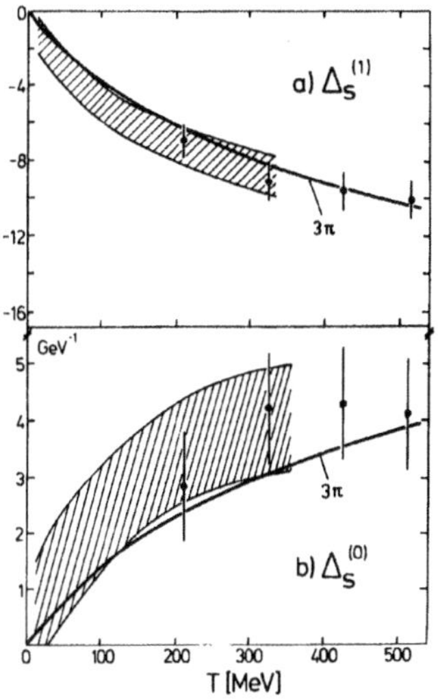

Figure 2 - The discrepancy functions $\Delta_s^{(1)}$ (a) and $\Delta_s^{(o)}$ (b) as a function of the kinetic energy. The hatched band indicates the experimental data (10), only the BASQUE results (11) are shown separately (•). The solid line is the 3π contribution (compare section 4)

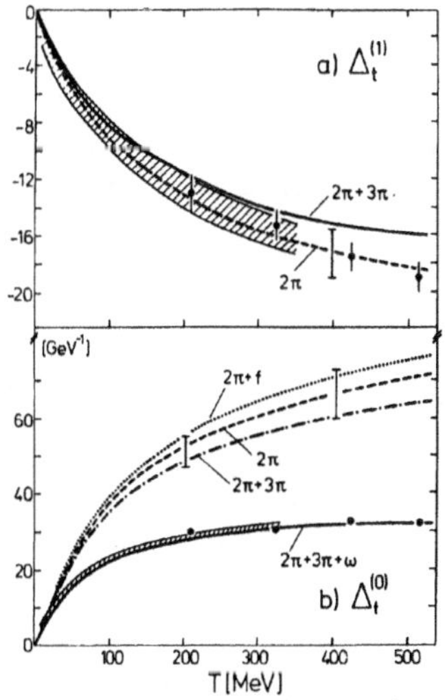

Figure 3 - The discrepancy functions $\Delta_t^{(1)}$ (a) and $\Delta_t^{(o)}$ (b) as a function of T. Various cut and OBE contributions are shown (compare section 4). The 2π contributions are calculated from $N\bar{N} \rightarrow \pi\pi$ partial waves of ref. 7. Typical (systematic) errors of them are indicated. The f couplings are those given in table 1. For further notations compare figure 2.

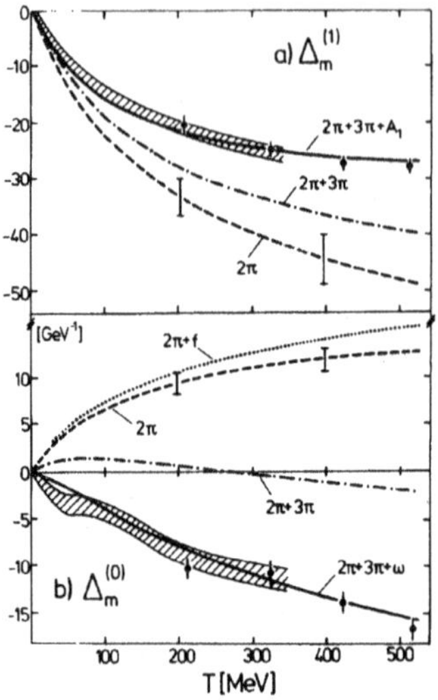

Figure 4 - The discrepancy functions $\Delta_m^{(1)}$ (a) and $\Delta_m^{(0)}$ (b) as a function of T. For notations compare figure 3.

iv) describing the difference between $\Delta_{t,m}^{(0)}$ and the 2π term solely by the ω pole, one obtains for the ω vector coupling the value $g_{V\omega}^2/4\pi = 7.7$ and the rather large ratio $g_{T\omega}/g_{V\omega} = 0.7$ in contrast to pure pole models where the tensor coupling comes out very small. The reason for the large tensor coupling in our description lies in the 2π contribution to $\Delta_m^{(0)}$ to which no low mass meson pole corresponds. However, part of the difference between $\Delta_{t,m}^{(0)}$ and the 2π term may come from the 3π continuum which starts with a $\pi\rho$ P wave.

4. A quantitative study of the LHC

From the preceding discussion it should be clear that we obtain a qualitative understanding of the NN discrepancy functions only if the 3π continuum is taken into account. A quantitative investigation, however, i.e. a check for consistency of the $N\bar{N} \to \pi\pi$ partial waves and a determination of coupling constants requires a detailed dynamical model for the 3π discontinuity which, of course, is hard to perform. As a first step the 3π continuum is assumed to be determined by $\pi\varepsilon$ and $\pi\rho$ and the dynamics of the reactions $N\bar{N} \to \pi\varepsilon$, $\pi\rho$ is approximated by nucleon exchange. Exploiting unitarity the 3π discontinuity can then be calculated in a similar fashion as that of the 2π cut from the $N\bar{N} \to \pi\pi$ amplitudes. In order to take into account the finite width of the ρ and the ε the discontinuities are averaged over the contribution of the $N\bar{N} \to \pi\pi$ S and P waves. Graphically the model is described by

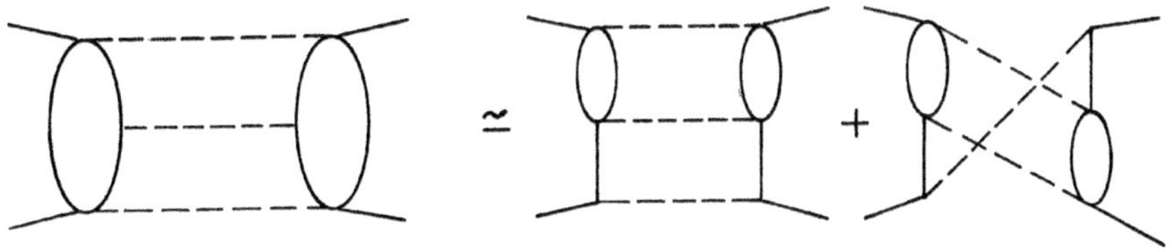

With this averaging procedure the model discontinuity really starts at the 3π threshold instead than at the $\pi\varepsilon(\rho)$ threshold. Furthermore, it has the advantage that the results do not depend on the approximations one has to make in going from the partial waves to masses and coupling constants.

At each vertex of the graphs shown above a form factor is introduced. In the actual calculation all form factors are set equal and the then only parameter of the model is adjusted such that $\Delta_s^{(1)}$ is well reproduced.

As can be seen from figures 2-4 the model removes most of the difficulties in the discrepancy functions discussed above. $\Delta_s^{(0)}$ is solely saturated by the 3π term, nothing is left for pole terms. One, therefore, has to conclude that the coupling constants of both the η and the η' are zero within the errors. On the other hand, in $\Delta_t^{(0)}$, $\Delta_m^{(0,1)}$ there are still gaps left to other contributions for which by inspection of table 2, the ω and the A_1 poles are the most important candidates. Fitting the pole terms to the data, one ends up with a good description of all NN discrepancy functions and with reasonable coupling constants (results are given in table 1). Also the high energy behaviour of our model for the LHC is correct as can be seen from figure 5 where Δ_{1pp} and Δ_{1pn} are plotted.

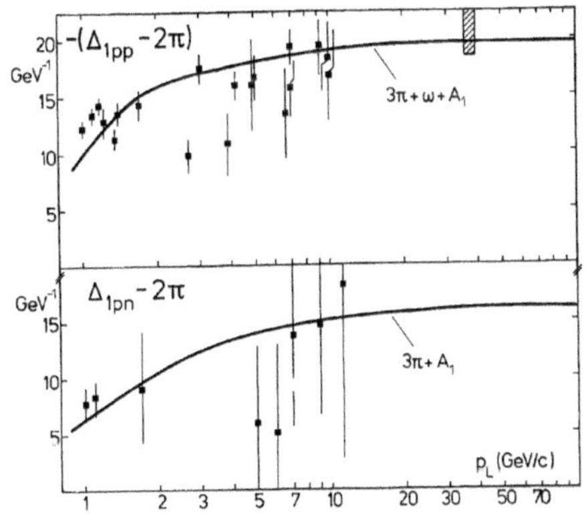

Figure 5 - The high energy behaviour of the discrepancy functions Δ_{1pp} and Δ_{1pn}. The 2π contribution has been subtracted out. The solid lines are our fits. The hatched bar is the average of the pp data from 15 to 2000 GeV/c.

I want to end this section with a few remarks concerning the contributions from other pole terms.

i) There is no way to separate a small ϕ contribution from that of the ω. The values of the ω couplings quoted in table 1 are obtained under the neglect of the ϕ.

ii) There are no hints at contributions from the δ and S^*. Nowadays these mesons are interpreted as cryptoexotics (12) with a $s\bar{s}$ quark content (compare table 1). Their weak coupling to nucleons may be then taken as evidence that the low energy NN interaction obeys the Zweig rule. This justifies the neglect of the ϕ. According to Lovas (13) the validity of Zweig's rule may be also the reason why there is no hint at contributions from the 4 (or more) π continuum in the data.

iii) Contributions from heavy mesons are negligible. In the case of the f meson for which the coupling constants have been computed from πN hyperbolic dispersions relations (14) (compare table 1) one can, however, demonstrate the effect of such a pole term. As is seen from figures 3 and 4 the contributions of the f lie within the error bars. Similar (or smaller) effects are expected from other heavy mesons.

5. Conclusions

Exploiting the experimentally available information on forward nucleon-nucleon scattering in terms of discrepancy functions one can clearly demonstrate the need for contributions from the 3π continuum which are not included in the usual pole approximation. The 3π term is sufficiently well described by a conventional nucleon exchange model. Having done this, one finds a surprisingly perfect agreement of the 2π contribution as calculated from the available $N\bar{N} \to \pi\pi$ partial waves and the NN scattering data.

In addition to the 2π and 3π contributions effects of the ω and the A_1 pole terms are seen in the discrepancy functions, allowing us to fix their coupling constants.

References and notes

1. Bugg, D.V.: 1968, Nucl. Phys. B5, 29.

2. Verwest, B., and Riska, D.O.: 1974, Phys. Lett. B48, 17.

3. Grein, W., and Kroll, P., to be published in Nucl. Phys. A.

4. The deuteron parameters are $B = 2.225$ MeV, $r_e = 1.786$ fm and ρ (D/S) $= 0.0278$.

5. Grein, W., and Kroll, P.: 1978, Nucl. Phys. B137, 173.

6. Grein, W., König,A., and Kroll, P.: 1980, preprint WU B 80-13, Wuppertal.

7. Gustafson, G., Nielsen H., and Oades, G.C.: 1979, published in M.M. Nagels et al., Nucl. Phys. B147, 189.

Oades, G.C.: 1978, private communication.

8. Höhler,G., and Pietarinen, E.: 1975, Nucl. Phys. B95, 210.

Pietarinen, E.: 1978, private communication.

9. Pietarinen, E.: 1979, published in Handbook of Pion-Nucleon Scattering, Fachinformationszentrum Karlsruhe, 12-1.

10. Arndt, R., Hackman, R.H., and Roper, L.D.: 1977, Phys. Rev. C15,1002.

Arndt, R., private communication.

Lacombe, M., et al.: 1978, preprint IPNO/TH 78-46, Orsay.

11. Bugg, D., et al.: 1978, Journ. of Physics G4, 1025.

Bugg, D., private communication.

12. Jaffe, R.L.: 1977, Phys. Rev. D15, 267.

13. Lovas, I.: 1980, Journal of Physics G6, 179.

14. Borie, E., and Kaiser, F.: 1977, Nucl. Phys. B126, 173.

DISCUSSION

- C.B. Dover:

I was struck by the implications of your remarks (on the role of the three-pion continuum) for conventional one-boson-exchange models of the nucleon-nucleon potential. Some of the OBE models, for instance the very successful one due to the Nijmegen group, have a sizeable coupling to η and η' as well as a large value $f/g \approx 0.7$ for the ω exchange. As I understand it, this could be an artifact of having treated three-pion-exchange improperly. Do you agree ?

- P. Kroll :

Yes, I agree. As an example, let me explain how one can arrive at a large η coupling in the dispersion theoretic approach. Let us assume that we would not have at our disposal the $\Delta\sigma_L(pn)$ data. Then we could only compute $\Delta_S(pp)$ but not separate the isospins. $\Delta_S(pp)$ is negative (compare figure 2) and we would take this as the η and not call for 3π contributions. Fitting a pole term to $\Delta_S(pp)$, we would obtain $g^2_{\eta}/4\pi \simeq 4.7$.

- G. Violini :

Could you make some comment on the quality of the input data for the three forward contributing amplitudes ? And did you make any attempt to study by your method also the two non-forward contributing ones ?

- P. Kroll :

Our analysis is based on σ_{tot}, $\Delta\sigma_L$ data for pp and pn scattering as well as $\Delta\sigma_T(pp)$ and the pn charge exchange forward $d\sigma/dt$. Apart from $\Delta\sigma_L(pn)$ the quality of the data is good, the statistical errors are less than 5%. $\Delta\sigma_L(pn)$, obtained from proton-deuteron data, has statistical errors of about 10% and in addition it suffers from the uncertainties of the Glauber corrections which amount to a systematical error of about 1 mb for P_L = 1 to 2 GeV/c. At higher energies the systematical error drops down ripidly.

We have not attempted an investigation of non-forward amplitudes since above the phase shift region there is no input available.

A REVIEW OF RECENT KN AND $\bar{K}N$ PHASE SHIFT ANALYSES*

G. C. Oades
Institute of Physics - Aarhus University

1. Introduction

The aim of this review is to give a status report on developments in the phase shift analysis of the KN and $\bar{K}N$ systems from around the time of the 1976 Oxford Baryon Resonance Conference until the present time just before 1980 Toronto Baryon Resonance Conference. I will try to describe the significance of new experimental results which are available and set the stage for the new results which will appear at Toronto. The emphasis will reflect my personal interests, other recent reviews covering parts of the same field have been given by Dover (1) at the 1979 Houston Conference and by Gopal (2) at the 1979 Warsaw Conference.

Phase shift analyses have two main aims, a determination of the complex scattering amplitudes as functions of energy and angle and a determination of the resonance spectrum. It is by now well known that many ambiguities complicate such a phase shift analysis and it is worth remembering how these are dealt with.

1.1. "Energy Independent" Analyses

The requirements that the amplitudes are analytic functions can be used to restrict the number of solutions at each particular energy and then pick a suitable path through these solution from one energy to another. These techniques have been developed in connection with πN phase shift analyses, the Karlsruhe group (3) making use of fixed-t and fixed-angle analyticity while the Carnegie-Mellon-LBL analysis (4) makes use of fixed-s analyticity and smooths by use of hyperbolic dispersion relations. The πN situation will be reviewed by Dr. Koch so I will not give more details here but only point out that similar methods can be used in KN and $\bar{K}N$ analyses. Up to now the only examples are the K^+p analysis of Cutkosky et al. (5) which makes use of fixed-s analyticity and smooths by use of partial wave dispersion relations and the K^+p and $\bar{K}N$ analyses of the Aarhus group (6), (7) which make use of Pietarinen's fixed-t amplitudes (8) to pick a "unique" solution. Further work along these lines is underway and it is hoped that new results will be ready for Toronto.

E. Ferrari and G. Violini (eds.), Low and Intermediate Energy Kaon-Nucleon Physics, 53–71.

1.2. "Energy Dependent" Analyses

In these type of analyses the amplitudes are represented by some suitably chosen parametric form and the parameters are varied to simultaneously fit the whole data set. In some cases only one solution is found and in other cases several solutions appear. Examples of this type of KN analysis are the old BGRT analysis (9) which parametrizes δ and η, the Martin analysis (10) which parametrizes the inverse partial wave amplitudes and the K^+p analysis of Arndt et al. (11) which uses a two channel K-matrix parametrization. In all of these KN cases there is no complicated resonance structure and the parametrizazions are simply chosen so as to be as flexible as possible without violating unitarity. Because of the many S = -1 resonances the situation in the $\bar{K}N$ case is a little different and here it is usual to chose a parametrization giving factorizing resonances plus some form of background. The RLIC analysis (12) fits $\bar{K}N$-$\bar{K}N$, $\bar{K}N$-$\pi\Sigma$ and $\bar{K}N$-$\pi\Lambda$ data with coupled channel Breit-Wigner resonances plus separately parametrized background amplitudes for each channel while the Martin and Pidcock analysis (13) fits the multichannel data with a K-matrix parametrization. Both of these analyses hope that the multichannel fit gives a better determination of the resonance spectrum but it is possible that the background amplitudes do not have enough freedom to reproduce all the details of the scattering amplitudes in each channel. The later $\bar{K}N$ LBL analysis (14) also uses Breit-Wigner resonances plus background amplitudes but by concentrating on $K^-p \rightarrow K^-p$ and $K^-p \rightarrow \bar{K}^o n$ hopes to get a better fit to the $\bar{K}N$-$\bar{K}N$ amplitudes.

As a general rule of thumb one can say that the energy independent analyses will give a better determination of the scattering amplitudes provided plentiful and accurate experimental data are available while energy dependent analyses are more suited to situation with limited data. In particular multichannel energy dependent analyses are well suited to determining the main features of the resonance spectrum but probably give a poorer fit to the scattering amplitudes in the individual channels.

I will now consider the situation for K^+p, I = 0 KN and I = 0 and 1 $\bar{K}N$-$\bar{K}N$ separately.

2. $K^+p \rightarrow K^+p$

Data which have appeared since the analyses of Fich et al. (6) and Cutkosky et al. (5) are

(i) $\frac{d\sigma}{d\Omega}$ at 12 momenta in the range 130-755 MeV/c measured in the CERN 81 cm hydrogen bubble chamber by Cameron et al. (15).

(ii) $\frac{d\sigma}{d\Omega}$ at 5 momenta in the range 178-588 MeV/c measured in the BNL-Columbia 30" hydrogen bubble chamber by Burnstein et al. (16).

(iii) $\frac{d\sigma}{d\Omega}$ (180°) at 11 momenta in the range 1.0 to 1.5 GeV/c measured in a counter experiment by Adams et al (17).

(iv) $\frac{d\sigma}{d\Omega}$ at 22 momenta in the range 0.865 to 2.125 GeV/c measured in a very high statistics counter experiment at ANL by Abe et al. (8).

(v) $\frac{d\sigma}{d\Omega}$ and polarization near the forward and backward directions at 5 momenta in the range 1.7 to 3.0 GeV/c measured in a counter experiment at BNL by Patton et al. (19).

(vi) Polarization at 4 momenta in the range 650-940 MeV/c measured in a counter experiment at BNL by Zeller et al. (20).

All this new data with exception of (iii) is included in the energy dependent analysis of Arndt et al. (11) which is an update of an earlier analysis by the same group (21). A two channel K-matrix parametrization is used in the range 0-2 GeV/c, inelasticity being assumed to be in the KΔ channel for all partial waves except S11 where the s-wave K^*p channel is used. No attempt is made to include any KN-KΔ and KN-K^*N data in the fit even though meausurements are available, the latest being those of the BGRT group (22). Excluding some old data, Arndt et al. end up with two very similar solutions H and L, with χ^2/(data point) of 1.73 and 1.72 respectively for a data set with 3006 points. After allowing renormalization of the data these values fall to 1.25 and 1.26 for H_r and L_r respectively. In view of the size of some of the renormalizations the authors prefer to discuss solutions H and L. The Argand diagrams of these solutions are shown in figure 1.

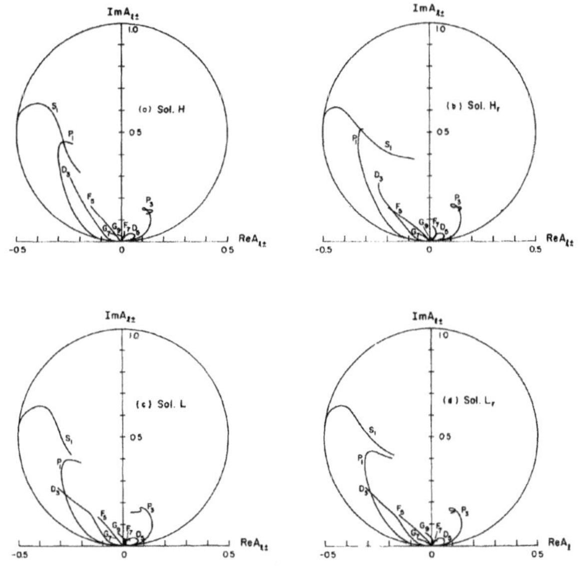

Figure 1 - Argand diagrams for the K^+p partial wave amplitudes of Arndt et al. (11).

The earlier solution of this group was criticised for the smallness of the low energy p-waves (23). Here the low energy polarization, which measures the interference of the p-waves with the dominant repulsive s-wave, plays a crucial role. Indeed, the inclusion of the new polarization data (vi) resulted in an increase in the low energy p-waves. Another general criticism is that allowing the possibility of renormalization is always dangerous in energy dependent fits. The large values of χ^2/(data point) before renormalization suggest that the parametrization is too inflexible and that some of the renormalizations are removing details which sould be fitted.

The combined I = 0 and 1 KN analysis of Martin (10) fits a selection of K^+p data in the range 0-1.5 GeV/c with an inverse amplitude parametrization. The data set includes the low energy data (i)and a selection of the high statistics data (iv) and the fit has a χ^2/(data point) of 1.08 for 1715 K^+p data points after making some small renormalizations. The fit does not include the low energy data (ii) nor does it use the low energy polarization data (vi). A comparison of the predictions of Martin's solution with this polarization data is shown in figure 2. The χ^2/(data point) is 1.21 so even without minimization the fit cannot be said to be bad. Argand diagrams are shown in figure 3.

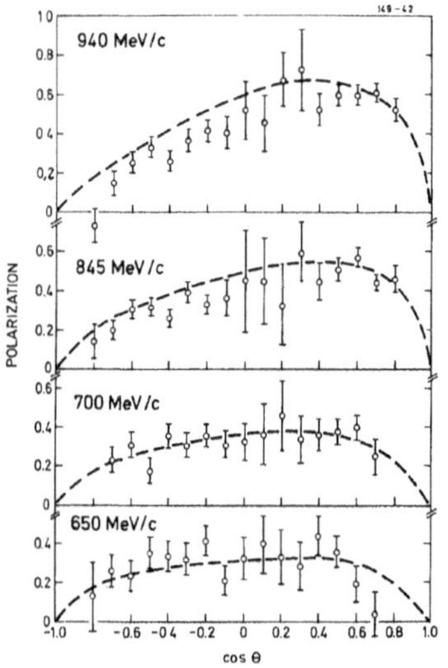

Figure 2 - The predictions of Martin's K^+p solution (10) compared to the polarization data (20).

A comparison of figs. 1 and 3 shows that in both cases the p_{13} amplitude shows an anticlockwise movement which has sometimes been interpreted as indicating the presence of a Z_1^* resonance even though there is no speed maximum. Indeed, the parametrization of Arndt et al. has an unphysical sheet pole at (1796-101i) MeV and that of Martin has an unphysical

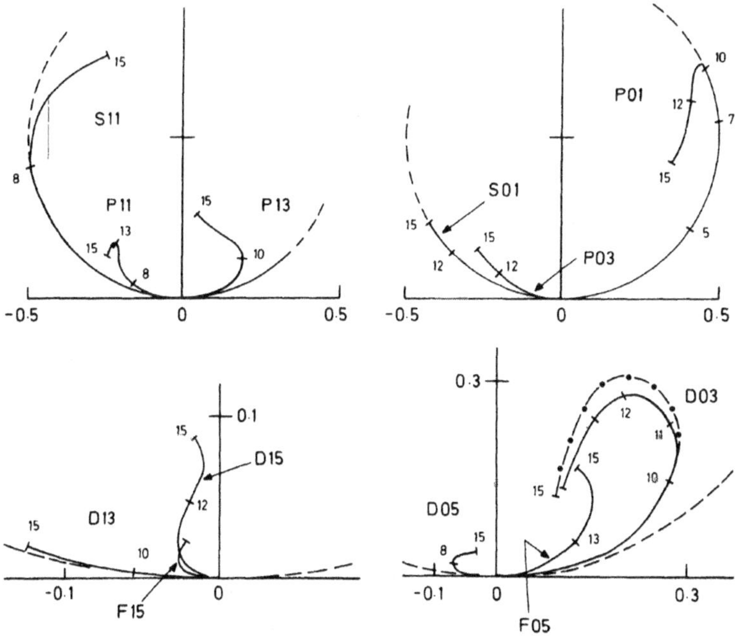

Figure 3 - Argand diagrams for Martin's KN amplitudes (10).

sheet pole at (1820-134i) MeV. If this is to be interpreted as a resonance rather than a consequence of the particular parametric forms used then the coupling to the KN channel is small and we would consequently expect a large coupling to some other channel, presumably KΔ. An analysis of KN-KΔ data by the BGRT group (24) also finds signs of an anticlockwise motion in the p_{13} KN-KΔ Argand diagram but again the coupling is not large and there is no speed maximum. In view of this, the general opinion is that the Argand diagram of the p_{13} KN-KN amplitude is simply due to the combination of a weak attraction with increasing inelasticity due to the opening of the KΔ channel.

3. I = 0 KN

Since the analysis of Martin (10), several experiments have been published reporting data which can be used to determine the I = 0 KN-KN scattering amplitudes. Experiments with deuteron targets :-

(i) $\frac{d\sigma}{d\Omega}$ (180°) for $K^+d \rightarrow K^+n(p)$ at 11 momenta in the range 1.0 to 1.5 GeV/c measured in a counter experiment by Adams et al. (17).

(ii) $\frac{d\sigma}{d\Omega}$ at 3 momenta in the range 342-587 MeV/c for the elastic process $K^+d \rightarrow K^+d$ and for the breakup process $K^+d \rightarrow K^+np$ and at 4 momenta in the range 252-587 MeV/c for the charge exchange breakup process $K^+d \to K^o pp$ measured in BNL-Columbia 30" deuterium bubble chamber by Glasser et al. (25).

(iii) $\frac{d\sigma}{d\Omega}$ at 3 momenta in the range 690-890 MeV/c for $K^+d \to K^o pp$ measured in the BNL-Columbia 30" deuterium bubble chamber by Sakitt et al.(26).

(iv) Polarization for $K^+d \to K^+n(p)$ and $K^+d \to K^o p(p)$ at 5 momenta in the range 851-1351 MeV/c measured using a polarized deuterium target by Robertson et al. (27).

Experiments using a K_L^o beam :-

(v) $\frac{d\sigma}{d\Omega}$ for $K_L^o p \to K^+n$ in 4 momentum bins in the range 500-1000 MeV/c measured in a counter experiment at ANL by Edelstein et al. (28).

(vi) $\frac{d\sigma}{d\Omega}$ for $K_L^o p \to K^+n$ in 9 momentum bins in the range 600-1500 MeV/c measured in a counter experiment at DNPL by Armitage et al. (29).

The I = 0 and 1 KN solution of Martin (10) fitted considerably more data than the older BGRT analysis (9) so I will mainly confine my comments and comparsions to Martin's solution, the Argand diagrams of which are shown in figure 3. The I = 0 amplitudes were determined from deuteron data only where, in the plane wave impulse approximation, the following formulae hold :-

$$\frac{d\sigma}{d\Omega}(K^+d \to K^+d) = \left[\,|f_p+f_n|^2 + \tfrac{2}{3}|g_p + g_n|^2\right] D$$

$$\frac{d\sigma}{d\Omega}(K^+d \to K^o pp) = \left[\,|f_{cx}|^2\right](I_o - J_o) + \left[\,|g_{cx}|^2\right](I_o - \tfrac{1}{3}J_o)$$

$$\frac{d\sigma}{d\Omega}(K^+d \to K^+np) = \left[\,|f_p|^2+|f_n|^2+|g_p|^2+|g_n|^2\right] I_o + 2\left[\mathrm{Re}f_n^* f_p + \tfrac{1}{3}\mathrm{Re}g_n^* g_p\right] J_o$$

$$\frac{d\sigma}{d\Omega}(K^+d \to K^+n(p)) = \left[\,|f_n|^2 + |g_n|^2\right] I_o$$

$$\frac{d\sigma}{d\Omega}(K^+d \to K^+p(n)) = \left[\,|f_p|^2 + |g_p|^2\right] I_o$$

where f_p and g_p are the spin non-flip and spin flip amplitudes for $K^+p \to K^+p$ (including the electro-magnetic amplitudes), f_n and g_n are the corresponding amplitudes for $K^+n \to K^+n$, and f_{cx} and g_{cx} are the corresponding amplitudes for $K^+n \to K^o p$. D is the elastic deuteron form factor and I_o and J_o are the inelastic deuteron form factors. They are functions of energy and angle and can be calculated from a given deuteron wave function. Martin uses the same form factors as BGRT, typical values for $I_o \pm J_o$ being shown in figure 4.

The elastic form factor D falls off quickly away from the forward direction so the elastic $K^+d \to K^+d$ data is limited to that region.Neither Martin nor BGRT attempt to use this data in their fit but check later

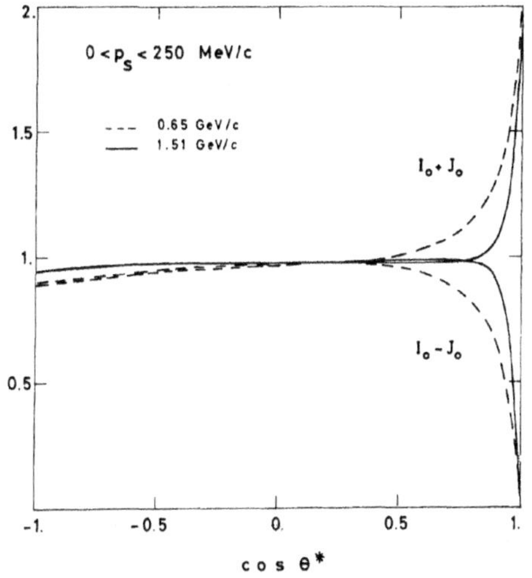

Figure 4 - The inelastic deuteron form factors $I_o \pm J_o$ as used by Martin (10) and BGRT (9).

that there is qualitative agreement. Martin quotes a χ^2/(data point) of 2.5 for 141 $K^+d \to K^+d$ data points which is considerably worse than the fits to the breakup or charge exchange data which have χ^2/(data point) of 1.65 for 311 data points and 1.26 for 470 data points respectively. He suggests that uncertainties on the very rapidly varying form factor can account for the poor fit. In the case of the breakup data, only the $K^+d \to K^+n(p)$ differential cross sections are included in the analysis since the genuine K^+p elastic data is available instead of $K^+ \to K^+p(n)$. In both cases the form factor I_o reduces the differential cross section compared to the genuine elastic value by at most 15% to 20%. Elastic $K^+p \to K^+p$ data have been used to check the values of I_o by both the Birmingham-RHEL group (30) and by BGRT (31) and the agreement seems satisfactory. Martin comments are an apparent problem with the Birmingham-RHEL values for the $K^+d \to K^+n(p)$ differential cross sections and if he removes them from his data set the χ^2/(data point) for the "elastic" K^+n data falls from 1.65 to 1.28. This is a problem which awaits further clarification.

In the case of the charge exchange data the form factors play a more important role since the Pauli exchange principle suppress the forward direction as compared to the charge exchange process on free neutrons. It is here that the new $K_L^o p \to K^+n$ data are important since no such effects are present in this case. A comparison of the predictions of Martin's solution with the data of the Daresbury-Manchester group (29) gives a χ^2/(data point) of 3.46. This large χ^2 is not confined to the forward direction but is distributed over the whole angular range.

Fits at 650 MeV/c, 750 MeV/c and 950 MeV/c are compared with the Daresbury-Manchester data and with the data of ref. (28) in figure 5. It will be seen that the predictions are showing a turn over near the forward direction which is not shown by the $K^o_Lp \rightarrow K^+n$ data and that possibly there are normalization problems associated with the K^o_L beam. Is this trouble due to the amplitudes trying to compensate for effects which should have been taken care of by the deuteron form factors or is there trouble with the K^o_Lp data ? Theoretical work along the lines described by Martin (32) could help to resolve this problem.

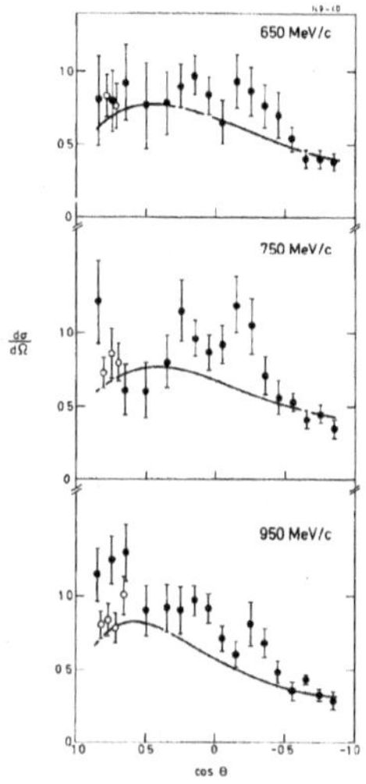

Figure 5 - The predictions for Martin's KN amplitudes (10) compared with the $K^o_Lp \rightarrow K^+n$ data of (28) (○) and of (29) (●).

Finally let us consider the polarization data. Up to now the selection of the type of KN I = 0 solution has been based on theoretical prejudice plus 5 polarization points at 600 MeV/c. For this reason the new RHEL-QMC polarized deuteron data (27) are extremely valuable. The predictions of Martin's solution are compared with the $K^+d \rightarrow K^+n(p)$ data in figure 6 and with the $K^+d \rightarrow K^op(p)$ data in figure 7. χ^2/(data points) is 1.37 for the "elastic" data and 1.13 for the charge exchange data. As can be seen from the figures the agreement is, at least qualitatively, good and it will be very interesting to see the results of a new fit including this polarization data.

The situation regarding the existence of a Z^*_o resonance has not changed since 1976. As can be seen from figure 3 the P_{01} KN-KN partial wave amplitude is quite large and the Argand diagram shows an anticlockwise motion. Is this due to the existence of an exotic Z^*_o resonance or is it due to a fairly strong attractive force in the elastic channel combined with the opening of inelastic channels ? Martin's P_{01} amplitude

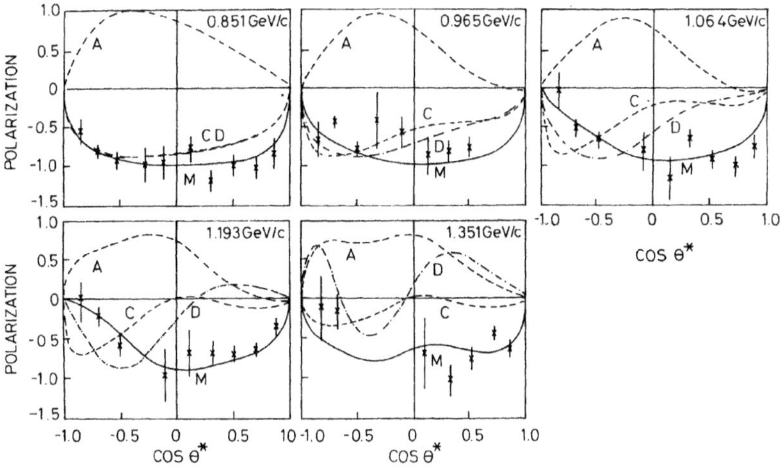

Figure 6 - The predictions of Martin's KN amplitudes (10) compared with $K^+d \rightarrow K^+n(p)$ polarization data (27)

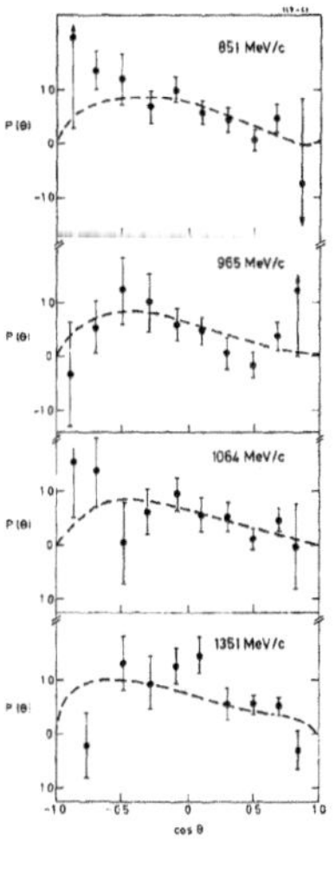

Figure 7 - The predictions of Martin's KN amplitudes (10) compared with preliminary $K^+d \rightarrow K^0p(p)$ polarization data (27)

does have an unphysical sheet pole but it is rather far from the real axis so it is impossible to say whether it is physically significant or whether it is simply a consequence of the particular parametrization used.

4. I = 0 and 1 $\bar{K}N$

I will confine my attention to the $\bar{K}N$ channels and will not attempt to review the situation for the other two body channels. Details of these can be found in Gopal's review (2). Recent experiments have provided the following data :-

(i) $\frac{d\sigma}{d\Omega}$ $(K^-p \rightarrow K^-p)$ at 4 momenta in the range 1.423 to 1.815 GeV/c measured at ANL in a high statistics counter experiment by Abe et al. (33).

(ii) $\sigma_{tot}(K^-p)$ and $\sigma_{tot}(K^-p)$ measured at 36 momenta in the range 410-1070 MeV/c at BNL by Carroll et al. (34).

(iii) $\frac{d\sigma}{d\Omega}$ $(K^-p \rightarrow K^-p)$ in 25 momentum bins and $\frac{d\sigma}{d\Omega}$ $(K^-p \rightarrow \bar{K}^on)$ in 23 momentum bins in the range 225 to 465 MeV/c measured in the Berkeley 25" hydrogen bubble chamber by Mast et al. (35).

(iv) $K^-p \rightarrow K^-p$ polarization at 8 momenta in the range 650-1071 MeV/c measured at BNL by Ehrlich et al. (36).

(v) $\frac{d\sigma}{d\Omega}$ $(K^-d \rightarrow K^-n(p))$ and $\frac{d\sigma}{d\Omega}$ $(K^-p \rightarrow K^-p(n))$ measured at 14 momenta in the range 610-940 MeV/c in a counter experiment at RHEL by Damerell et al. (37).

(vi) $\sigma(K^-p \rightarrow \bar{K}^on)$ at 48 momenta in the range 515-1065 MeV/c measured in a counter experiment at BNL by Alston-Garnjost et al. (38).

(vii) $\frac{d\sigma}{d\Omega}$ $(K^-p \rightarrow \bar{K}^on)$ at 23 momenta in the range 515-956 MeV/c measured in a counter experiment at BNL by Alston-Garnjost et al. (39).

(viii) $\frac{d\sigma}{d\Omega}_{180^o}$ $(K^-p \rightarrow K^-p)$ at 49 momenta in the range 476-1084 MeV/c measured in a counter experiment at BNL by Alston-Garnjost et al. (40).

(ix) $K^-n \rightarrow K^-p$ polarization at 9 momenta in the range 955-1272 MeV/c measured by Bryant et al. (41).

Most of this new data (but not (i), (v), (viii) and (ix)) have been included in a new energy dependent partial wave analysis in the range 360-1320 MeV/c by Alston-Garnjost et al. (14). This LBL analysis concentrates on the $\bar{K}N$-$\bar{K}N$ I = 0 and 1 channels and uses partial wave parametrizations consisting of non-resonant background amplitudes plus Breit-Wigner resonances and makes a special point of including the $\eta\Lambda$ and $\eta\Sigma$ s-wave threshold cusps. Argand diagrams comparing the LBL s and p-wave amplitudes with the RLIC amplitudes are shown in figure 8, which gives some idea of the uncertainties still attached to the individual amplitudes.

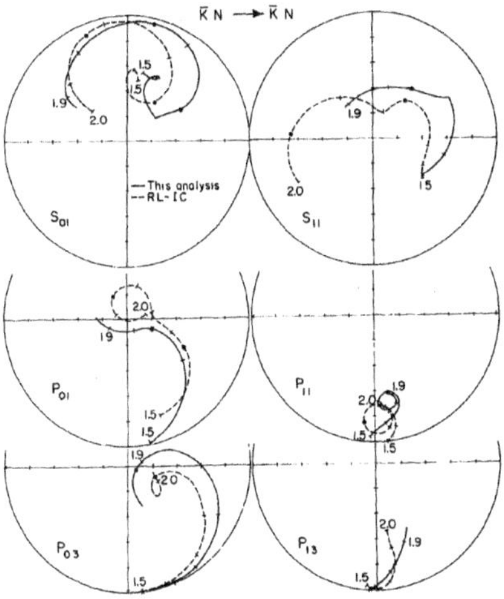

Figure 8 - Argand diagrams for the RLIC and LBL s and p-wave $\bar{K}N$-$\bar{K}N$ amplitudes (14).

The accurate K^-p and K^-d total cross section measurements (34) claimed evidence for 4 new Y_1^* states and 2 new Y_0^* states in the isospin separated total cross sections. These new states are not confirmed by this analysis which fits the K^-p σ_{tot} values very well but which suggests that there may be some trouble with extracting the K^-n σ_{tot} values from the K^-d measurements.

An interesting point of this analysis is that forward real part data for $K^-p \rightarrow K^-p$ and $K^-n \rightarrow K^-n$ calculated from the fit of A. Martin (42) is included in the fit. The input and the results of the fit are shown in figure 9.

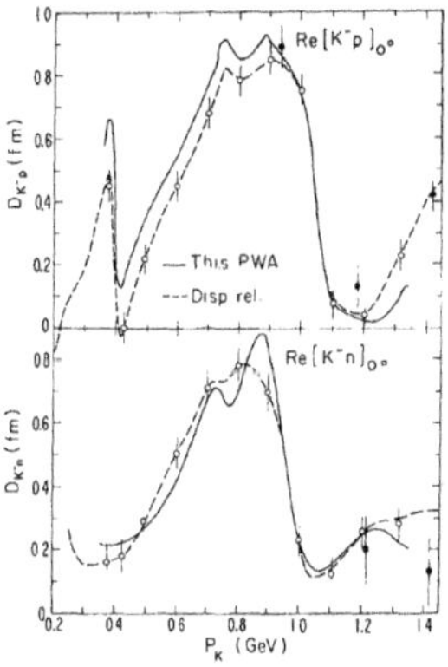

Figure 9 - LBL fit to the $K^-p \rightarrow K^-p$ and $K^-n \rightarrow K^-n$ forward real part predictions of (42).

The Birmingham-RHEL deuteron data (37) were not included in this fit which is a pity since their $K^-d \rightarrow K^-n(p)$ measurements provide pure I = 1 angular distributions. Allowing for 10% normalization uncertainties both the $K^-d \rightarrow K^-n(p)$ and $K^-d \rightarrow K^-p(n)$ data seem well fitted by the older RLIC amplitudes.

The new backward K^-p elastic measurements have been used to update the LBL analysis (40). Figure 10 shows the data compared to the predictions of a new fit where some of the higher energy background amplitudes are increased. The structure between 800 and 850 MeV/c coincides with one of Carroll et al.'s suggested new Y_1^* states. It can be fitted a little better by including an extra D_{13} resonance but the decrease in χ^2 is not significant.

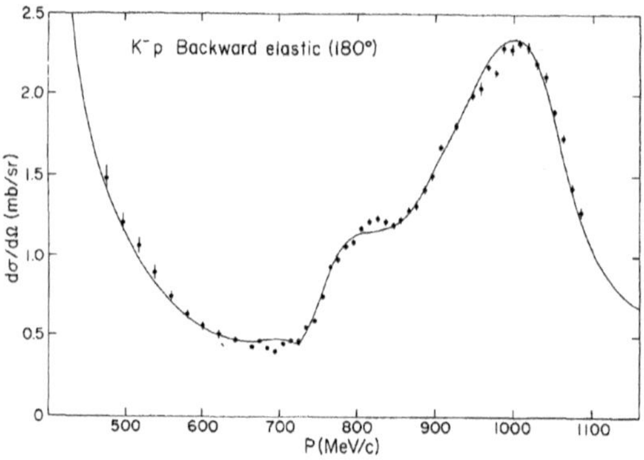

Figure 10 - LBL data and fit to the K^-p backward elastic cross section (40).

The polarization data from 955 to 1272 MeV/c (41) are not included in either the LBL analysis or its update but the LBL predictions seem to reproduce the data very well as can be seen from figure 11 so no surprises should be expected here.

In all analyses of the $\bar{K}N$-$\bar{K}N$ system the only polarization information so far available is for $K^-p \rightarrow K^-p$. Until some measurements of $K^-p \rightarrow \bar{K}^o n$ are made on a polarized proton target or of $K^-d \rightarrow K^-n(p)$ are made on a polarized deuteron target there will always be an unpleasant uncertainty associated with the amplitudes. Another problem concerns the region very close to the K^-p threshold where mass difference and other electromagnetic effects complicate the analysis. Figure 9 shows that the LBL $K^-p \rightarrow K^-p$ forward amplitude lies considerably over the dispersion relation values of A. Martin (42). Should we take this as a sign that the K^-p s-wave scattering length combination is not as negative as the low energy fits suggest and is this associated with the surprising results of the K-mesic atom measurements in hydrogen (43)? I hope we will hear various views on this point at this workshop.

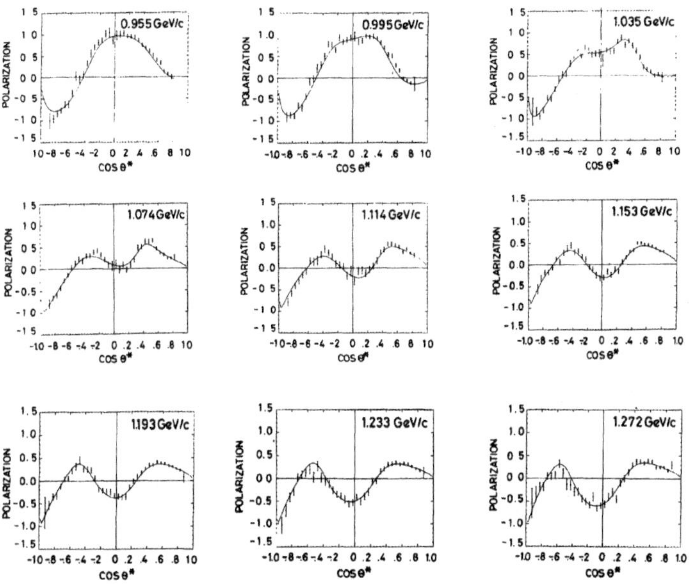

Figure 11 - $K^-p \rightarrow K^-p$ polarization data (41) compared to the LBL predictions.

I will not attempt to give a detailed review of the resonance spectrum of Λ and Σ states since this relies heavily on information from the other two body and quasi two body channels as well as from the $\bar{K}$N channels. Details can be found in the 1978 Review of Particle Properties and in many reviews. In summary the situation is :-

S_{01} As well as the below threshold Λ(1405), there is general agreement on a rather narrow Λ(1670) close to the ηΛ threshold and there are claims for a wider state with a more uncertain mass denoted in the tables by Λ(1870).

P_{01} The tables list two states Λ(1600) and Λ(1800). RLIC found two rather wide states at 1573 MeV and 1853 MeV while LBL claims a single very wide state at 1703 MeV.

P_{03} The Λ(1860) state is well established although its mass is probably close to 1900 MeV.

$\underline{D}_{03}$ The two states Λ(1520) and Λ(1690) are well established. The third D_{03} state required in the 70, 1^- SU(6) multiplet is expected to couple only very weakly to the $\bar{K}N$ channel and is only claimed by the older Aarhus analysis.

$\underline{D}_{05}$ The Λ(1815) has been well established for many years and a higher Λ(2110) state is claimed by some analyses.

$\underline{F}_{07}$ There is some evidence for a Λ(2020) state.

$\underline{G}_{07}$ There are claims for a very wide Λ(2100) state.

$\underline{S}_{11}$ The situation is very confused. The tables list three states. Σ(1620), Σ(1750) and Σ(2000) but the lowest of these is seen neither by RLIC nor by LBL.

$\underline{P}_{11}$ Also a very confused situation, three states Σ(1660), Σ(1770) and Σ(1880) being listed in the tables. LBL claims a single state at 1679 MeV and RLIC claim two states at 1676 MeV and 1738 MeV.

$\underline{P}_{13}$ Apart from the below threshold Σ(1385) there are some claims for a Σ(1840) state although neither RLIC nor LBL see such a resonance.

$\underline{D}_{13}$ The Σ(1670) is well established and there are some claims for a high mass Σ(1940).

$\underline{D}_{15}$ The Σ(1775) is well established.

$\underline{F}_{15}$ The Σ(1915) is well established.

$\underline{F}_{17}$ The Σ(2030) is rather well established state.

$\underline{G}_{17}$ There are some claims for a Σ(2100) state.

5. $K^o_L p \to K^o_s p$

Finally let me briefly mention new data on the regeneration process $K^o_L p \to K^o_s p$:-

(i) $\frac{d\sigma}{d\Omega}$ in 12 energy bins in the c.m. energy range 1540-1710 MeV, measured in the CERN 2m. hydrogen bubble chamber by Cameron et al. (44).

(ii) $\frac{d\sigma}{d\Omega}$ in the c.m. energy bin 1540-1610 MeV, measured in the ANL 12' hydrogen bubble chamber by Engler et al. (45).

(iii) $\frac{d\sigma}{d\Omega}$ in 9 energy bins in the c.m. energy range 1605-1910 MeV measured in the CERN 2m. hydrogen bubble chamber by Cordon et al. (46).

The regeneration process is very interesting since the amplitude is given by

$$T(K^o_L p \rightarrow K^o_s p) = \frac{1}{2}(T(K^+n \rightarrow K^+n) - T(K^-n \rightarrow K^-n))$$

i.e. it provides a direct measure of the antisymmetric $K^{\pm}n$ scattering amplitude. (Similar measurements of regeneration on neutrons would provide a direct measure of the $K^{\pm}p$ antisymmetric scattering amplitude). The problem is, of course, that any analysis must deal with both the K^+n and K^-n amplitudes. The usual procedure is to compare the data with the predictions of combinations of K^+n and K^-n amplitudes, the usual choice being RLIC K^-n and Martin or BGRT K^+n. The general conclusions are that RLIC + Martin reproduce the low energy data very well, as can be seen in figure 12, but that there is more uncertainty at the high energy end, as shown in figure 13.

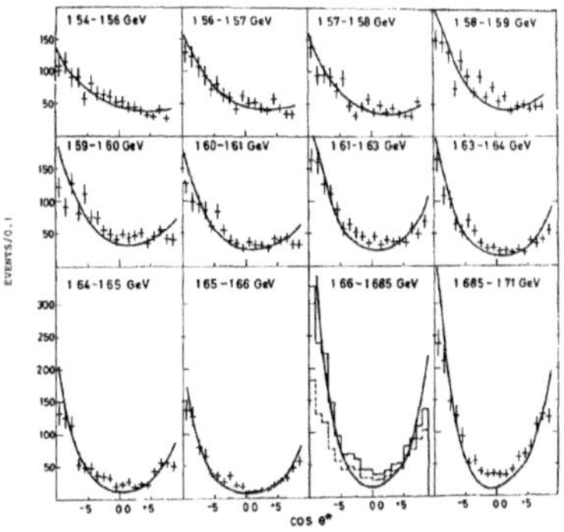

Figure 12 - Comparison of the BEGPR data for $K^o_L p \rightarrow K^o_s p$ (44) with the predictions of Martin (10) plus RLIC (12).

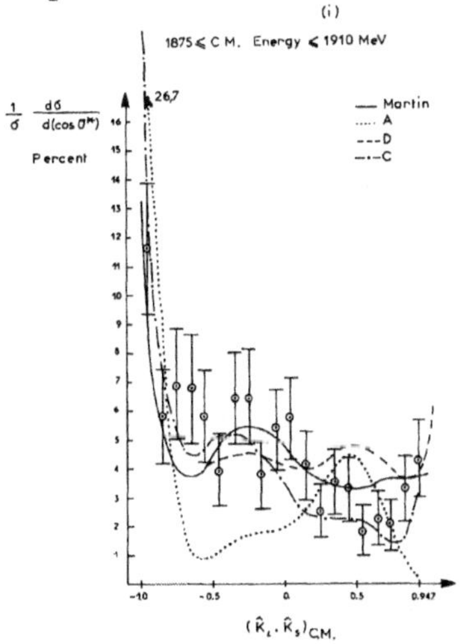

Figure 13 - Comparison of the highests energy Birmingham-Paris data (46) with the predictions of Martin (10) plus RLIC (12) or BGRT (9) solutions A, C and D plus RLIC (12).

6. Conclusions

To briefly sum up, the outstanding experimental problems are :-

(i) The conflict between the $K^+d \to K^o pp$ and $K^o_L p \to K^+n$ differential cross sections.

(ii) The lack of $K^-p \to \bar{K}^o n$ or $K^-n \to K^-n$ polarization measurements.

(iii) The join onto the very low energy and threshold $\bar{K}N$ data.

On the theory side there is much which could be done to help the analysis of the scattering data. Many techniques have been developed to analyse πN scattering and these should be adapted and applied to the coupled KN and $\bar{K}N$ systems.

References and notes

*. Work supported by a grant from the Danish Natural Science Research Council.

1. Dover, C.B.: 1979, Brookhaven preprint, BNL-25961.
2. Gopal, G.P.: 1979, RHEL preprint, RL-79-071.
3. Höhler,G., Kaiser, F., Koch, R., and Pietarinen, E.: 1979, "Handbook of Pion-Nucleon Scattering", Physics Data 12-1.
4. Cutkosky, R.E., et al.: 1979, Phys. Rev. D20, 2804.
5. Cutkosky, R.E., et al.: 1976, Nucl. Phys. B102, 139.
6. Fich, O., Oades, G.C., and Pietarinen, E.: 1974, Proceedings of the International Conference on High Energy Physics.
7. Hansen, P.N., Moss, J., and Oades, G.C.: 1976, Proceedings of the Oxford Baryon Resonance Conference, 275.
8. Pietarinen,E.: 1974, Nucl. Phys. B76, 231.
9. Giacomelli, G., et al.: 1974, Nucl. Phys. B71, 142.
10. Martin, B.R.: 1975, Nucl. Phys. B94, 426.
11. Arndt, R.A., Roper, L.D., and Steinberg, P.H.: 1978, Phys. Rev. D18, 3283.
12. Gopal, G.P., et al.: 1977, Nucl. Phys. B119, 362.
13. Martin, B.R., and Pidcock, M.K.: 1977, Nucl. Phys. B126, 266.
14. Alston-Garnjost, M., et al.: 1978, Phys. Rev. D18, 186.
15. Cameron, W., et al.: 1974, Nucl. Phys. B78, 93.
16. Burnstein, R.A., et al.: 1974, Phys. Rev. D10, 2767.
17. Adams, U., et al.: 1975, Nucl. Phys. B87, 41.
18. Abe, K., et al.: 1975, Phys. Rev. D11, 1719.

19. Patton, R., et al.: 1975, Phys. Rev. Letters, 34, 975.

20. Zeller, M.: 1978, private communication to R. Kelly, reported in Kelly's 1978 review,Berkeley preprint LBL-7976.

21. Arndt, R.A., et al.: 1974, Phys. Rev. Letters 33, 987.

22. Giacomelli, G., et al.: 1976, Nucl. Phys. B111, 365.

23. Martin, B.R.: 1976, Proceedings of the Oxford Baryon Resonance Conference, 409.

24. Giacomelli, G., et al.: 1976, Nucl. Phys. B110, 67.

25. Glasser, R.G., et al.: 1977, Phys. Rev., D15, 1200.

26. Sakitt, M., et al.: 1977, Phys. Rev. D15, 1846.

27. Robertson, A.W., et al.: 1980, RHEL preprint, RL-80-001, and private communication from D.V. Bugg; 1980, Phys. Letters 91B, 469.

28. Edelstein, R.M., et al.: 1976, Phys. Rev. D14, 702.

29. Armitage, J.C.M., et al.: 1977, Nucl. Phys. B123, 11.

30. Damerell, C.J.S., et al.: 1975, Nucl. Phys. B94, 374.

31. Wilson, B.C.: 1972, in K-Meson Physics Below 5 GeV/c RHEL report RHEL/R 245.

32. Martin, B.R.: 1977, J. Phys. G3, 1.

33. Abe, K., et al.: 1975, Phys. Rev. D12, 6.

34. Carroll, A.S., et al.: 1976, Phys. Rev. Letters 37, 806.

35. Mast, T.C., et al.: 1976, Phys. Rev. D14, 13.

36. Ehrlich, R.D., et al.: 1977, Phys. Letters 71B, 455.

37. Damerell, C.J.S., et al.: 1977, Nucl. Phys. B129, 397.

38. Alston-Garnjost, M., et al.: 1978, Phys. Rev. D17, 2216.

39. Alston-Garnjost, M., et al.: 1978, Phys. Rev. D17, 2226.

40. Alston-Garnjost, M., et al.: 1979, Berkeley preprint LBL-9083; 1980, Phys. Rev. D21, 1203.

41. Bryant, H.C., et al.: 1979, RHEL preprint RL-79-078.

42. Martin, A.: 1976, Phys. Letters 65B, 346; 1978,Phys. Rev. D18, 194.

43. Davies, J.D., et al.: 1979, Phys. Letters 83B, 55.

44. Cameron, W., et al.: 1978, Nucl. Phys. B132, 197.

45. Engler, A., et al.: 1978, Phys. Rev. D18, 3061.

46. Corden, M.J., et al.: 1979, Nucl. Phys. B155, 36.

DISCUSSION

- G. Violini :

I have one question and one comment. First about the discrepancy between $K^o p \to K^+ n$ data and B. Martin's analysis. Would it not be possible that this data, if included in the fit, could be fitted through some renormalization? Second, about the inclusion of dispersion relations as constraints in phase shift analysis. It seems to me that even if this is a very popular procedure one should be very cautious. In fact it is certainly highly meaningful the point-to-point comparison of $D(\omega)$, but this will mainly depend on σ_{tot} in the region which is in general already included in the fit. Alternatively one can wish to include this information from far aways cross-sections but then just a few values of $D(\omega)$ would be included giving them their correct weight (this means without factors which reduce it).

- G. Oades :

The $K^o_L p \to K^+ n$ data even if renormalized by up to 25% are still not very well fitted by B. Martin's phases. There seems to be problems both with the shape and with the normalization.

There are several reasons for including dispersion relation predictions as constraints on phase shift analyses. I agree that only a few values for the forward real part D need be included provided σ_{tot} is also fitted. This is, however, not the whole story. Lack of polarization data and of A and R measurements give the possibility of considerable ambiguity which can be reduced by including dispersion relation predictions for both invariant amplitudes at several angles and energies.

- M. Svec :

At present there exists no data on $K^+ n \to K^+ n$ or $K^+ n$ CEX reactions for $P_{LAB}=1{\cdot}5 \to 2{\cdot}5$ GeV/c. Do you know of any proposals to make measurements in this region? Obviously such data would help to extend kaon-nucleon phase shift analysis for isospin I=0 into the same energy region as that for isospin I=1.

- G. Oades :

I do not know of any plans to extend the $K^+ n \to K^+ n$ or $K^+ n \to K^o p$ data to higher energies.

- F. Takasaki :

At KEK we plan to make K^+n experiments up to P_{LAB}=2.2 GeV/c.

- C. B. Dover :

Cottingham proposed a change of sign of the $S_{\frac{1}{2}}$ K^+N phase shift at low momentum. Is this consistent with the data on dispersion relations?

- G. Oades :

The fixed-t dispersion relations predict a negative I = 0 $S_{\frac{1}{2}}$ KN scattering amplitude and could not accept the Cottingham et al. suggestion of a low energy sign change and a large positive scattering length.

- C. B. Dover :

Why are higher standards applied to Z^* resonances than to N^* resonances? The loop in the I = 0, $p_{\frac{1}{2}}$ Argand plot for K^+n phase shifts seems more suggestive of a resonance than some very wide N^* structures which are regarded as well established.

- G. Oades :

The standards applied to resonances are of course vey subjective. The KN I = 0 $p_{\frac{1}{2}}$ amplitude describes an anticlockwise motion in the Argand plot as expected of a resonance but it does not show the same speed variation as shown by similar N^* structures. This suggests that the second sheed structure is more complicated than a pair of resonance poles.

THE LOW ENERGY FIXED-t KN AND $\bar{K}$N AMPLITUDES*

G.C. Oades
Institute of Physics, Aarhus University

1. Introduction

The use of fixed-t dispersion relations has greatly helped the analysis of πN scattering data and it is clear that such dispersion relations can be of similar help in analysing the KN and $\bar{K}$N data. Pietarinen (1) has given a convenient method for constructing fixed-t amplitudes which have the correct analyticity properties and which reproduce the experimental scattering data. He shows how the problem can be formulated as the determination of two functions $h_\pm(z)$ which are real analytic in the z plane cut along the real axis and where $|h_\pm(z)|$ is given in terms of $\frac{d\sigma}{d\Omega}$ (1±P) for z > 1 (s-channel) and z < -1 (u-channel). For negative t values the cut due to the s-channel process extends from 1 down to some threshold value z_s and in this region $\mathrm{Im}h_\pm(z)$ can be calculated by continuation from physical scattering angles provided t is not too negative. Similarly the cut due to the u-channel process extends from -1 to z_u and again $\mathrm{Im}h_\pm(z)$ can in principle be calculated by continuation. Poles due to s-channel stable particles lie to the left of z_s and those due to u-channel particles lie to the right of z_u. As t becomes more negative z_s moves to the left and z_u moves to the right and the cuts eventually overlap.

Pietarinen and I have attempted to apply this method to the KN and $\bar{K}$N systems where we use existing phase shift solutions to interpolate $\frac{d\sigma}{d\Omega}$ and to interpolate or predict P so as to calculate $|h_\pm(z)|$ for a particular fixed value of t. This information is supplemented by high energy data where available. The main problem comes in contructing values of $\mathrm{Im}h_\pm(z)$ for $z_s < z < 1$ and $-1 < z < z_u$ where the scattering angle is unphysical. In the s-channel (K^+p or K^+n) case we can use existing partial wave amplitudes to continue to the required unphysical values. In the u-channel (K^-p or K^-n) case a similar procedure is possible down to the $\bar{K}$N threshold but in the region between the $\bar{K}$N and ηΛ thresholds some model must be used.

It soon became clear that at present the method is subject to far too much uncertainty due to the badly known low energy and unphysical

E. Ferrari and G. Violini (eds.), Low and Intermediate Energy Kaon-Nucleon Physics, 73–95.

region amplitudes and for this reason we have postponed the analysis until better low energy information can be obtained. In the πN case the low energy amplitudes were reliably determined by the use of finite contour fixed-t dispersion relations (2) and so it was decided to try the same method for the KN and $\bar{K}N$ amplitudes. This report describes the first results of such calculations, examines the special case of the forward direction, makes some comments on the values of σ^{KK}_{pp} and σ^{KK}_{nn} and looks at some of the predictions for the low energy partial wave amplitudes.

2. Method

As well as the invariant amplitudes A and B we will also use the amplitude C defined in the usual way by

$$C = A + \frac{\nu B}{1-t/4M_N^2} \tag{2.1}$$

where ν is given by

$$\nu = \frac{s-u}{4M_N} \tag{2.2}$$

Symmetric and antisymmetric combinations for the $K^{\pm}p$ or $K^{\pm}n$ amplitudes are defined by

$$A^{(\pm)} = \frac{1}{2}\left(A_{\bar{K}N} \pm A_{KN}\right) \tag{2.3}$$

with similar definitions for $B^{(\pm)}$ and $C^{(\pm)}$. The amplitudes $A^{(+)}$, $B^{(-)}$ and $C^{(+)}$ are symmetric under $s \leftrightarrow u$ crossing while $A^{(-)}$, $B^{(+)}$ and $C^{(-)}$ are antisymmetric.

For a fixed value of t the symmetric amplitude $A^{(+)}$ obeys a fixed-t dispersion relation of the form

$$\mathrm{Re}A^{(+)}(\nu,t) = A_B^{(+)}(\nu,t) + \frac{2P}{\pi}\int_{\nu_{\pi\Lambda}}^{\infty} \frac{\nu' \mathrm{Im}A^{(+)}(\nu',t)}{\nu'^2-\nu^2}\, d\nu' \tag{2.4}$$

with similar forms for $B^{(-)}$ and $C^{(+)}$. The antisymmetric amplitude $A^{(-)}$ obeys a fixed-t dispersion relation of the form

$$\mathrm{Re}A^{(-)}(\nu,t) = A_B^{(-)}(\nu,t) + \frac{2\nu P}{\pi}\int_{\nu_{\pi\Lambda}}^{\infty} \frac{\mathrm{Im}A^{(+)}(\nu',t)}{\nu'^2-\nu^2}\, d\nu' \tag{2.5}$$

with similar forms for $B^{(+)}$ and $C^{(-)}$. The lower integration limit $\nu_{\pi\Lambda}$ corresponds to the $\pi\Lambda$ threshold

$$\nu_{\pi\Lambda} = \left((M_\Lambda+m_\pi)^2 - M_N^2 - m_K^2 + t/2\right)/(2M_N) \tag{2.6}$$

and $A_B^{(\pm)}$, $B_B^{(\pm)}$ and $C_B^{(\pm)}$ are the hyperon pole terms. For the $K^{\pm}p$ case we combine the Σ and Λ poles to a common pole at the Λ position and write

$$A_B^{(\pm)}(\nu,t) = \frac{M_N-M_\Lambda}{4M_N} \cdot G^2_{YNK} \left(\frac{1}{\nu_\Lambda-\nu} \pm \frac{1}{\nu_\Lambda+\nu}\right) \tag{2.7}$$

$$B_B^{(\pm)}(\nu,t) = \frac{G^2_{YNK}}{4M_N} \left(\frac{1}{\nu_\Lambda-\nu} \mp \frac{1}{\nu_\Lambda+\nu}\right) \tag{2.8}$$

$$C_B^{(\pm)}(\nu,t) = \frac{M_N-M_\Lambda+\nu_\Lambda/(1-t/4M_N^2)}{4M_N} \cdot G^2_{YNK} \left(\frac{1}{\nu_\Lambda-\nu} \pm \frac{1}{\nu_\Lambda+\nu}\right) \tag{2.9}$$

where

$$G^2_{YNK} = G^2_{\Lambda NK} + 0.84G^2_{\Sigma NK} \tag{2.10}$$

and

$$\nu_\Lambda = (M^2_\Lambda - M^2_N - m^2_K + t/2)/(2M_N) \tag{2.11}$$

The corresponding $K^{\pm}n$ results are obtained by replacing G^2_{YNK} by $2G^2_{\Sigma NK}$ and by changing the pole position to ν_Σ where

$$\nu_\Sigma = (M^2_\Sigma - M^2_N - m^2_K + t/2)/(2M_N) \tag{2.12}$$

In order to make use of the fixed-t dispersion relation we first remove the Born term and, in the case of the antisymmetric amplitudes, we also divide by ν, so that we have the 6 symmetric Born stripped amplitudes

$$A^{(+)} - A_B^{(+)},\ B^{(-)} - B_B^{(-)},\ C^{(+)} - C_B^{(+)}$$

$$(A^{(-)} - A_B^{(-)})/\nu,\ (B^{(+)} - B_B^{(+)})/\nu,\ (C^{(-)} - C_B^{(-)})/\nu$$

If X is one of these amplitudes then (2.4) becomes

$$\mathrm{Re}X(\nu,t) = \frac{2P}{\pi}\int_{\nu_{\pi\Lambda}}^{\infty} \frac{\nu'\,\mathrm{Im}X(\nu',t)}{\nu'^2-\nu^2}\,d\nu' \tag{2.13}$$

The problem in using (2.13) is that $\mathrm{Im}X(\nu',t)$ is not directly calculable from experimental data (except for the special case of $\mathrm{Im}C^{(\pm)}(\nu',t=0)$) but must be contructed from the results of some phase shift analysis together with a model for the region between the $\pi\Lambda$ and $\bar{K}N$ thresholds. For this reason we subtract at $\nu = 0$ and rewrite (2.13) as

$$\mathrm{Re}X(\nu,t) = \frac{2\nu^2P}{\pi}\int_{\nu_{\pi\Lambda}}^{\nu_{max}(t)} \frac{\mathrm{Im}X(\nu',\ t)}{\nu'(\nu'^2-\nu^2)}\,d\nu' + \Delta_X(\nu,t) \tag{2.14}$$

where

$$\Delta_x(\nu,t) = X(0,t) + \frac{2\nu^2 P}{\pi}\int_{\nu_{max}(t)}^{\infty} \frac{ImX(\nu',t)}{\nu'(\nu'^2-\nu^2)} d\nu' \tag{2.15}$$

and $\nu_{max}(t)$ is determined by the highest energy for which we can reconstruct $ImX(\nu',t)$ from phase shifts.

If we define $\nu_{min}(t)$ by

$$\nu_{min}(t) = (s(q^2=-t/4) - M_N^2 - m_K^2 + t/2)/2M_N) \tag{2.16}$$

then for $\nu \geq \nu_{min}(t)$, the scattering angle is physical and we can also use phase shifts to reconstruct $ReX(\nu,t)$. This means for $\nu_{min}(t) < \nu < \nu_{max}(t)$ we can calculate "experimental" values for $\Delta_x(\nu,t)$ from

$$\Delta_x(\nu,t) = ReX(\nu,t) - \frac{2\nu^2}{\pi} P \int_{\nu_{\pi\Lambda}}^{\nu_{max}(t)} \frac{ImX(\nu',t)}{\nu'(\nu'^2-\nu^2)} d\nu' \tag{2.17}$$

We know from (2.15) that $\Delta_x(\nu,t)$ is an even function of ν which is analytic in the whole ν plane apart from the real axis cut $\nu_{max}(t) \leq \nu < \infty$. This means that $\Delta_x(\nu,t)$ should be a smooth function of ν provided we are not too close to ν_{max} and so we should be able to fit the "experimental" values with a simple parametric form which (i) smooths the input "experimental" values and (ii) allows extrapolation to the region $0 \leq \nu \leq \nu_{min}(t)$.

Once we have the fitted values of $\Delta_x(\nu, t)$ we can use (2.14) to calculate values of $ReX(\nu,t)$. These output values should be an improvement on the initial phase shift values in the region $\nu_{min}(t) < \nu < \nu_{max}(t)$ and they are predictions in the region $0 \leq \nu \leq \nu_{min}(t)$ where phase shifts cannot be used to calculate the real parts of the invariant amplitudes because of convergence problems.

3. Input

(a) Phase Shifts

The phase shift inputs used in these calculations were :-

K^+p The phases of Cutkosky et al. (3) above 780 MeV/c supplemented by some Aarhus values in the range 400-780 MeV/c (4). The highest energy corresponds to $\nu = 18.5$ for $t = 0$ and $\nu = 17.0$ for $t = 40$. (All numerical values will be given in m_π units unless otherwise stated).

K^-p Two sets were used, the phases of RLIC (5) extending up to $\nu = 13.7$

for t = 0 and 12.2 for t = -40 while the phases of LBL (6) extend up to 10.2 for t = 0 and 8.7 for t = -40.

K^+n The phases of B. Martin (7) were used. These extend up to ν = 11.3 for t = 0 and 9.8 for t = -40.

K^-n The RLIC and LBL sets of phases were used as in the K^-p case.

The partial waves of these solutions do not continue right down to threshold and so the phases were continued with a suitable scattering length approximation. In the case of K^-p and K^-n the s-wave continuations were made in such a way as to fit onto the threshold values of A. Martin (8).

(b) Below the $\bar{K}N$ threshold

The imaginary parts of the invariant amplitudes were approximated in the region between the $\pi\Lambda$ and $\bar{K}N$ thresholds by the s-wave K-matrix parametrization of A. Martin (8) supplemented by a $p_{3/2}$ Breit-Wigner type $Y^*(1385)$ contribution with a variable coupling. This was chosen to have the form

$$\mathrm{Im} f_{1+}^{(1)} = \frac{G_R^2}{4\pi} \cdot \frac{q_{\bar{K}N}^2 \cdot \gamma}{(M_R^2-s)^2 + M_R^2\gamma^2} \tag{3.1}$$

where

$$\begin{aligned} \gamma &= \Gamma_{tot} \cdot \left(\frac{q_{\pi\Lambda}}{q_{R_{\pi\Lambda}}}\right)^3 \qquad & s < M_R^2 \\ &= \Gamma_{tot} \left(\frac{|q_{\bar{K}N}|}{|q_{R_{\bar{K}N}}|}\right)^3 \qquad & s > M_R^2 \end{aligned} \tag{3.2}$$

and where the width and energy used were

$$\Gamma_{tot} = 37.5 \text{ MeV}, \qquad M_R = 1384 \text{ MeV}$$

The Su(3) value taken for the coupling was (9)

$$\frac{G_R^2}{4\pi} = \frac{2}{3} \; \frac{G^2_{Y_1^*\Lambda\pi}}{4\pi} = 4.7 \tag{3.3}$$

Calculations were carried out both with this Su(3) value and with 25% increased and decreased values.

4. The Forward Scattering Amplitude

In the case of the forward scattering amplitude $C^{(\pm)}(\nu,t=0)$ the results of this method can be compared with the results of conventional forward dispersion relations. If we first consider the antisymmetric forward amplitude and use the optical theorem result.

$$\mathrm{Im}C_{K^{\pm}N}(\nu,t=0) = p_{lab} \cdot \sigma_{tot}^{K^{\pm}N} \tag{4.1}$$

We can write the forward dispersion relation for $C^{(-)}(\nu,t=0)/\nu$ in the form

$$\frac{\mathrm{Re}\tilde{C}^{(-)}(\nu,t=0)}{\nu} = \frac{2P}{\pi}\int_{\nu_{\pi\Lambda}}^{\nu_{\bar{K}N}} \frac{\mathrm{Im}C^{(-)}(\nu',t=0)}{\nu'^2-\nu^2}\,d\nu' + \frac{2P}{\pi}\int_{\nu_{\bar{K}N}}^{\infty} \frac{p'_{lab}\,\sigma_{tot}^{(-)}(\nu')}{\nu'^2-\nu^2}\,d\nu' \tag{4.2}$$

where

$$\mathrm{Re}\tilde{C}^{(-)}(\nu,t=0) = \mathrm{Re}C^{(-)}(\nu,t=0) - C_B^{(-)}(\nu,t=0) \tag{4.3}$$

In particular we then have the sum rule

$$\left.\frac{\tilde{C}^{(-)}(\nu,t=0)}{\nu}\right|_{\nu=0} = \frac{2}{\pi}\int_{\nu_{\pi\Lambda}}^{\nu_{\bar{K}N}} \frac{\mathrm{Im}C^{(-)}(\nu',t=0)}{\nu'^2}\,d\nu' + \frac{2}{\pi}\int_{\nu_{\bar{K}N}}^{\infty} \frac{p'_{lab}\,\sigma_{tot}^{(-)}(\nu')}{\nu'^2}\,d\nu' \tag{4.4}$$

However from (2.15) we see that

$$\Delta_c^{(-)}(\nu=0,\ t=0) = \left.\frac{\tilde{C}^{(-)}(\nu,t=0)}{\nu}\right|_{\nu=0} \tag{4.5}$$

so extrapolating $\Delta_c^{(-)}(\nu,\ t=0)$ to $\nu = 0$ gives a value which can be compared with the direct sum rule calculation using (4.4). In this calculation the model described in section 3 was used for the unphysical region, the various phase shift solutions were used to calculate $\sigma_{tot}^{(-)}$ up to just over 1GeV/c lab. momentum and above this momentum $\sigma_{tot}^{(-)}$ was obtained directly from the experimental total cross section values. The resulting values are :-

Table 1 - Values of $\tilde{C}^{(-)}(\nu,\ t=0)/\nu\Big|_{\nu=0}$ calculated from (4.4). m_π units

Y_1^* coupling	Su(3) x 0.75	Su(3)	Su(3) x 1.25
$K^\pm p$	1.39	1.37	1.35
$K^\pm n$	0.53	0.48	0.44

Note that the sum rule (4.4) does not depend on the coupling of the hyperon pole whereas the values of $\Delta_c^{(-)}(\nu,t=0)$ do depend on the choice of $G^2/4\pi$. Fig. 1 shows a typical set of $K^\pm p$ values for $\Delta_c^{(-)}$ together with the result of fitting with a form linear in ν^2. A quadratic form in ν^2 gives an almost identical fit. Figure 2 shows a similar set of $K^\pm n$ values where there is now a small difference between a linear and quadratic fit.

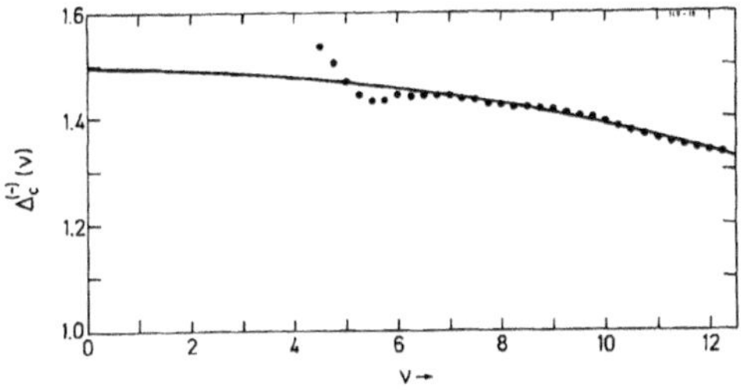

Figure 1 - Values of $K^\pm p$ $\Delta_c^{(-)}(\nu)$ for the set (K^+p: Cutkosky, K^-p: RLIC, Y_1^*: Su(3) × 1.0, $\frac{G^2_{YNK}}{4\pi}$: 16.0) and t = 0 are shown by dots. The solid line shows a fit of the form $a + b\nu^2$.

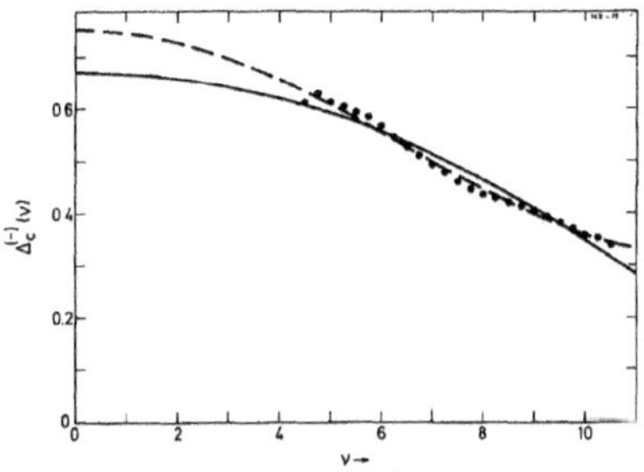

Figure 2 - Values of $K^\pm n$ $\Delta_c^{(-)}(\nu)$ for the set (K^+n: Martin, K^-n: RLIC, Y_1^*: Su(3)×1.0, $\frac{2G^2_{\Sigma NK}}{4\pi}$: 2.0) and t = 0 are shown by dots. The solid line shows a fit of the form $a + b\nu^2$ and the dashed line a fit of the form $a + b\nu^2 + c\nu^4$.

Extrapolating these and similar fits to $\nu = 0$ gives values of $\tilde{C}^{(-)}(\nu, t=0)/\nu\big|_{\nu=0}$ which are compared with the sum rule values in figure 3 for $K^{\pm}p$ and in figure 4 for $K^{\pm}n$.

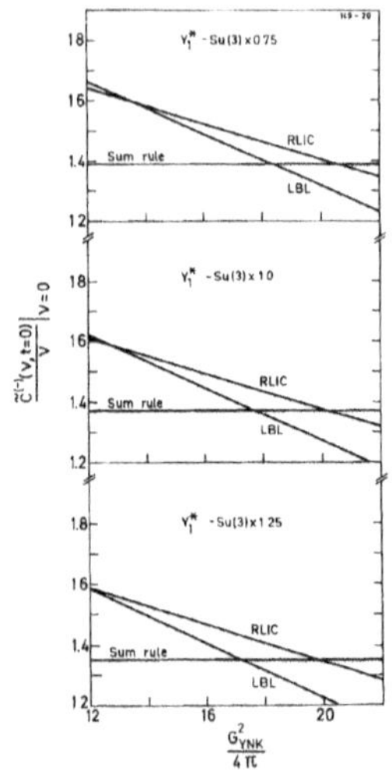

Figure 3 - Comparison of $K^{\pm}p$ values for $\tilde{C}^{(-)}(\nu, t=0)/\nu\big|_{\nu=0}$

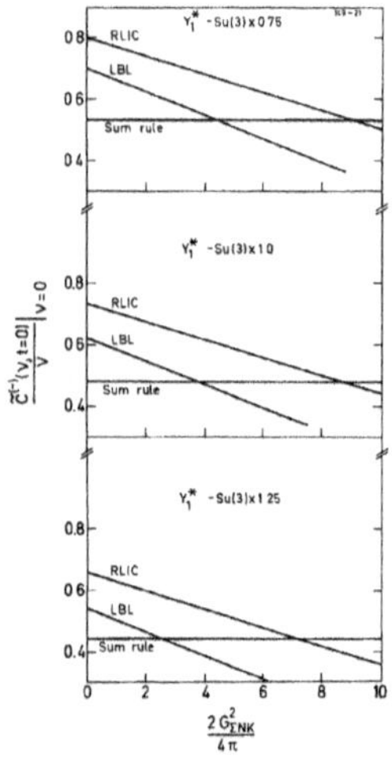

Figure 4 - Comparison of $K^{\pm}n$ values for $\tilde{C}^{(-)}(\nu, t=0)/\nu\big|_{\nu=0}$

Consistent values are obtained for :-

Table 2 - Couplings giving consistent values of $\tilde{C}^{(-)}(\nu, t=0)/\nu\Big|_{\nu=0}$

	Y_1^* / Phases	Su(3) x 0.75	Su(3)	Su(3) x 1.25
$K^{\pm}p$	Cutkosky RLIC	$\frac{G^2_{YNK}}{4\pi} = 20.6$	20.2	19.8
	Cutkosky LBL	$\frac{G^2_{YNK}}{4\pi} = 18.3$	17.7	17.2
$K^{\pm}n$	Martin RLIC	$\frac{2G^2_{\Sigma NK}}{4\pi} = 9.2$	8.8	7.4
	Martin LBL	$\frac{2G^2_{\Sigma NK}}{4\pi} = 4.4$	3.8	2.6

More work hat to be done before realistic errors can be assigned to these values but the trend is clear. In the $K^{\pm}p$ case the difference between the LBL and RLIC values is smaller than in the $K^{\pm}n$ case but in both cases the LBL values require a smaller value of the hyperon coupling. It is interesting to note that if we use (2.10) to extract $G^2_{\Lambda NK}/4\pi$ then the RLIC and LBL values agree rather well, the values for the Su(3) value of the Y_1^* coupling being 16.5 and 16.1 respectively.

There is no reason why the comparison between the results of our method and of relation (4.2) should be restricted to $\nu = 0$. The comparison is possible in the range $0 \leq \nu < \nu_{max}(t=0)$ which in practice means $0 \leq \nu \lesssim 10$. Such a comparison has been made, not directly, but via the use of logarithmic dispersion relations, the details being given by Dr. Dumbrajs in his contribution. The values of $G^2/4\pi$ giving the best agreement over the whole range are reported in table 3.

A comparison with other determinations of these couplings (8), (9), (10), shows that these values are very reasonable and suggests that the method is working rather well at least in the case of $C^{(-)}(\nu, t=0)$.

The situation in the case of the symmetric forward amplitude $C^{(+)}(\nu, t=0)$ is a little different since the integral over $\sigma^{(+)}_{tot}$ does not converge without a subtraction.

The equivalent relation to (4.2) is :

Table 3 - "Best fit" values of the couplings

	Y_1^* / Phases	Su(3) x 0.75	Su(3)	Su(3) x 1.25
$K^{\pm}p$	Cutkosky RLIC	$\frac{G^2_{YNK}}{4\pi} = 17.8$	17.5	17.1
	Cutkosky LBL	$\frac{G^2_{YNK}}{4\pi} = 15.8$	15.4	15.0
$K^{\pm}n$	Martin RLIC	$\frac{2G^2_{\Sigma NK}}{4\pi} = 9.7$	9.0	8.5
	Martin LBL	$\frac{2G^2_{\Sigma NK}}{4\pi} = 5.8$	5.3	4.8

$$\mathrm{Re}\tilde{C}^{(+)}(\nu,t=0) = \tilde{C}^{(+)}(\nu{=}0,t{=}0) + \frac{2\nu^2 P}{\pi}\int_{\nu_{\pi\Lambda}}^{\nu_{\bar{K}N}} \frac{\mathrm{Im}C^{(+)}(\nu',t=0)}{\nu'(\nu'^2-\nu^2)}\,d\nu'$$

$$+ \frac{2\nu^2 P}{\pi}\int_{\nu_{\bar{K}N}}^{\infty} \frac{p'_{lab}\,\sigma^{(+)}_{tot}(\nu')}{\nu'(\nu'^2-\nu^2)}\,d\nu' \qquad (4.6)$$

where

$$\tilde{C}^{(+)}(\nu,t=0) = C^{(+)}(\nu,t=0) - C_B^{(+)}(\nu,t=0) \qquad (4.7)$$

From (2.15) we see that

$$\Delta_C^{(+)}(\nu=0,t=0) = \tilde{C}^{(+)}(\nu=0,t=0) \qquad (4.8)$$

but now there is no sum rule equivalent to (4.4) and we cannot make a direct comparison with the value of $\tilde{C}^{(+)}(\nu=0, t=0)$ obtained by extrapolating to $\nu = 0$.

Information is available from conventional forward dispersion relation fits to experimental forward real part data with D(o) and $G^2/4\pi$ as free parameters. If D(o) is measured in fermi and $\tilde{C}^{(+)}(\nu=0, t=0)$ in m_π units then

$$D_{K^\pm p}(o) = 0.1125\ \tilde{C}^{(+)}(\nu = 0,\ t = 0) - 0.1854\ \frac{G^2_{YNK}}{4\pi} \qquad (4.9)$$

$$D_{K^\pm n}(o) = 0.1125\ \tilde{C}^{(+)}(\nu = 0,\ t = 0) - 0.06263\ \frac{2G^2_{\Sigma NK}}{4\pi} \qquad (4.9')$$

A typical set of $K^\pm p$ values for $\Delta_C^{(+)}$ and of $K^\pm n$ values for $\Delta_C^{(+)}$ are shown in figures 5 and 6.

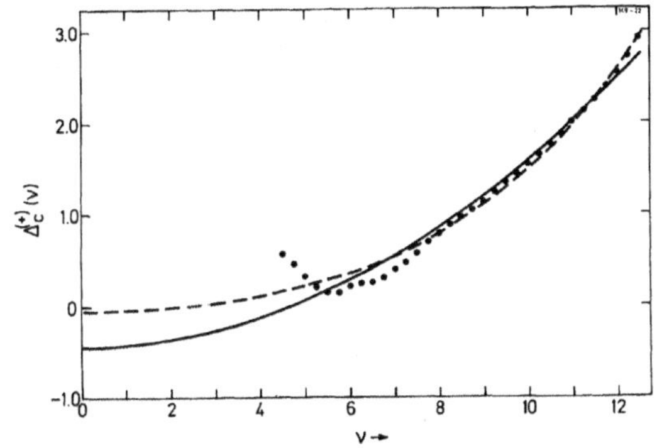

Figure 5 - Values of $K^\pm p\ \Delta_C^{(+)}(\nu)$ for the same set as figure 1 are shown by dots. The solid line shows a fit of the form $a + b\nu^2$ and the dashed line a fit of the form $a + b\nu^2 + c\nu^4$.

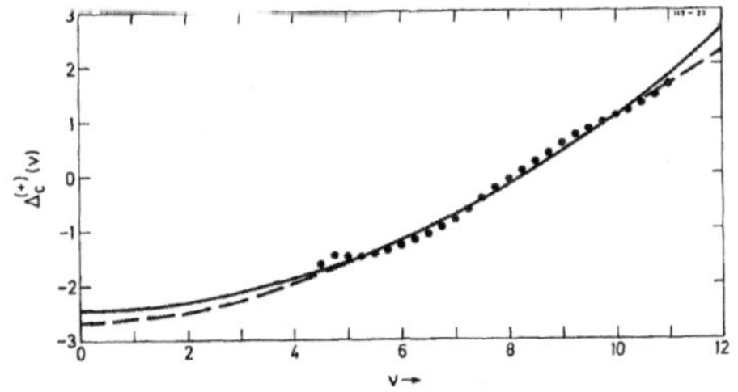

Figure 6 - Values of $K^\pm n\ \Delta_C^{(+)}(\nu)$ for the same set as figure 2 are shown by dots. The solid line shows a fit of the form $a + b\nu^2$ and the dashed line a fit of the form $a + b\nu^2 + c\nu^4$.

Extrapolating to $\nu = 0$ with a form linear in ν^2 and using (4.9) or (4.9') then gives the following values for $D(o)$: -

Table 4 - Values of D(o) from extrapolating $\Delta_C^{(+)}(\nu,\ t = 0)$

		Y_1^* \ $G^2_{YNK}/4\pi$	12.0	16.0	20.0
$K^{\pm}p$	Cutkosky RLIC	Su(3) x 0.75	- 2.26	- 3.01	- 3.76
		Su(3)	- 2.27	- 3.02	- 3.77
		Su(3) x 1.25	- 2.28	- 3.03	- 3.78
	Cutkosky LBL	Su(3) x 0.75	- 2.24	- 2.99	- 3.74
		Su(3)	- 2.25	- 3.00	- 3.76
		Su(3) x 1.25	- 2.26	- 3.02	- 3.77
		Y_1^* \ $2G^2_{\Sigma NK}/4\pi$	0.0	4.0	8.0
$K^{\pm}n$	Martin RLIC	Su(3) x 0.75	- 0.25	- 0.52	- 0.79
		Su(3)	- 0.27	- 0.54	- 0.81
		Su(3) x 1.25	- 0.29	- 0.56	- 0.83
	Martin LBL	Su(3) x 0.75	- 0.33	- 0.61	- 0.89
		Su(3)	- 0.36	- 0.64	- 0.92
		Su(3) x 1.25	- 0.38	- 0.66	- 0.94

Figure 7 shows these $K^{\pm}n$ values for D(o) compared to the values obtained by A. Martin (8) and by Baillon et al. (11) (the dashed lines show the expected correlation between $G^2_{\Sigma NK}$ and D(o)). Taking into account the uncertainty on the extrapolation of $\Delta_C^{(+)}(\nu,\ t = 0)$ the agreement is very good. Figure 8 shows the corresponding $K^{\pm}p$ values together with the values of A. Martin (8), Di Claudio et al. (10) and Baillon et al. (12). Here the situation is not so satisfactory. Inverting (4.9) we see that

$$\tilde{C}^{(+)}(\nu = 0,\ t = 0) = 8.89\left(D(o) + 0.1854\ \frac{G^2_{YNK}}{4\pi}\right) \qquad (4.10)$$

so a change of 0.2 fm in D(o) corresponds to a change of 1.8 in $\tilde{C}^{(+)}(\nu = 0,\ t = 0)$. From figure 5 we see that such an uncertainty on the extrapolation of $\Delta_C^{(+)}(\nu,\ t = 0)$ is already very unlikely so there is no way in which the values of Baillon et al. or of Di Claudio et al. can be reconciled with our calculation (15).

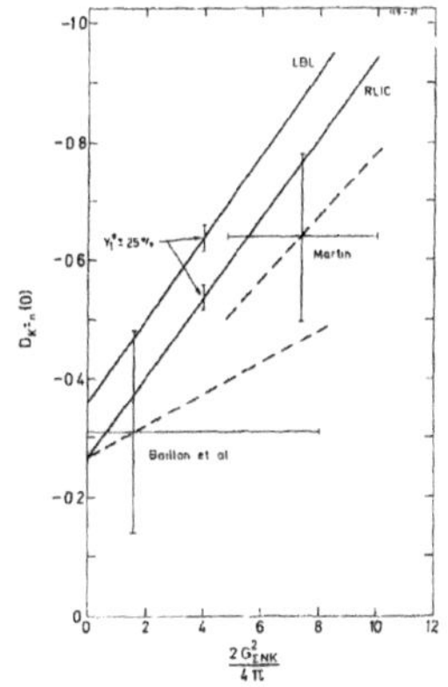

Figure 7 - D(o), $K^{\pm}n$ as a function of $2G^2_{\Sigma NK}/4\pi$

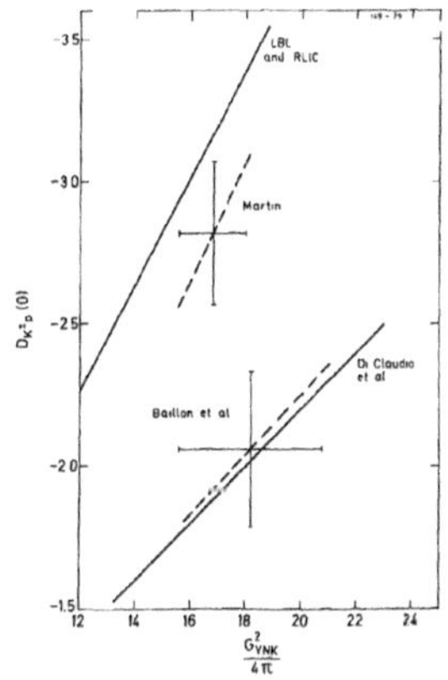

Figure 8 - D(o), $K^{\pm}p$ as a function of $G^2_{YNK}/4\pi$

This is another example of the problem mentioned by Di Claudio et al. (13). If (4.6) is rewritten with subtractions at the KN and $\bar{K}N$ thresholds then D(o) can be calculated in terms of these two subtraction constants, and integrals over the unphysical region and over σ_{tot}. Di Claudio et al. (13) make such a calculation and obtain a value close to that of A. Martin (8). They conclude that fits including the low energy data tend to give more negative values of D(o) than those which only fit high energy data. Clearly the results of our method which concentrates on the phase shift region agree with this trend.

5. The KN σ term

If we consider the meson-nucleon scattering process

$$m(q) + N \rightarrow m(q') + N \qquad (5.1)$$

and define an amplitude $T^{(+)}$ by

$$T^{(+)}(\nu, t; q^2, q'^2) = A^{(+)} + \nu B^{(+)} \quad (5.2)$$

then the σ-commutator is given by (14)

$$\sigma_{NN}^{mm} = -f_m^2 \, T^{(+)}(0,0; 0,0) \quad (5.3)$$

where f_m is the meson decay constant (i.e. f_π or f_K). Assuming linearity about (0,0; 0,0) and making use of the Adler consistency conditions, the off mass shell value of $T^{(+)}$ can be related to an on mass shell value at the point $\nu = 0$, $t = 2m^2$, giving

$$\sigma_{nn}^{mm} = f_m^2 \, T^{(+)}(\nu = 0, t = 2m^2; q^2 = -m^2, q'^2 = -m^2) + O(\frac{m}{M_n})^4$$

$$= f_m^2 \, C^{(+)}(\nu = 0, t = 2m^2) + O(\frac{m}{M_N})^4 \quad (5.4)$$

There have been several calculations of $\sigma_{NN}^{\pi\pi}$ where $C^{(+)}(\nu = 0, t)$ is calcualted for a series of negative t-values which are then continued to $t = 2m_\pi^2$ (2), (9). In principle such calculations are also possible in the KN case, the only problem being that $t = 2m_K^2$ is considerably farther away from t = 0 than is $t = 2m_\pi^2$. Indeed $C^{(+)}$ develops an absorption part due to the process $K\bar{K}$-$N\bar{N}$ already at $t = 4m_\pi^2$ so the continuation must be made into $K\bar{K}$-$N\bar{N}$ region.

If for the moment we concentrate on the negative t-values then (2.15) gives

$$C^{(+)}(\nu = 0, t) = \Delta_C^{(+)}(\nu = 0,t) \quad (5.5)$$

and so extrapolation of $\Delta_C^{(+)}(\nu,t)$ to ν=0 for a series of t values gives values for $\tilde{C}^{(+)}(\nu=0, t)$. Figure 9 shows $K^\pm p$ values obtained using a linear form in ν^2 to extrapolate to ν = 0 while figure 10 shows similar $K^\pm n$ values in the range $-40 \leq t \leq 0$. If $C^{(+)}(\nu = 0, t)$ is expressed in m_π units

$$\sigma_{pp}^{KK} = 106(\tilde{C}^{(+)}(\nu = 0, t = 25.03) + 0.231 \frac{G_{YNK}^2}{4\pi}) \text{ MeV} \quad (5.6)$$

$$\sigma_{nn}^{KK} = 106(\tilde{C}^{(+)}(\nu = 0, t = 25.03) + 0.262 \frac{2G_{\Sigma NK}^2}{4\pi}) \text{ MeV} \quad (5.7)$$

The vertical lines in figures 9 and 10 at $t = 2m_K^2$ show the scale for σ_{pp}^{KK} and σ_{nn}^{KK} as calculated from (5.6) and (5.7).

The extrapolation to $t = 2m_K^2$ will be influenced by the $K\bar{K}$-$N\bar{N}$ absorptive part. This can be taken into account by writing a fixed-ν disper-

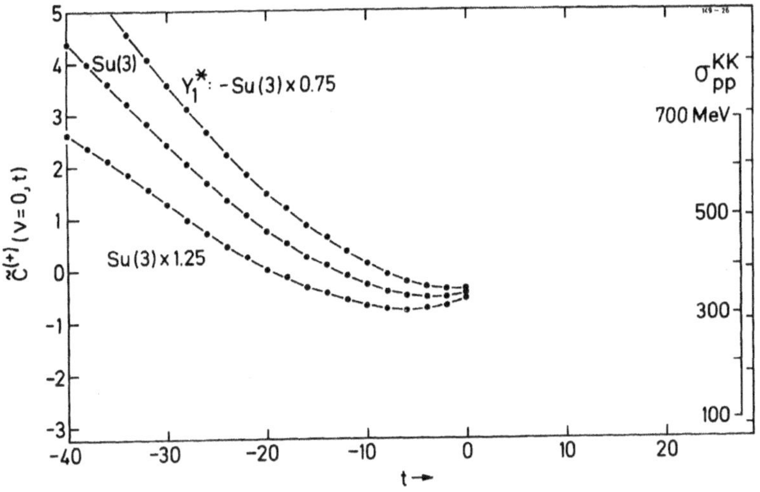

Figure 9 - $\tilde{C}^{(+)}(\nu=0,t)$ for $K^{\pm}p$ using Cutkosky and RLIC.
The reader is left to make the extrapolation to $t = 2m_K^2$ where the vertical line shows the scale for σ_{pp}^{KK}.

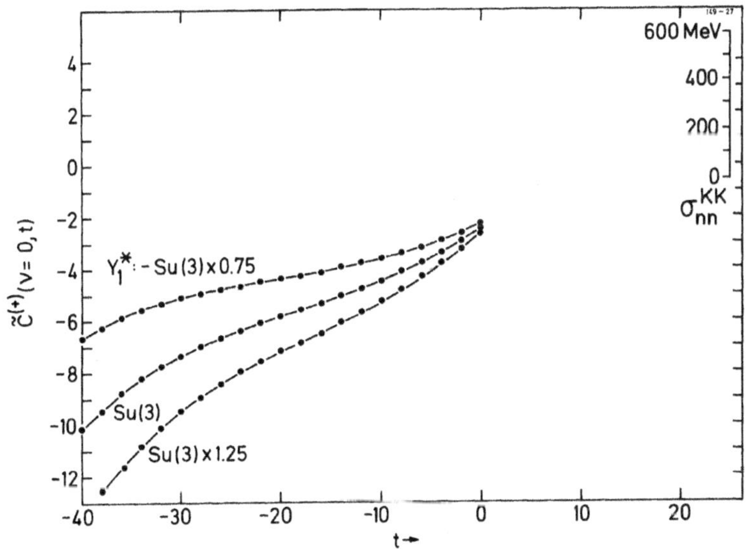

Figure 10 - $\tilde{C}^{(+)}(\nu=0,t)$ for $K^{\pm}n$ using Martin and RLIC.
The reader is left to make the extrapolation to $t = 2m_K^2$ where the vertical line shows the scale for σ_{nn}^{KK}

sion relation in the form

$$\tilde{C}^{(+)}(\nu = 0,\ t) = \frac{1}{\pi}\int_{4m_\pi^2}^{t_{max}} \frac{\mathrm{Im}C^{(+)}(\nu = 0,\ t')}{t'-t}\,dt' + \Delta(t) \qquad (5.8)$$

where $\Delta(t)$ has cuts for $t_{max} \leq t < \infty$ and for $-\infty < t \leq t_{min}$. The left hand cut is determined by double spectral functions and lies to the left of -40. On the right hand cut using extended unitary and neglecting states with $J \gtrsim 2$,

$$\mathrm{Im}C^{(+)}(\nu = 0,\ t) = -\frac{3}{4M_N^2-t}\sqrt{\frac{t-4m_\pi^2}{4t}}\; f_+^0(t)\,g_0^{(+)*}(t) \qquad (5.9)$$

where $f_+^0(t)$ is the $J = 0$ $\pi\pi$-$N\bar{N}$ amplitude and $g_0^{(+)}(t)$ is the s-wave $\pi\pi$-$K\bar{K}$ amplitude. If we choose $t_{max} \simeq 4m_K^2$ then we can use known values of $f_+^0(t)$ and $g_0^{(+)}(t)$ (9) to calculate $\mathrm{Im}C^{(+)}(\nu = 0,\ t)$. We can now use (5.8) to calculate $\Delta(t)$ for $-40 \leq t \leq 0$ and then extrapolate these values to $t = 2m_K^2$ giving

$$\mathrm{Re}\tilde{C}^{(+)}(\nu = 0,\ t = 2m_K^2) = \Delta(t = 2m_K^2) + \frac{P}{\pi}\int_{4m_\pi^2}^{t_{max}} \frac{\mathrm{Im}C^{(+)}(\nu = 0,\ t)}{t'-t}\,dt' \quad (5.10)$$

At present no detailed attempts at extrapolation have been made but first estimates suggest that including the $K\bar{K}$-$N\bar{N}$ effects in this way decreases the extrapolated value of $\tilde{C}^{(+)}(\nu = 0,\ t = 2m_K^2)$ by roughly 2 units thus lowering the values of σ_{pp}^{KK} and σ_{nn}^{KK} by 200 MeV compared to extrapolations neglecting these effects.

Both σ_{pp}^{KK} and σ_{nn}^{KK} have been recently determined by another method (13), (14) giving the results

$$\sigma_{pp}^{KK} = 493 \pm 716 \text{ MeV (for } \frac{G_{\Sigma NK}^2}{4\pi} = 2.0)$$

$$\sigma_{nn}^{KK} = 638 \pm 438 \text{ MeV}$$

Comparing these values with figures 9 and 10 it will be seen that the values obtained by extrapolation are likely to be lower than these values.

6. The Low Energy Partial Wave Amplitudes

In the previous two sections we have concentrated on the amplitudes $C^{(\pm)}(\nu,\ t)$. The method has also been used for $A^{(\pm)}(\nu,\ t)$ and $B^{(\pm)}(\nu,\ t)$; it might seem unnecessary to use both $A^{(\pm)}$ and $B^{(\pm)}$ since only one of them is independent of C but that does not necessarily guarantee that all three have smooth Δ's. In the case of the (−) amplitudes, provided we fit all three Δ's with the same order expansion in ν^2, the output amplitudes automatically satisfy the defining relation (2.1) and it is only necessary to work with either $A^{(-)}$ or $B^{(-)}$ in addition to $C^{(-)}$. This is not the case for the (+) amplitudes and here the requirement that the output amplitudes should satisfy the relation

$$C^{(+)}_{out}(\nu,\ t) \;=\; A^{(+)}_{out}(\nu,\ t) + \frac{\nu}{1-t/4M_N^2}\, B^{(+)}_{out}(\nu,\ t) \qquad (6.1)$$

must be imposed as an extra constraint when fitting the $\Delta^{(+)}$'s, this constraint being imposed for all ν.

The output values of the real parts of the amplitudes can be unfolded to give $\mathrm{Re}A_{K^\pm p}$, $\mathrm{Re}B_{K^\pm p}$ and $\mathrm{Re}C_{K^\pm p}$ and these values will now be used as extra data in a new cycle of phase shift analyses. In order to get some idea of the likely consequences of this recycling we can project out the real parts of the partial wave amplitudes from our output amplitudes. The calculations are restricted to the range $-40 \lesssim t \lesssim 0$ which means that the output amplitudes only cover the full angular range for $p_{lab} \simeq 800$ MeV/c corresponding to $E_{cm} \lesssim 1700$ MeV. As the energy falls below this value the most negative t values start to move out of the physical angular range and so the partial waves are projected out from fewer and fewer t-values as we move to lower energies. This means that the higher ℓ values which require the evaluation of integrals involving high P_ℓ values become more uncertain at the lowest energies. Some typical values are shown in figures 11 and 12 for $K^\pm p$ and figures 13 and 14 for $K^\pm n$. At present work is underway to see which partial waves are particularly sensitive to variations on the couplings of the Λ, Σ and Y_1^* and to the choice of phase shift input.

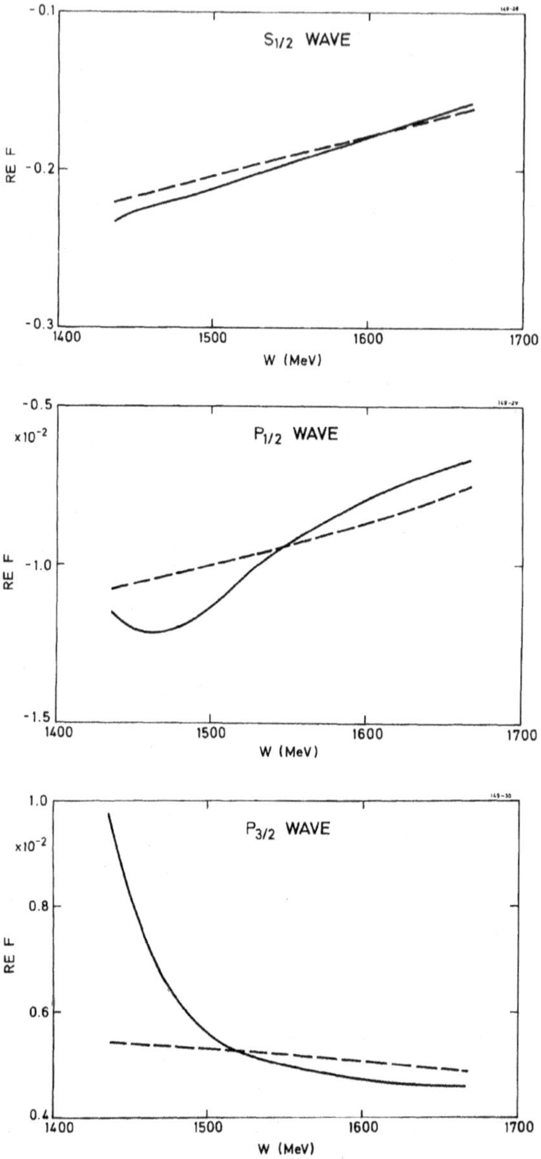

Figure 11 - Real parts of the K^+p s-wave and reduced p-waves for the set (K^+p: Cutkosky, K^-p: LBL, Y_1^* : Su(3)×1.0, $\frac{G^2_{YNK}}{4\pi}$: 16.0). The dashed line is the input and the solid line is the output.

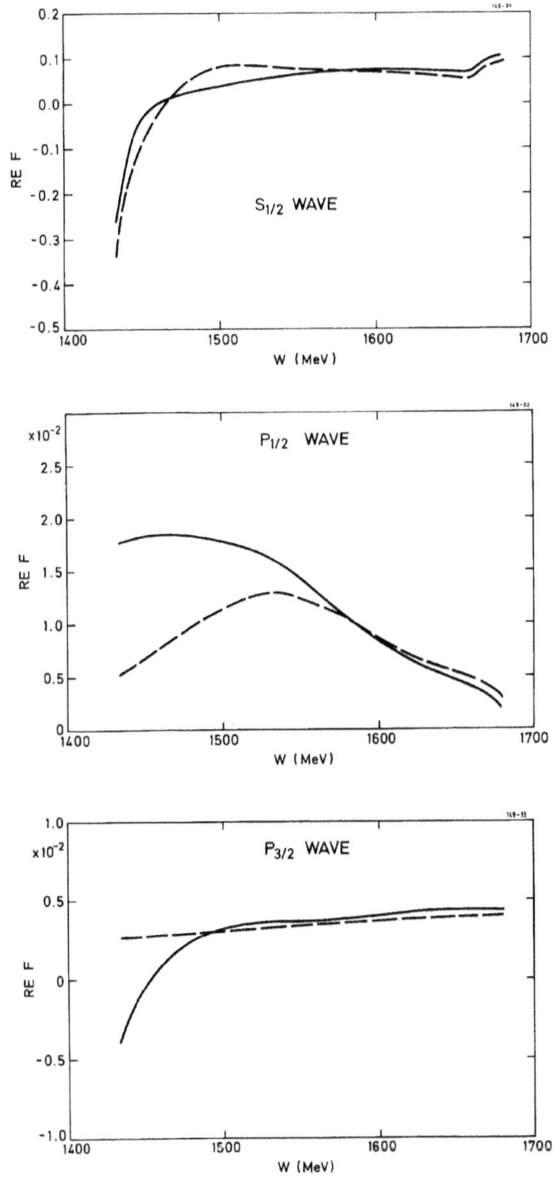

Figure 12 - As figure 11 but for K^-p.

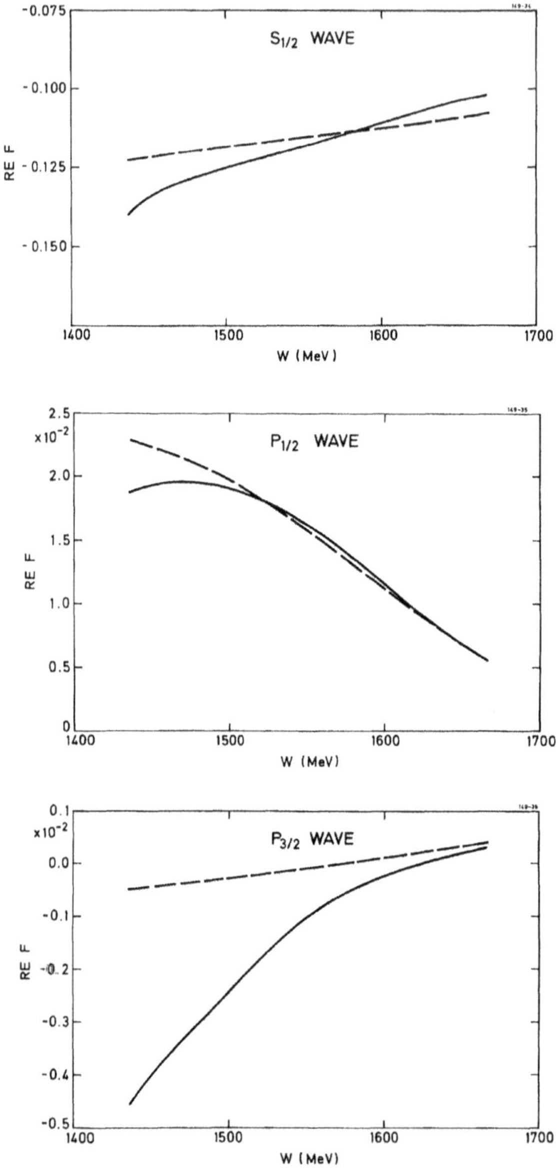

Figure 13 - Real parts of the K^+n s-wave and reduced p-waves for the set (K^+n: Martin, K^-n: LBL, Y_1^*: Su(3)$\times$1.0, $\frac{2G^2_{\Sigma NK}}{4\pi}$: 2.0). The dashed line is the input and the solid line is the output.

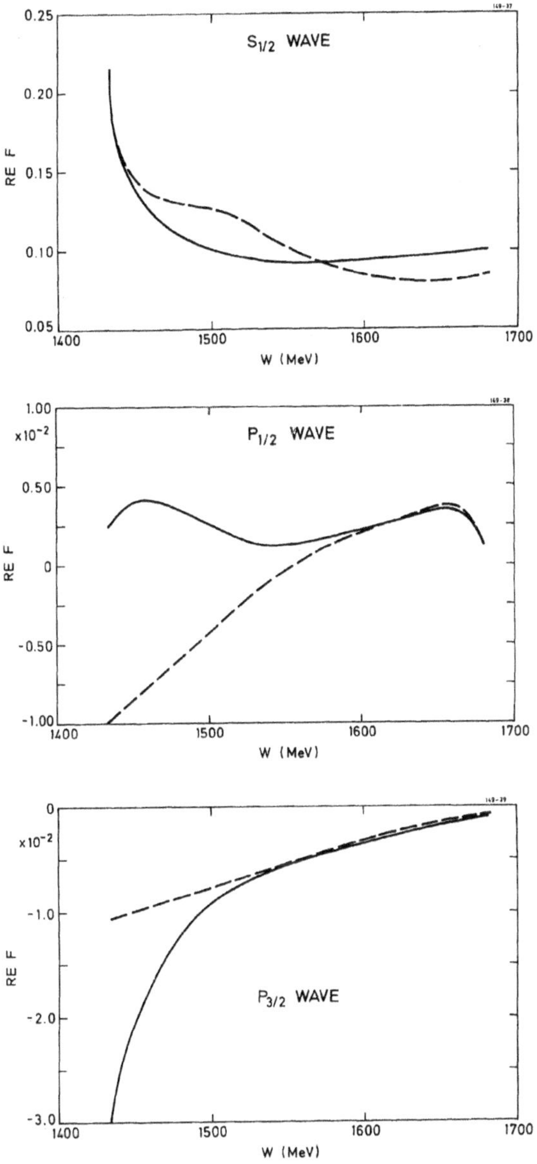

Figure 14 - As figure 13 but for K^-n.

References and notes

*. Work supported by a grant from the Danish Natural Science Research Council.

1. Pietarinen, E.: 1976, Physica Scripta 14, 11.

2. Nielsen H., and Oades, G.C.: 1974, Nucl. Phys. B72, 310.

3. Cutkosky, R.E., et al.: 1976, Nucl. Phys. B102, 139.

4. Hansen, P.N., Moss, J., and Oades, G.C.: 1976, Proceedings of the Oxford Baryon Resonance Conference, 275.

5. Gopal, G.P., et al.: 1977, Nucl. Phys. B119, 362.

6. Alston-Garnjost, A., et al. :1978, Phys. Rev. D18, 182.

7. Martin, B.R.: 1975, Nucl. Phys. B94, 413.

8. Martin, A.D.: 1976, Phys. Letters 65B, 346.

9. Compilation of Coupling Constants and Low-Energy Parameters: 1979, Nucl. Phys. B147, 189.

10. Di Claudio, B., Violini, G., and Queen, N.M.: 1979, Nucl. Phys. B161, 238.

11. Baillon, P., et al.: 1976, Phys. Letters 61B, 171.

12. Baillon, P., et al.: 1974, Phys.Letters 50B, 383.

13. Di Claudio, B., Rodriguez-Vargas, A.M., and Violini, G.: 1979, Z. Physik C3, 75.

14. Rodriguez-Vargas, A.M., and Violini, G.: 1979, preprint ANDES 13(P).

15. During the workshop a discussion with A. Martin cleared up this problem. Baillon et al. and Di Claudio et al. do not put their pole at the Λ position. Moving the pole to the Λ position changes the value of D(o) to around -3.

DISCUSSION

- C. B. Dover :

What is the currently accepted value for the πN σ-term? What is your opinion of the recent determination of this quantity by Banerjee and Cammarata?

- G. Oades :

If we confine ourselves to the Cheng-Dashen method of determining the σ-commutator then the currently accepted value is around 60 MeV with an uncertainty of about 15%. Banerjee and Cammarata, using their model to fit the low energy πN s-waves, find that they need a value of $\sigma_{NN}^{\pi\pi}$ = 25.5 MeV. Personally I feel that this is a problem for their model. They must first show that the model reproduces the full details of the low energy πN scattering amplitudes before it can be accepted as more than a qualitative model for the low energy πN interaction.

- G. Violini :

I think that from this analysis and from other results like for instance those I will present tomorrow on the g_{Σ}^2 calculations one gets two lessons. First it is needed a simultaneous analysis of all the data, second they must have the same weight in the analysis, which does not occur for the current ones.

- G. Oades :

I agree completely that a simultaneous analysis of all the data is required. The question of the weight to be attached to the various pieces of data is very difficult and often requires a detailed knowlegde of the various experiments. One should often beware of taking the experimental errors too literally.

THE KAON-NUCLEON INTERACTION NEAR THRESHOLD

A.D. Martin
Department of Physics - University of Durham

1. Introduction

Data for the complete set of K^-p channels below 250 MeV/c became available about twelve or more years ago, and several independent K-matrix analyses of these data were presented shortly afterwards. The data were inadequate for a complete solution to this multichannel problem and so the analyses were far from unique. We might have hoped that the $\bar{K}N \to \bar{K}N$ amplitudes were reasonably well determined, but unfortunately the situation is complicated by the presence of the $\Lambda(1405)$ S wave resonance just below the $\bar{K}N$ threshold. We summarize the problem in section 2.

In the following years little extra information became available, unitl the accurate measurements of the $K^{\pm}p$ and K^-d scattering in the Coulomb-nuclear interference region rekindled interest in $K^{\pm}N$ forward dispersion relations. These were found to impose powerful constraints on the low energy amplitudes and so improve the determination of the low energy $\bar{K}N$ parameters. We describe this development in section 3.

More recently further data have become available; but none, unfortunately, for the unmeasured $\pi Y \to \pi Y$ channels, where $Y = \Sigma$ or Λ. As we discuss, in section 4, the new data mainly help to tighten the dispersion relation constraints. In section 5 we present an up-dated determination of the $\bar{K}N$ parameters taking the new data into account. Finally we summarize the present situation and include a few remarks concerning the $\Lambda(1405)$.

2. Early K matrix analyses

Our knowledge of the threshold properties of the $\bar{K}N$ interaction is based on bubble chamber data in the laboratory momentum range below 300 MeV/c for the reactions (1-4)

E. Ferrari and G. Violini (eds.), Low and Intermediate Energy Kaon-Nucleon Physics, 97–114.

$$K^-p \to K^-p, \quad \bar{K}^0 n, \quad \pi^\pm \Sigma^\mp, \quad \pi^0 \Sigma^0, \quad \pi^0 \Lambda, \tag{1}$$

together with a few at-rest branching ratio measurements, and some data for $K_L^0 p$ channels (5). The K^-p cross section data (1-4) are shown in figure 1.

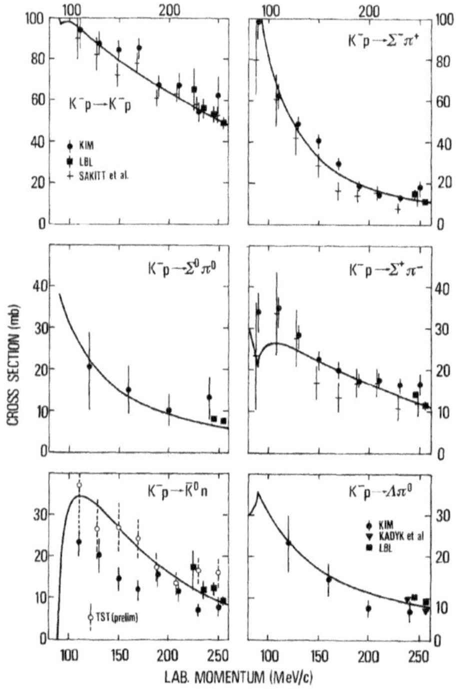

Figure 1 - The low energy K^-p data from ref. 1-4. The curves correspond to the M matrix fit, with analyticity constraints, described in section 5.

Below about 200 MeV/c the data show that the S wave interaction is dominant and show no significant three-body production, such as $\Lambda\pi\pi$. Finally the Coulomb-nuclear interference observed (2) in the low energy elastic K^-p differential cross section indicates a repulsive K^-p interaction.

The original K-matrix analyses (6) performed before and around 1970 were based primarily on the data of Kim (2) and of Sakitt et al. (1). A K (or $M \equiv K^{-1}$) matrix approach is appropriate since we have a multichannel problem. For isospin I = 0 we have two relevant channels ($\bar{K}N$, $\pi\Sigma$) and for I = 1 three channels ($\bar{K}N$, $\pi\Sigma$, $\pi\Lambda$). Unitarity requires the S wave am plitudes, T^I, to satisfy

$$(T^I)^{-1} = M^I - iQ$$

where the M^I are real (7), symmetric matrices and Q is a diagonal matrix of the channel centre-of-mass momenta. When confronting the data corrections are included for electromagnetic mass differences and for Coulomb effects in the K^-p channel as originally proposed by Dalitz and Tuan (8). The simplest parametrization consistent with unitarity is to take the K (or M) matrices to be constant, a total of nine independent parameters.

A crucial problem is the absence of data for the $\pi Y \rightarrow \pi Y$ channels, where Y is Λ or Σ. This means that the low energy K^-p and K^o_Lp data may be described at each energy by six S wave parameters ($A_I \equiv a_I + ib_I$ with $I = 0,1$; ε and ϕ) and therefore only these combinations of M matrix elements are determined. The six quantities arise as follows. A_I are complex scattering lengths which specify the $\bar{K}N \rightarrow \bar{K}N$ amplitudes,

$$T^I_{KK} = \frac{A_I}{1-ikA_I} , \qquad (2)$$

and determine the (S wave) inelastic cross section, $\bar{K}N \rightarrow \pi Y$, for isospin I; k is the $\bar{K}N$ c.m. momentum. Since both $\pi\Lambda$ and $\pi\Sigma$ production occur for $I = 1$ we need to specify the fraction, ε, of $\pi\Lambda$ production. Finally to determine the $K^-p \rightarrow \pi^{\mp}\Sigma^{\pm}$ cross sections we need to know the relative phase, ϕ, of the $I = 0$ and $I = 1$ $\bar{K}N \rightarrow \pi\Sigma$ amplitudes. Accurate data for the six reactions of eq. (1) at a given energy determine the six parameters at that energy. The K^o_Lp data further constrain the $I = 1$ parameters. All nine M matrix elements can only be fitted because, via the πY channel momenta, the six quantities have a definite energy dependence. For example

$$A_o^{-1} = \alpha_{KK} - \alpha^2_{K\Sigma}/(\alpha_{\Sigma\Sigma} - iq) \qquad (3)$$

where q is the $\pi\Sigma$ c.m. momentum and the α's are the $I = 0$ M matrix elements.

Although the M matrix solutions are not unique we do expect them to give similar values for A_o, A_1, ε and ϕ. For example, the scattering lengths at the K^-p threshold are all within about 0.1 fm of the following values (6)

$$A_o = -1.65 + i\,0.7 \text{ fm}$$
$$A_1 = 0.0 + i\,0.6 \text{ fm}. \qquad (4)$$

However, even here we can anticipate uncertainties greater than 0.1 fm, since the determinations all rely on the same data for six different reactions some of which are not well known.

If the I = 0 amplitudes are extrapolated below the $\bar{K}N$ threshold they contain a factor $1/(1+\kappa A_o)$ where $\kappa_- \equiv -ik$ is real. For a_o large and negative, see eq. (4), we obtain a $\bar{K}N$ 'bound state' resonance in the neighbourhood of $\kappa = -1/a_o$. This is the $\Lambda(1405)$. In the absence of $\pi\Sigma \rightarrow \pi\Sigma$ data the properties of the $\Lambda(1405)$ are ill-determined, and depend sensitively on the extrapolation of A_o below threshold. The problem is well illustrated by figure 2 which is taken from the analysis of Chao et al. (6).

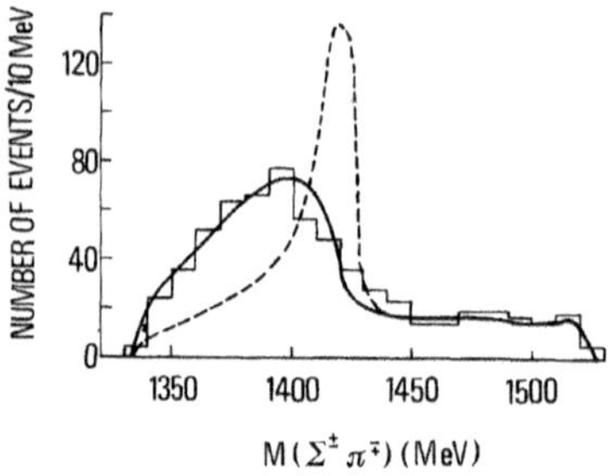

Figure 2 - The $\pi\Sigma$ mass spectrum observed in the reaction $\pi^- p \rightarrow \Sigma\pi K^o$ at 1.69 GeV/c. The dashed curve is the prediction of solution A and the continuous curve is the fit of solution B of the M matrix analysis of Chao et al. (6). The figure is taken from Chao et al.

It shows the $\pi\Sigma$ spectrum observed in the 1.69 GeV/c $\pi^- p \rightarrow \pi\Sigma K^o$ data, compared with curves calculated using the I = 0 $\pi\Sigma \rightarrow \pi\Sigma$ of two different M matrix solutions of the low energy $K^- p$ and $K^o_L p$ data. Both solutions describe the low energy $\bar{K}N$ data equally well, although solution B, which corresponds to the continuous curve, is also required to fit (9) the $\pi\Sigma$ spectrum. Although this constraint from such low energy $\pi\Sigma$ production data is at the best very approximate, it is clearly a step in the right direction.

3. Inclusion of forward dispersion relation constraints

A new ingredient in the determination of low energy $\bar{K}N$ parameters is the constraints arising from dispersion relations for the $K^{\pm}N$ forward scattering amplitudes, denoted by $F^{\pm}_N = D^{\pm}_N + iA^{\pm}_N$ in the laboratory frame. For a long time KN forward dispersion relations have been ill-determined, due mainly to the uncertainties in the sizeable $\Lambda(1405)$ contribution (10). However the situation changed with the Coulomb-nuclear interference measurements (11) of the real parts, $D^{\pm}_N(\omega)$, of the forward amplitudes at several values of ω, where ω is the kaon energy in the laboratory frame. This together with the improved knowledge of the $K^{\pm}N$ total cross sections, leaves much less freedom for the $\bar{K}N$ unphysical region contribution.

The relevance to the determination of the $\bar{K}N$ parameters can be seen from the dispersion relation results of the CERN-Caen group (11). In order to use their measurements of $D^{\pm}(\omega)$ they rewrote the once-subtracted relation

$$D^{+}(\omega) = D(0) + \sum_{Y=\Lambda,\Sigma} \frac{\omega R_Y g_Y^2}{\omega_Y(\omega-\omega_Y)} + \frac{\omega}{\pi}\int_{\omega_{th}}^{\infty}\left[\frac{A^{+}(\omega')}{\omega'(\omega'-\omega)} - \frac{A^{-}(\omega')}{\omega'(\omega'+\omega)}\right] d\omega' \quad (5)$$

in the form

$$\left[D^{+}(\omega) - I(\omega)\right](\omega-\omega_Y) = (\omega-\omega_Y)\, D^{+}(0) + \frac{R_\Lambda G^2}{\omega_Y}\,\omega, \quad (6)$$

where $I(\omega)$ denotes the contribution of the integral term in eq. (5). We first consider their results for the $K^{\pm}p$ relation and then discuss the $K^{\pm}n$ relation. In the $K^{\pm}p$ case, they replaced the Λ and Σ poles by a single effective pole, with a coupling $G^2 = g_\Lambda^2 + 0.844\, g_\Sigma^2$, at the average of the Λ and Σ pole positions (12). R_Y is a known kinematical factor and $g_Y^2 = g^2(YK^-p)$. $I(\omega)$ is calculated from the known total cross sections and low energy parametrizations. Thus the left hand side of eq. (6) is known at those values of ω at which the K^+p real part, $D^+(\omega)$, is measured, and at values $-\omega$ if the K^-p real part $D^-(\omega) = D^+(-\omega)$ is known. The known left hand side of eq. (6) is called a 'discrepancy function' and is denoted by $\Delta(\omega)$. Note that the right hand side is linear in ω. Thus by fitting the known values of $\Delta(\omega)$ to this linear form they determined G^2 and $D_p(0)$. In fact where this linear form intersects $\omega=\omega_Y$ we have simply $G^2 = \Delta(\omega_Y)/R_\Lambda$. They evaluated the $K^{\pm}p$ relation using several different low energy $\bar{K}N$ parametrizations; the results of two such calculations are shown in figure 3.

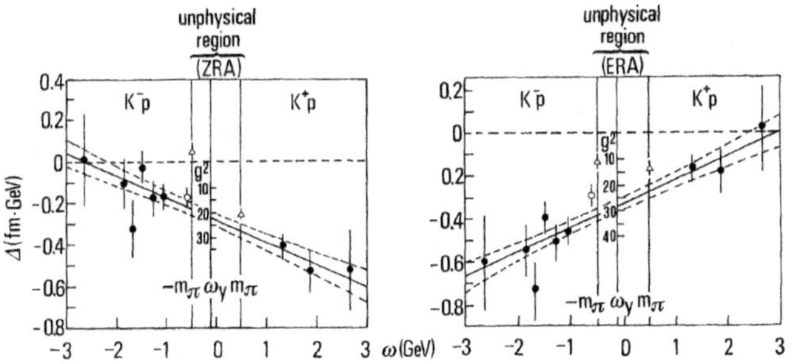

Figure 3 – $\Delta_p(\omega)$ as a function of ω, taken from ref. 11, calculated using two different $\bar{K}N$ parametrizations: (a) Martin and Sakitt (6), (b) Berley et al. (6). The full circles are the values of $\Delta_p(\omega)$ obtained using the measured $D^{\pm}p$, and the triangles come from the $K^{\pm}p$ scattering lengths. In our notation $g^2 = G^2$. When they include their more recent data they find (14) $G^2 = 18.6 \pm 2.6$.

For the present discussion the point to note is the inconsistency at the K^-p threshold, that is a discrepancy between the triangle and the linear form at $\omega=-m_K$. In other words the 'output' value of $D^-(m_K)$ obtained from the dispersion relation is very different from the 'input' value given by the $\bar{K}N$ parametrization used in the evaluation of the dispersion integrals. This is not surprising because the evaluation of the dispersion integral over the $\bar{K}N$ unphysical region involves an extrapolation of $A^-(\omega)$ below the K^-p threshold and across the S wave $\Lambda(1405)$ resonance. However it does mean that imposing self-consistency on the dispersion relation (in the sense of requiring equality between the input and output values of $D^-(m_K)$) will be an important constraint on any low energy parametrization.

The CERN-Caen results (15) for the $K^{\pm}n$ dispersion relation, obtained form their D_n^- measurements, are shown in figure 4.

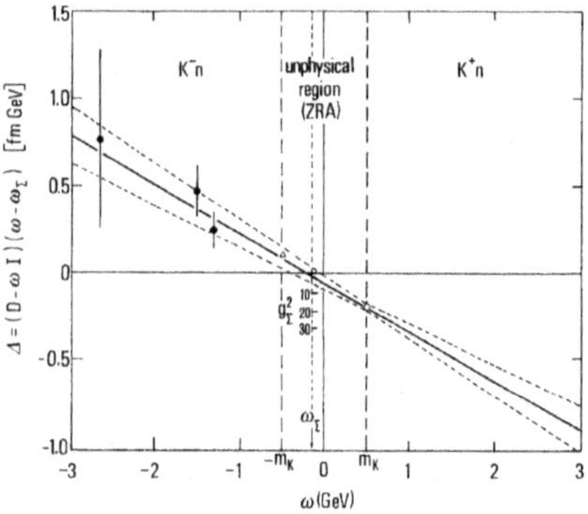

Figure 4 - $\Delta_n(\omega)$ versus ω for $K^{\pm}n$ scattering taken from ref. 15. In our notation the scale for g_Σ^2 is half the value shown.

Here only the Σ pole enters (and the $\Sigma(1385)$ in the unphysical region). We note the small g_Σ^2 coupling and the absence of any apparent discrepancy, but in this case there are no K^+ data. However a check of the results can be obtained from data for the charge exchange reaction $K^-p \to \bar{K}^0n$. In terms of the K^-p and K^-n amplitudes the forward charge exchange cross section is

$$\frac{d\sigma}{d\Omega_L}(K^-p \to \bar{K}^0n) = |D_p^- - D_n^-|^2 + |A_p^- - A_n^-|^2 \qquad (7)$$

and is therefore predicted knowing $D_{p,n}^-$ from dispersion relations and $A_{p,n}^-$ from the total cross sections.

The comparison (16), using the CERN-Caen results for $D_{p,n}^-$, shows the prediction does not reproduce the dip observed in the data near 300 MeV/c. The dip requires $D_p^- \simeq D_n^-$ at this momentum and the dispersion relation results do not satisfy this approximate equality.

Motivated by these discrepancies, a more ambitious M matrix analysis of the low energy K^-p and $K_L^o p$ data was attempted (17), which simultaneously used these same parameters in the evaluation of the low energy dispersion integrals over A_- and required that the $K^\pm p$ and $K^\pm n$ relations be satisfied at all energies at which reliable information on the real parts, $D^\pm$, exists. Specifically this combined analysis fitted

(a) the low energy K^-p and $K_L^o p$ data and branching ratios,

(b) the observed values of $D_p^\pm$ and D_n^-,

(c) the known KN scattering lengths,

$$\alpha_o = 0.04 \pm 0.10 \text{ fm}, \qquad \alpha_1 = -0.32 \pm 0.02 \text{ fm},$$

(d) the self-consistency of the $\bar{K}N$ amplitudes at threshold,

(e) the forward $\bar{K}^o p \rightarrow K^o n$ cross section data.

It was found impossible to obtain a satisfactory fit using energy independent M matrices (18). This contrasts with the constant M matrix analyses without analyticity constraints, where the low energy $\bar{K}N$ data can be well described, and moreover where the solution was not unique. Due to the proximity of the $\Lambda(1405)$ it was necessary to introduce extra parameters in the I = 0 M matrix. An effective range form was used $\underline{M} = \underline{A} + \underline{R}\, k^2$. The detailed results are given in ref. 17. Here we note that the corresponding $\bar{K}N$ scattering lengths, the values of A_I at the K^-p threshold, are

$$A_o = -1.66 + i\, 0.75 \text{ fm}$$

$$A_1 = 0.35 + i\, 0.66 \text{ fm}.$$

An interesting feature of imposing analyticity constraints is the increase in a_1 from 0 to 0.35 fm (20). This was to be expected as self-consistency at the K^-p threshold implies an increase in $(a_o + a_1)$, whereas the dip in the charge exchange cross section at 300 MeV/c requires a decrease in $(D_{I=0}^- - D_{I=1}^-)$. The I=0 $\pi\Sigma \rightarrow \pi\Sigma$ spectrum in the $\Lambda(1405)$ region was similar to solution B of Chao et al. (6), which is encouraging as no below threshold $\pi\Sigma$ constraints were included. However there is still some freedom in the extrapolation (17).

4. Recent information relevant to the determination of $\bar{K}N$ parameters

Since the above analysis was performed new information has become available, or is anticipated, from a variety of sources which should improve our knowledge of the kaon-nucleon interaction. We discuss these new data below and describe an up-dated analysis which takes them into account.

4.1 LBL study of K^-p charge exchange and elastic scattering

High statistics counter measurements of $K^-p \to \bar{K}^on$ have been made in the 0.5 - 1 GeV/c momentum range (21). The results for the forward cross section are shown in figure 5. To use such high precision data to constrain the dispersion relation value of $|D_p^- - D_n^-|$ requires care. First over much of the energy range the second term, $|A_p^- - A_n^-|^2$, of eq. (7) dominates the $K^-p \to \bar{K}^on$ forward cross section, and so the results depend sensitively on the value used for the K^-N total cross sections.

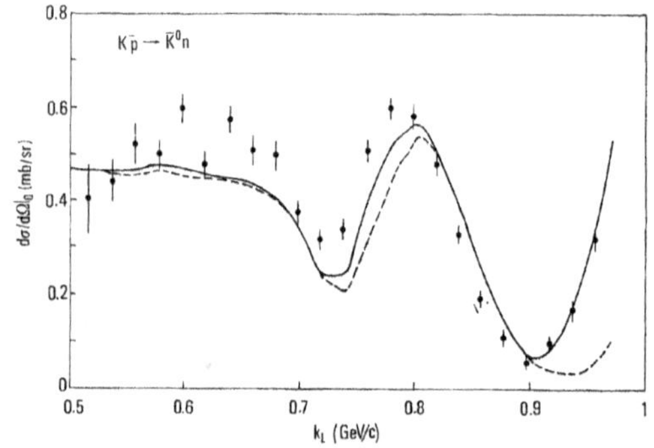

Figure 5 - The charge exchange differential cross section at 0°. The data are from ref. 21. The dashed curve is the prediction using only the imaginary part of the forward amplitude as given by the K^-p and K^-n total cross sections. The continuous curve is the result of including the real part as given by the $K^\pm p$ and $K^\pm n$ dispersion relation analysis of section 5. Only the data above 0.89 GeV/c are fitted in this analysis for the reason explained in the text.

The LBL partial wave analysis (22) of $K^-p \to K^-p$, K^on data in the region 0.36 - 1.32 GeV/c leads to values of $\sigma_T(K^-n)$ which have some significant differences from the measured values. We regard the partial wave predictions for $\sigma_T(K^-n)$ as more reliable. The dashed line in figure 5 shows the 'total cross section' contribution, $|A_p^- - A_n^-|^2$, to the forward $K^-p \to \bar{K}^on$. We conclude that it is only in the region 0.9 GeV/c, and above, that these data should be used to constrain $|D_p^- - D_n^-|$. The final prediction, including the dispersion relation real part contribution of section 5, is shown by the continuous curve. The residual disagreement between the curve and the data in the 0.6 - 0.8 GeV/c region can be accounted for by small changes in either the K^-n or the K^-p total cross section.

4.2 Importance of regeneration data

Recently the process $K_L^op \to K_s^op$ has been measured in the 0.5 - 1.2

GeV/c region by three independent groups (23). These data are particularly useful, as the regeneration amplitude is proportional to the difference of the $K^{\pm}n$ amplitudes. The forward cross section is given by

$$4\frac{d\sigma}{d\Omega_L}(K_L^o p \to K_S^o p) = |D_n^+ - D_n^-|^2 + |A_n^+ - A_n^-|^2 \qquad (8)$$

and the data are displayed in figure 6. Again the 'total cross section' contribution is shown by the dashed curve, and indicates that the data in the region 0.5 - 0.9 GeV/c will give a good determination of $|D_n^+ - D_n^-|$. The continuous curve shows the final result, including the dispersion relation contribution, for the cross section.

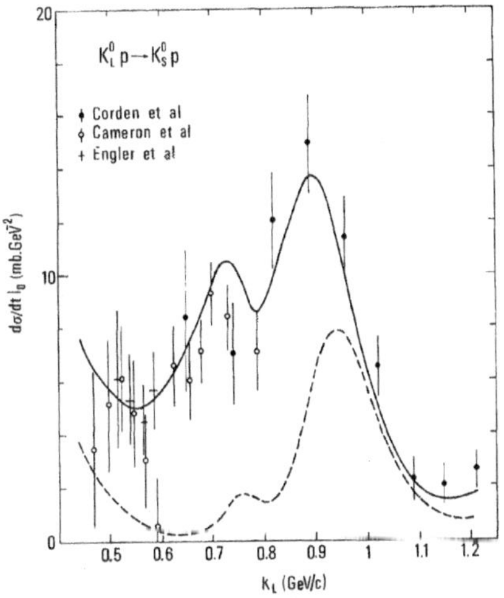

Figure 6 - The regeneration differential cross section at 0°. The data are from refs. 23. The dashed curve is the prediction using only the imaginary part of the forward amplitude as given by the $K^{\pm}n$ total cross section. The continuous curve is the result of including the real part as given by the $K^{\pm}n$ dispersion relation analysis of section 5.

The particular importance of these data is that they act as a constraint on D_n^+ in a region where no other experimental information exists (see section 4.3).

4.3 The I = 0 K^+N scattering length

The value of the I = 0 K^+N scattering length usually used in $K^{\pm}n$ dispersion relations is α_o = 0.04 fm (24). This was obtained from $K^{\pm}d$ data. Recently the uncertainties in this determination have been emphasized and a value α_o = -0.23 ± 0.18 fm proposed (25). The implications of this change can be deduced from figure 4; it would lead to an increased g_Σ^2 coupling constant. However with the regeneration data we now have a constraint on D_n^+. Indeed the overall analysis of section 5 used the new value of α_o, but found the $K^\pm n$ dispersion relations predicted $\alpha_o = 0.03$ (see figure 9).

4.4 Results of the K^-p TST collaboration

The progress in the analysis of the track sensitive target bubble chamber data has been reported by Miller (26). The collaboration plan to provide a complete set of K^-p data up to about 250 MeV/c with statistics comparable to or better than those of Kim (2), with a more careful study of the possible sources of bias than hitherto. So far (4) at-rest branching ratios are available, but only preliminary results for the scattering data. The preliminary $K^-p \to \bar{K}^o n$ cross section data are found to be significantly larger than those of Kim (2), see figure 1. The confirmation of this result is particularly important because such a change is required by the imposition of analyticity constraints on the M matrix analysis (17).

With the improved data it may be possible to include P waves in the low energy analysis. There are indications in the observed asymmetry of the LBL $\Lambda\pi^o$ production data (3) of an I = 1 P wave contribution at 250 MeV/c. Dispersion relations give further self-consistency conditions that have to be satisfied if P waves are included.

4.5 K^- hydrogen atom results

The measurement of the energy shift and width of the 2p → 1s X-ray transition in kaonic atoms of hydrogen are related to the K^-p scattering length at threshold. The existing data have been discussed by Batty (27) at this meeting. If the forthcoming experiment confirms the preliminary data (28) and if the method of interpretation of the data is essentially correct (29), then there will be a serious conflict with the results of the analysis of the low energy scattering data. It is found impossible to fit the scattering data with the additional constraint of the small K^-p scattering length implied by the mesic atom results. In the analysis described below this constraint is not imposed. Definitive kaonic hydrogen results are awaited with interest.

5. Up-dated $\bar{K}N$ parameters and amplitudes

Here we present the results of an M matrix analysis, with dispersion relation constraints, which includes the new information discussed in sections 4.1 - 4.4. The parameters of the combined fit to all types of data are listed in table 1, and the description of the low energy K^-p scattering data is shown by the curves on figure 1.

Table 1 - Parameter values for the combined fit to the kaon nucleon data

I = 0 M matrix parameters,		$M = A(fm^{-1}) + R(fm)\,k^2$:	
A_{KK}	= - 0.07	$A_{K\Sigma}$	= - 1.02
$A_{\Sigma\Sigma}$	= 1.94	R_{KK}	= 0.18
$R_{K\Sigma}$	= 0.19	$R_{\Sigma\Sigma}$	= - 1.09
I = 1 $K \equiv M^{-1}$ matrix elements in fm :			
K_{KK}	= 1.07	$K_{K\Sigma}$	= - 1.32
$K_{K\Lambda}$	= - 0.30	$K_{\Sigma\Sigma}$	= 0.27
$K_{\Sigma\Lambda}$	= 1.54	$K_{\Lambda\Lambda}$	= - 1.02
Subtraction and coupling constants (see eq.(5))			
$D_p(0)$	= - 2.83 fm	$D_n(0)$	= - 0.64 fm
g_Λ^2	= 13.9	g_Σ^2	= 3.4[a]

a) The value of g_Σ^2 is an upper bound. If a $\Sigma(1385)$ pole is included the best fit has $g^2_{\Sigma(1385)}$ consistent with zero, but it is hard to separate the Σ and $\Sigma(1385)$ contributions. However, if the $\Sigma(1385)$ pole is included with a residue fixed by SU(3), as given in ref. 38, then, for example, $g_\Sigma^2 \sim 0$, $D_n(0) = -0.32$ and $A_1 = 0.53 + 0.60i$; but the fit is not so satisfactory.

The additional $K_L^0 p \to K_S^0 p$ (and $K^- p \to \bar{K}^0 n$) forward data should lead to more reliable $K^\pm n$ amplitudes. The real parts of the forward amplitudes resulting from fit are shown in figures 7 and 8.

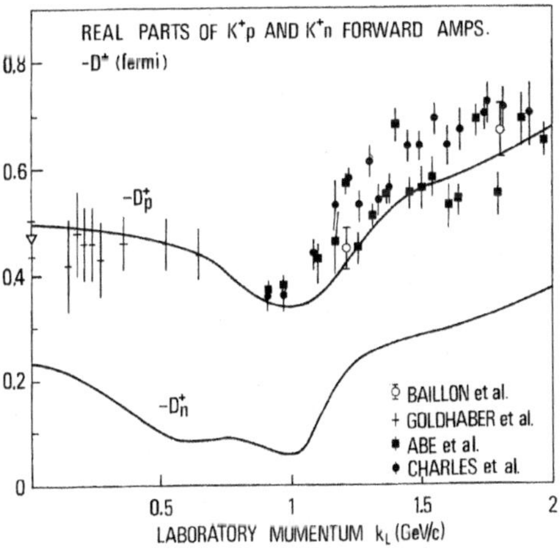

Figure 7 - The curves are the real parts of the K^+p and K^+n forward scattering amplitudes found in the analysis of section 5. Only the open points are used in the analysis; the remaining ones are obtained by extrapolating the K^+p elastic differential cross section data of refs. 32 to 0° and are taken from Queen's compilation (33).

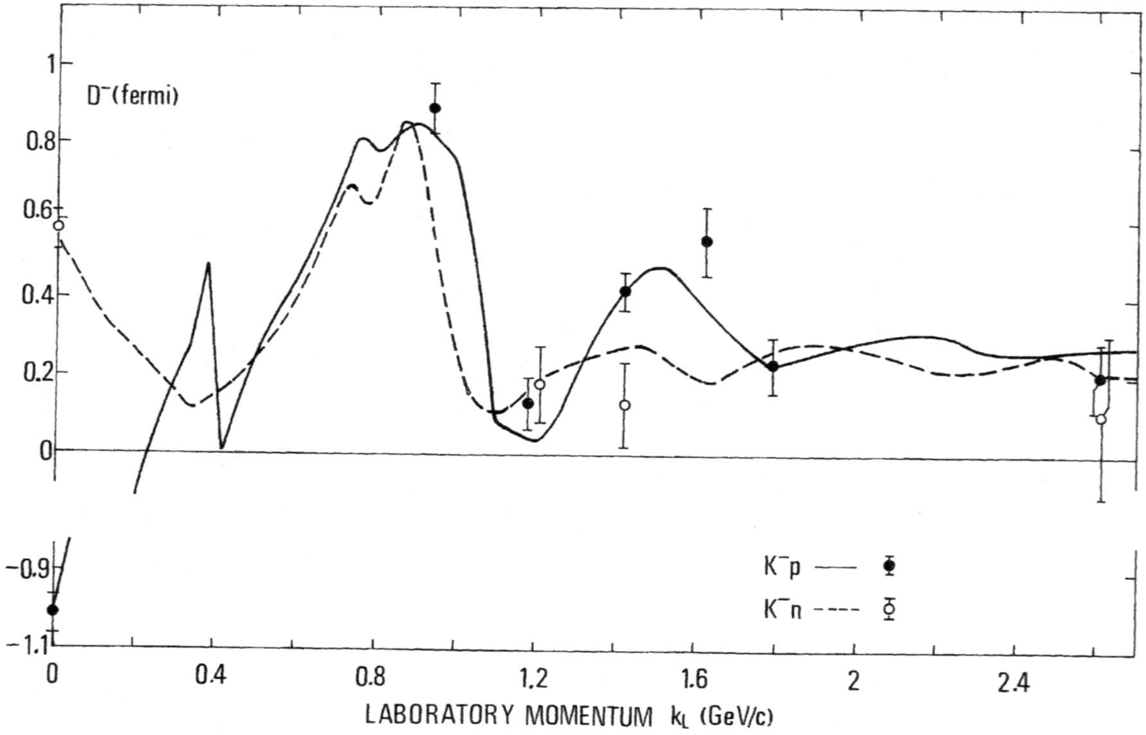

Figure 8 - The real parts of the K^-p and K^-n scattering amplitudes D^-_p and D^-_n respectively.

For the purposes of comparison we show in figure 9 the dispersion relation results in terms of the discrepancy functions, $\Delta(\omega)$, defined by the left hand side of eq.(6).

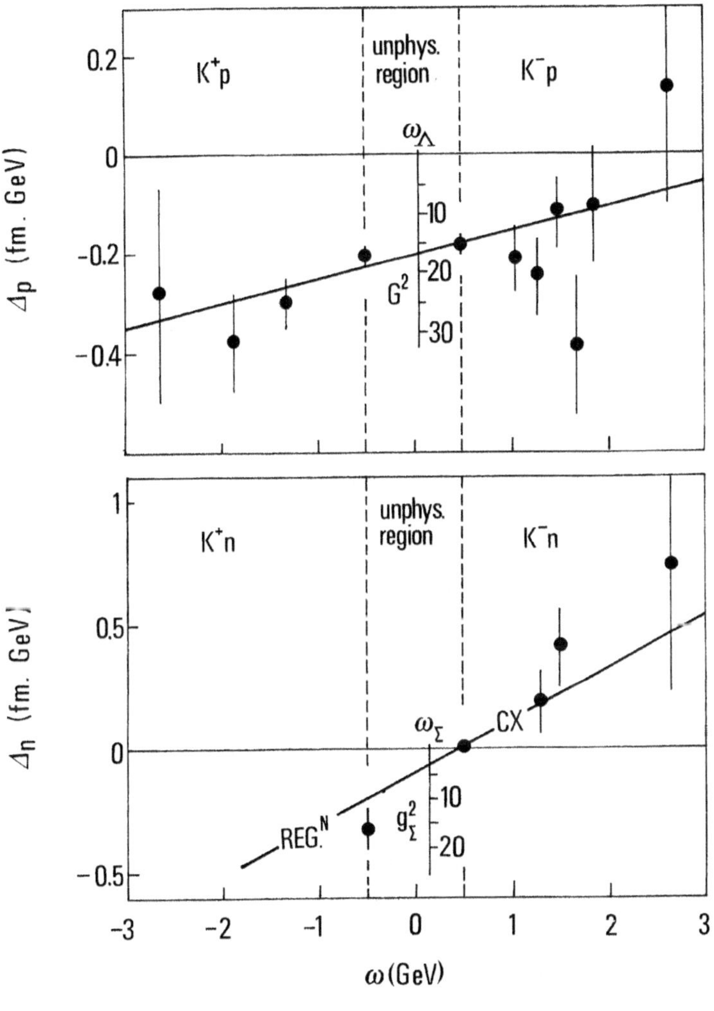

Figure 9 - Plots of Δ_p and Δ_n versus ω obtained from the $K^{\pm}p$ and $K^{\pm}n$ dispersion relations of section 5. Note that here Δ has been defined in terms of D^- so that to compare with figures 3 and 4 we must reflect about $\omega=0$.

However the analysis is performed using eqs. (5) with $D_p(0)$, $D_n(0)$, g_Λ^2 and g_Σ^2 as free parameters, and so near the pole positions we do not have a linear form for $\Delta_p(\omega)$. Moreover the lines are not simply the best fits to the data points shown; there are many other constraints.

In figure 10 we show the energy dependence of the $\bar{K}N$ scattering lengths, $A_I = a_I + ib_I$ of eq. (2). The values at the K^-p threshold are

$$A_o = -1.70 + i\,0.68 \text{ fm}$$

$$A_1 = 0.37 + i\,0.60 \text{ fm}.$$

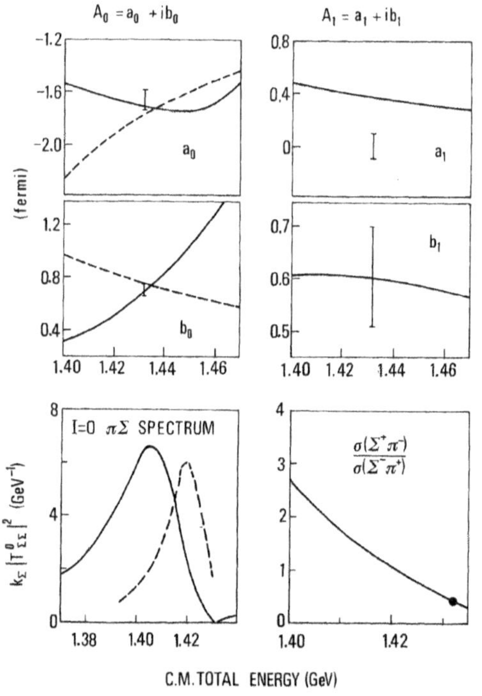

Figure 10 – The continuous curves show the energy dependence of the $\bar{K}N$ scattering lengths, $A_I \equiv a_I + ib_I$, corresponding to the parameters of table 1. The error bars give the range of A_I found in the early K matrix analyses (6). The use of a Breit-Wigner parametrization for the I = 0 S wave, which leads to the dashed curves, does not give a satisfactory description of the data. We also show the S wave contribution to the I = 0 $\pi\Sigma$ spectrum and the ratio of Σ^+/Σ^- production below threshold

6 The nature of the Λ(1405) resonance

In the above analyses the Λ(1405) arises as a $\bar{K}N$ 'bound state' resonance, whereas in the quark model we would expect a CDD pole interpretation (30). An alternative simple parametrization is therefore to try a Breit-Wigner description of the Λ(1405). In this description all the I = 0 K matrix elements (and also A_o) have a pole, but unfortunately considerably displaced from 1405 MeV, and distant from the $\bar{K}N$ physical region (36)

$$K_{ij} = \frac{g_i g_j}{E_o - E}, \qquad A_o = \frac{g_1^2}{E_o - E - i q_\Sigma g_2^2} .$$

Also if the I = 0 S wave is parametrized in terms of g_1, g_2 and E_o no adequate description of the data is possible. The parametrization leads to an energy dependence of A_o shown by the dashed curves in figure 10, which is very different from the behaviour required by the data (37).

To be fair no adequate fit was possible with a three parameter I = 0 constant M matrix either; effective range parameters were necessary. Thus we are led to include a background contribution

$$K_{ij} = \frac{g_1 g_j}{E_o - E} + (K_o)_{ij} .$$

Unfortunately in such a fit the pole residues become very small and K_o is such that we revert to the $\bar{K}N$ bound state description. A similar result is obtained by McGinley (19). In this work he also performs model calculations (34) of the Λ(1405) and finds, for instance, if the CDD pole is broad enough the resulting K matrix may have no nearby pole. The quark model multiplet interpretation of the Λ(1405) remains an open question.

6. Conclusion

With the exception of the preliminary kaonic hydrogen atom results, it is possible to obtain a self-consistent M matrix description of all the low energy $\bar{K}N$ data, simultaneously satisfying the dispersion relation constraints coming partly from the measured real parts of the forward $K^{\pm}N$ amplitudes and partly from the forward $K^- p \to K^o n$ and $K_L^o p \to K_S^o p$ data. The dispersion relation parameters and Λ and Σ couplings are stable to the introduction of the recent data into the analysis.

The imposition of the analyticity constraints enables the energy dependence of the I = 0 S wave $\bar{K}N$ amplitude to be determined and it is found to extrapolate to a Λ(1405) resonance which is compatible with the peak in the spectrum for Σπ production. However the dynamical origin of the Λ(1405) is not revealed, and a better determination of the Λ(1405) must await $\Sigma\pi \to \Sigma\pi$ data. With the advent of high energy hyperon beams, such information can now be contemplated, namely the study of π exchange reactions of the type $\Sigma N \to \pi\Sigma N$ with the produced $\pi\Sigma$ system going forward.

References and notes

1. Humphrey,W.E., and Ross, R.R.: 1962, Phys. Rev. 127, 1305.
   Sakitt, M., et al.: 1965, Phys. Rev. 139B, 719.
   Abrams, G.S., and Sechi-Zorn, B.: 1965, Phys. Rev. 139B, 454.
   Kittel, W., Otter, G., and Wacek, I.: 1966, Phys. Lett. 21, 349.
   Tovee, D., et al.: 1971, Nucl. Phys. B33, 493.
2. Kim, J.K.: 1966, Columbia University Report, Nevis 149.
3. LBL collaboration:
   Mast, T.S., et al.: 1975, Phys.Rev. D11, 3078; ibid: 1976, D14, 13.
   Tripp, R.D., private communication.
4. TST collaboration :
   Nowak, R., et al.: 1978, Nucl. Phys. B139, 61.
   Goossens, M.: 1978, Brussels University thesis.
   Bedford, N.H., et al.: 1979, Contribution to Conf. on Hyperfragment and K^-Physics, Jablonna, Warsaw.
   Bedford, N.H.: 1979, Durham University thesis.
   Goossens, M., et al., contribution to this Meeting.
   Miller, D.J., these Proceedings.
5. Kadyk, J.A., et al.: 1966, Phys. Rev. Lett. 17, 599; 1968, UCRI 18325.
   Sayer, G.A., et al.: 1968, Phys. Rev. 169, 1045.
6. Kim, J.K.: 1967, Phys. Rev. Lett. 19, 1074.
   Martin, B.R., and Sakitt, M.: 1969, Phys. Rev. 183, 1345.
   Martin, A.D., and Ross, G.G.: 1970, Nucl. Phys. B16, 479.
   Berley, D.,et al.: 1970, Phys. Rev. D1, 1996.
   Chao, Y.A., et al.: 1973, Nucl. Phys. B56, 46.
7. Although the M matrix elements are free of the 'right-hand' $\bar{K}N$, πY threshold singularities, they do contain the 'left-hand' crossed channel singularities of the S wave amplitudes. It may not be safe to ignore these if we extrapolate too far below threshold.
8. Dalitz, R.H., and Tuan, S.F.: 1960, Ann. Phys. 10, 307.

9. The fit uses an incoherent sum of intensities arising from the I =0 $\pi\Sigma \to \pi\Sigma$ amplitude, phase space, and small $\Sigma(1385)$ and $\Lambda(1520)$ resonance contributions.

10. Queen, N.M., Restignoli, M., and Violini G.: 1969, Fort. der Physik 17, 467; ibid : 1973, 21, 569.

11. Baillon, P., et al.: 1976, Nucl. Phys. B105, 365.

Baillon, P., et al.: 1974, Phys. Lett. 50B, 377.

Jenni, P., et al.: 1976, Nucl. Phys. B105, 1.

12. Note that although the value of G^2 is not sensitive to the precise choice of the effective pole position, the same is not true for $D_p(0)$. To determine $D_p(0)$ reliably we need to known both g_Λ^2 and g_Σ^2, and so have to consider both $K^\pm p$ and $K^\pm n$ relations. This is the origin of the discrepancy in $D_p(0)$ noted in ref. 13.

13. Di Claudio, B., Queen, N.M., and Violini, G.: 1979, Nucl. Phys.B161, 238.

14. Baillon, P.,et al.: 1976, Nucl. Phys. B107, 189.

15. Baillon, P., et al.: 1976, Phys. Lett. 61B, 171.

16. Tripp, R.D.: 1976, Proc. of Oxford Conf. on Baryon Resonances, Eds. R.T. Ross and D.H. Saxon, 496.

17. Martin, A.D.,: 1976, Phys. Lett. 65B, 346.

18. McGinley (19) notes that this is not entirely due to the dispersion relation constraints, but is partly due to the inclusion of the high precision LBL data (3) in the analysis.

19. McGinley, J.C.: 1979, Oxford University thesis.

20. After the workshop we received the results of a similar analysis by McGinley (19) in which this change is not found. This analysis gives less emphasis to the dispersion relation constraints and so, in particular, gives a better description of the $K^-p \to \bar{K}^0n$ data of Kim (2). However the recent indications are that this charge-exchange cross section is significantly larger (4) and in agreement with the results of ref. 17. The I = 0 $\pi\Sigma$ spectrum of the parametrization of ref. 19 does not have a conventional $\Lambda(1405)$ structure.

21. Alston-Garnjost, M., et al.: 1978, Phys. Rev. D17, 2226.

22. Alston-Garnjost, M., et al.: 1978, Phys. Rev. D18, 183.

23. Cameron, W., et al.: 1978, Nucl. Phys. B132, 189.

Engler, A., et al.: 1978, Phys. Rev. D18, 3061.

Corden, M.J., et al. :1979, Nucl. Phys. B155, 13.

24. Stenger, V.J., et al.: 1964, Phys. Rev. 134B, 1111.

25. Martin, B.R.: 1978, J. of Phys. G4, 335.

26. Miller, D.J., these Proceedings.

27. Batty, C.J., these Proceedings.

28. Davies, J.D., et al.: 1979, Phys. Lett. 83B, 55.

29. The substantial modifications to the K mesic atom analysis so far proposed (31) are not compelling, see ref. 19.

30. This was originally discussed by Dalitz (34). A detailed theoretical investigation has been made recently by McGinley (19), see ref. 35.

31. Deloff, A., and Law, J.: 1979, Phys. Rev. C20, 1597.

32. Goldhaber G., et al.: 1962, Phys. Rev. Lett. 9, 135.

Abe, K. et al.: 1975, Phys. Rev. D11, 1719.

Charles, B.J., et al.: 1977, Nucl. Phys. B131, 7.

33. Queen, N.M.: 1978, University of Birmingham Report UB-KP-1-78.

34. Dalitz, R.H.: 1970, HR-70-03, presented at the Duke University Hyperon Conference.

35. Dalitz, R.H., these Proceedings.

36. Martin, A.D.: 1977, Nukleonika 22, 857.

37. It is interesting to note that the I = 0 S wave determined in the LBL partial wave analysis (22) corresponds to $A_0 \approx 1{\cdot}0 + i2{\cdot}7$ fm at 360 MeV/c, supporting the rapidly increasing b_0 of figure 10.

38. Nagels, M.M., et al.: 1976, Nucl. Phys. B109, 1.

DISCUSSION

- G. Oades :

Do you have any comments on other determinations of $D_{K^{\pm}p}(0)$ which suggest a value closer to -2 fm?

- A. D. Martin :

The value found for $D_p(0)$ depends sensitively on the treatment of the Λ and Σ pole contributions. I believe a value of about -2 fm corresponds to taking an effective pole at the average of the Λ and Σ positions. With an exact treatment of the poles the value of $D_p(0)$ is near -3 fm.

KN DISPERSION THEORY

N.M. Queen
Department of Mathematical Physics - University of Birmingham, England

1. Historical introduction

As is evident from previous reviews (1-4), the techniques of dispersion theory have been applied very extensively to kaon-nucleon interactions, and the subject has a long and chequered history. The objective of the present review is to survey the present state of the art, with special emphasis on some of the developments of the past few years, which are not covered by the previous reviews. We make no claim to completeness, and attempt merely to indicate several of the current major trends. We shall also avoid excessive technical detail, and we refer the reader who desires further information to the cited references.

When the first applications of dispersion relations to strong interactions appeared about 25 years ago, their importance was quickly realized. The hypothesis of analyticity of the scattering amplitudes as functions of their kinematic variables imposes strong constraints on their possible behaviour and implies intimate relationships between their values in different kinematic regions. The utilization of those few general principles which could be regarded as reasonably well established was of great significance in the absence of any viable dynamical theory of the strong interactions. Even now, with the advent of QCD, which purports to be a fundamental theory of the strong interactions, dispersion theory remains one of the most powerful tools at our disposal for the phenomenological analysis of scattering data on strong-interaction processes; from a practical point of view, we seem to be no nearer the goal of predicting (or even understanding) the detailed behaviour of specific scattering amplitudes from first principles than we were 25 years ago.

The earliest applications of dispersion relations were made to pion-nucleon scattering, for which the experimental data were most abundant, and the success of the dispersion approach for this process has been quite impressive.

The πNN coupling constant has now been determined (5) with great accuracy,

E. Ferrari and G. Violini (eds.), Low and Intermediate Energy Kaon-Nucleon Physics, 115–130.

$$g^2_{\pi NN} = 13.90 \pm 0.36 \qquad (1)$$

and the forward dispersion relations for πN scattering have been found to be satisfied up to the highest accelerator energies. The successful applications of non-forward fixed-t dispersion relations and partial-wave dispersion relations for πN scattering are also well known.

Historically, the development of KN dispersion relations followed that of πN dispersion relations, using the latter as a model. Attempts were made to deduce the magnitude of the $KN\Lambda$ and $KN\Sigma$ coupling constants, $g_{KN\Lambda}$ and $g_{KN\Sigma}$, by analyzing forward KN dispersion relations. This problem was of particular interest, since SU(3) symmetry (6) yields the definite predictions

$$\begin{aligned} g^2_{KN\Lambda} &= \tfrac{1}{3}(1 + 2\alpha)^2 \, g^2_{\pi NN} \\ g^2_{KN\Sigma} &= \tfrac{1}{3}(1 - 2\alpha)^2 \, g^2_{\pi NN} \end{aligned} \qquad (2)$$

where $\alpha = f/(f + d)$ is the fraction of F-type coupling. Higher symmetries lead to more stringent predictions. Thus, SU(6) symmetry (7) would require that $\alpha = 0.4$, i.e., $g^2_{KN\Lambda} = 1.08 g^2_{\pi NN}$ and $g^2_{KN\Sigma} = 0.04 g^2_{\pi NN}$.

In the early applications of KN dispersion relations, it quickly became apparent that their analysis was far more problematical than in the case of πN scattering. The main reason for this is the existence of the unphysical energy region between the $\pi\Lambda$ and $\bar{K}N$ thresholds in the $\bar{K}N$ scattering amplitude, where there is an important contribution from the Y_0^* (1405) resonance in the I = 0 S-wave state. In the absence of a complete dynamical theory, the structure of the amplitude in this region can in principle be established only by analytic continuation from the low-energy physical region. The required extrapolation, which is usually carried out by means of some definite parametrization of the low-energy $\bar{K}N$ scattering data, proves to be rather unstable: different parametrizations give different structures in the unphysical region, which in turn lead to different predictions when inserted in the dispersion relations. A large part of the effort in KN dispersion theory in the past two decades has been devoted to the problem of pinning down the structure in the unphysical region or, alternatively, suppressing its importance in the analysis of the data.

The imaginary parts of the forward scattering amplitudes are known with great accuracy, via the optical theorem, up to the highest accelerator energies. Therefore, once the model for the unphysical region is fixed and the values of the coupling constants have been determined, the dispersion relations give direct predictions for the real parts of the amplitudes. Such predictions have been of special interest, since the experimental data on the real parts are much more limited, both in quantity and in energy. In particular, most of the extrapolations of the differential cross sections to the forward direction have provided information on only the modulus, but not the sign, of the real part of the amplitude, the exceptions being the limited number of energies at which careful observations of the Coulomb interference have been made. In certain energy regions, there are also significant systematic discrepancies between the real parts obtained from different independent experiments, which appear to be due to errors in the normalization of the cross sections. Notable ambiguities also exist in phase-shift analyses at low energies. Since dispersion relations severaly constrain the possible energy dependences of scattering amplitudes, they provide a powerful means of eliminating many of these ambiguities.

2. Experimental information

Our experimental knowledge of forward KN scattering can be summarized briefly as follows.

The $K^{\pm}p$ and $K^{\pm}n$ total cross sections have been measured at energies up to 240 GeV (8) and are known with good accuracy over the entire range of lower energies. Of course, the Kn data are somewhat less reliable than the Kp data, since they must be extracted from experiments on scattering by the deuteron. A rather complete tabulation of the experimental data on the $K^{\pm}p$ total cross sections, together with a careful treatment of the errors in these data, can be found in Ref. (9). Various Regge models may be used to estimate the asymptotic behaviour of the cross sections.

Our experimental knowledge of the real parts of the $K^{\pm}p$ forward scattering amplitudes has improved dramatically during the past decade, although relatively few experiments of this type have been performed in the past few years. A compilation of all known real parts published in 1971 (10) included 49 $K^{+}p$ data points and 70 $K^{-}p$ data points. That compilation has subsequently been brought up to date periodically; the latest edition (11), in which a number of older dubious data points have been discarded and many more accurate points have been added, contains 113 $K^{+}p$ data points at kaon laboratory momenta between 0.175 and 150 GeV/c, and 94 $K^{-}p$ data points between 0.3 and 150 GeV/c. The sign of the $K^{+}p$ real part has been determined in Coulomb-interference measurelents at 15 momenta, and the sign of the $K^{-}p$ real part has been determined at 18 momenta.

Experimental data on the real parts of the $K^{\pm}n$ forward scattering amplitudes (12) are extremely scarce, and this limits the accuracy of analyses based on dispersion relations for Kn elastic scattering.

In many of the early analyses (1), the dispersion relations were evaluated at a single energy, which was often taken to be the threshold energy. As more and more real-part data were accumulated, it became increasingly advantageous to perform global fits to the entire set of data, and this approach is now used almost exclusively.

3. The once-substracted dispersion relation

One of the simplest methods which permits full use of all the real-part data was first applied to Kp scattering by Perrin and Woolcock (13). This method is based on a standard once - subtracted dispersion relation, which in the case of Kp scattering takes the form (1)

$$D_{\pm}(\omega) = C \pm \sum_{Y} \frac{X(Y)}{\omega \pm \omega_Y} \mp \frac{\omega}{\pi} \int_{\omega_{\pi\Lambda}}^{m_k} \frac{A_-(\omega')\, d\omega'}{\omega' (\omega' \pm \omega)}$$

$$+ \frac{\omega}{4\pi^2} P \int_{m_k}^{\infty} \frac{k'}{\omega'} \left[\frac{\sigma_{\pm}(\omega')}{\omega' - \omega} - \frac{\sigma_{\mp}(\omega')}{\omega' + \omega} \right] d\omega', \qquad (3)$$

where, following the conventional notation, $F_{\pm} \equiv D_{\pm} + iA_{\pm}$ is the $K^{\pm}p$ forward scattering amplitude in the laboratory system, k and ω are the laboratory momentum and energy of the kaon, $\sigma_{\pm}$ are the $K^{\pm}p$ total cross sections,

$$X(Y) = \frac{(m_Y - m_p)^2 - m_k^2}{4m_p^2} g_{KNY}^2$$

are the residues of the hyperon poles ($Y = \Lambda, \Sigma$), and

$$\omega_Y = \frac{m_Y^2 - m_p^2 - m_k^2}{2m_p}$$

are their positions; the constant C is related to the zero-energy subtraction constant $D_{\pm}(0)$ by

$$C = D_{\pm}(0) - \sum_{Y} \frac{X(Y)}{\omega_Y} .$$

If it is assumed that the pair of hyperon poles can be replaced by a single effective pole near the position of the Λ, the dispersion relation (3) contains only two undetermined constants, namely the effective residue $X(\Lambda) + X(\Sigma)$, which is proportional to the effective coupling constant

$$G^2 \equiv g_{KN\Lambda}^2 + 0.84 g_{KN\Sigma}^2, \qquad (4)$$

and the subtraction constant $D_{\pm}(0)$.

The method consists in finding the values of these parameters which, when inserted in the dispersion relation (3), give the best fit to all the experimental data on $D_{\pm}$.

As we have already pointed out, the sign of $D_{\pm}$ is unkonwn at most energies. This problem can be overcome by simply finding that set of signs which results in the best overall fit, i.e. the set of signs which is most consistent with the dispersion relation. This procedure was followed in many of the calculations described in detail below.

The most recent analysis of this type was carried out in Ref. (14), using the 207 real-part data points from the compilation of Ref. (11). It was found that 13 points gave individual contributions to χ^2 greater than 9 (i.e., they deviated from the best fit by more than three standard deviations), and a final fit was obtained after rejecting these points (the same rejection criterion is adopted in many other calculations). Fits were found with two alternative parametrization for the imaginary part of the K^-p scattering amplitude in the unphysical region: the constant K-matrix parametrization of Martin and Sakitt (15) and the K-matrix parametrization of Martin (16), which includes dispersion-relation constraints. These fits gave the following values for the effective KNY coupling:

$$G^2 = 14.3 \pm 0.5, \quad \chi^2/\text{NDF} = 246/192 \quad \text{(Martin and Sakitt: 15)}, \qquad (5a)$$

$$G^2 = 16.6 \pm 0.7, \quad \chi^2/\text{NDF} = 242/192 \quad \text{(Martin: 16)}. \qquad (5b)$$

The values of the parameter $D_{\pm}(0)$ for these fits are -1.68 fm and -1.87 fm, respectively. We note that the second of these values is somewhat less negative than the value $D_{\pm}(0) = (-2.82 \pm 0.25)$ fm found by Martin (16) from his simultaneous determination of the K-matrix elements and the parameters G^2 and $D_{\pm}(0)$, although his value for G^2, namely $G^2=16.8 \pm 1.2$, is consistent with the value given in (5b). The reason for the discrepancy in $D_{\pm}(0)$ has now been understood (see the concluding talk by G. Violini at this Workshop).

4. The Haber-Schaim method

The advantages of exploiting different types of dispersion relations for a given process are well known. Different variants of the dispersion relations would give identical results only if the input data exactly satisfied the required analyticity properties. In the general case of error-affected data, a comparison of the results of different alternative methods provides a valuable check of the overall consistency of the input data and thus makes it possible to draw conclusions with greater confidence.

An alternative method which has received some attention is based on a technique originally proposed by Haber-Schaim (17) for πN scattering. By writing down an unsubtracted dispersion relation for $F_-(\omega) - F_+(\omega)$, the difference between the K^-p and K^+p forward scattering amplitude, and approximating the two hyperon poles by a single effective pole at $\omega = \omega_Y$, it can be shown (1) that the function

$$L(\omega) \equiv (\omega_Y^2 - \omega^2) \left[\frac{D_-(\omega) - D_+(\omega)}{2\omega} - \frac{1}{\pi} \int_{\omega_{\pi\Lambda}}^{m_k} \frac{A_-(\omega')\, d\omega'}{\omega'^2 - \omega^2} - \frac{\omega^2}{4\pi^2} \int_{m_k}^{\infty} \frac{k' \left[\sigma_-(\omega') - \sigma_+(\omega') \right] d\omega'}{\omega'^2 (\omega'^2 - \omega^2)} \right] \quad (6)$$

is equal to

$$L(\omega) = \sum_Y X(Y) - \frac{\omega^2 - \omega_Y^2}{4\pi^2} \int_{m_K}^{\infty} \frac{k' \left[\sigma_-(\omega') - \sigma_+(\omega') \right] d\omega'}{\omega'^2} \quad (7).$$

The method consists in evaluating (6) from experimental data at each energy ω at which both $D_+(\omega)$ and $D_-(\omega)$ are known and then finding the best fit of the resulting "experimental" values of $L(\omega)$ to the general functional form required by (7), namely

$$L(\omega) = A_0 - A_1(\omega^2 - \omega_Y^2). \quad (8)$$

From (7) we can make the obvious identifications

$$A_0 = X(\Lambda) + X(\Sigma), \quad (9)$$

$$A_1 = \frac{1}{4\pi^2} \int_{m_K}^{\infty} \frac{k' \left[\sigma_-(\omega') - \sigma_+(\omega') \right] d\omega'}{\omega'^2} . \quad (10)$$

Thus A_0 in proportional to the effective coupling G^2, while Eq. (10) constitutes a sum rule for A_1, which can be used either as a consistency test of the fit or as a constraint in the fit itself.

Of course, the fact that $D_+(\omega)$ and $D_-(\omega)$ are required at the same set of energies is a limitation of the method. But this is not a very serious defect, since the K^+p real parts can be safely interpolated in many energy regions where there is little structure in the amplitude.

In the most recent analysis of this method (14), 40 values of $L(\omega)$ at energies between 0.6 and 150 GeV were constructed and fitted. Using the low-energy parametrization of Refs. (15, 16), the unconstrained fits gave the following results (after rejection of one point):

$$G^2 = 14.2 \pm 1.9, \quad \chi^2/NDF = 45.1/37 \quad \text{(Martin and Sakitt: 15)}, \quad (11a)$$

$$G^2 = 16.4 \pm 1.9, \quad \chi^2/NDF = 44.3/37 \quad \text{(Martin: 16)}. \quad (11b)$$

The best-fit value of the parameter A_1 in either case was in almost perfect agreement with the sum rule (10).

More recently, the Haber-Schaim tecnique was used to analyze the $K^o_L p \rightarrow K^o_S p$ regeneration amplitude (18). By isospin invariance, that amplitude is proportional to the difference between the $K^{\pm}n$ elastic scattering amplitudes and therefore contains only the Σ pole. As had already been emphasized by others authors (19, 20), the kaon nucleon regeneration amplitudes (and also charge-exchange amplitudes) contain valuable information about the real parts which had not previously been exploited.

There exist only several data points for the real parts of the $K^{\pm}n$ elastic scattering amplitudes, and an analysis of a once-substracted Kn dispersion relation using these data (12) gave a value for the $KN\Sigma$ coupling constant with a rather large uncertainly:

$$g^2_{KN\Sigma} = 0.8 \pm 3.2 \quad (12)$$

It was shown in Ref. (18) that a substantial improvement in the accuracy can be achieved by exploiting the much larger set of experimental data on the regeneration phase. An analysis of the Haber-Schaim type using 67 data points gave the result

$$g^2_{KN\Sigma} = 0.9 \pm 0.4 \quad (13)$$

5. Analysis of the modulus of the amplitude

It is well known that the phase of an amplitude is completely determined by its modulus on the cuts, provided that the positions of the zeros and poles are known. One of the most direct expressions of this relationship is a dispersion relation for the logarithm of the amplitude. The first serious attempt to study such a logarithmic dispersion relation for Kp scattering was made in Ref. (21). The values of $D_{\pm}(\omega)$ predicted by this dispersion relation ware found to be in general qualitative agreement whit the experimental data, although some notable discrepancies were observed. A particular feature of this method is that the results depend strongly on the position of the zeros of the amplitude in the complex energy plane (instead of the residues of the poles, as in the ordinary dispersion relations).

The importance of the zeros in determining the structure of the amplitude was also stressed in another work (22), where an analysis was made of the modulus of the scattering amplitude. In particular, it was demonstrated that there is a strong correlation between the positions of the zeros and the residues of the poles, so that a knowledge of the zeros leads to definite constraints on the coupling constants, or vice versa.

More recently, in a somewhat related work (23), a determination was made of the domain of values of the $KN\Lambda$ and $KN\Sigma$ coupling constants consistent with experimental information on the modulus of the $K^{\pm}p$ forward scattering amplitude. The method is as follows.

The complex ω-plane is conformally mapped into the unit circle in a new variable z in such a way that $z(-m_K) = -1$, $z(\omega_\Lambda) = 0$ and $z(\omega_{\pi\Lambda}) = 1$; these properties determine the transformation $z(\omega)$ uniquely. In particular, the cuts in the energy plane are mapped onto the unit circle $|z| = 1$, and the cut-free region (which includes the hyperon poles) goes onto the interval $-1 < z < 1$. One then defines the function

$$g(z) \equiv zF_-(z) \left(\frac{z - z_\Sigma}{1 - zz_\Sigma} \right), \tag{14}$$

were $F_-(z)$ is the K^-p scattering amplitude as a function of the new variable, and $z_\Sigma \equiv z(\omega_\Sigma)$. It is easy to see that $g(z)$ is analytic inside the unit circle and has the property $|g(z)| = |F_-(z)|$ on the circle $|z| = 1$.

The function $g(z)$ is then represented in the form

$$g(z) = \chi(z) \ C(z), \tag{15}$$

where

$$\chi(z) = \prod_{i=1}^{N} \frac{z - z_i}{1 - zz_i^*} \tag{16}$$

is a product of Blaschke factors corresponding to the N zeros of the amplitude, and $C(z)$ is an analytic function which has no zeros inside the circle. Under certain general conditions, $C(z)$ is uniquely determined in terms of the values of $|g(z)|$ on the unit circle through the expression

$$C(z) = \exp\left\{ \frac{1}{2\pi} \oint \left[\log |g(z')| \right] \frac{z' + z}{z' - z} \frac{dz'}{iz'} \right\}, \tag{17}$$

where the integration is over the circle $|z| = 1$.

If the positions of the zeros are known, the Blaschke product (16) can be constructed explicitly; in general, the function $C(z)$ can be computed numerically from (17) in terms of $|g(z)|$ on the circle, which in turn is given by $|F_-(\omega)|$ on the cuts. Thus, when the zeros are known, the complete amplitude $F_-(\omega)$ can be reconstructed by means of Eqs. (14) and (15); in particular, the couplings $g^2_{KN\Lambda}$ and $g^2_{KN\Sigma}$ are determined.

In the more realistic case when the zeros are unknown, or when one assumes only partial knowledge of them, one finds, as in Ref. (22), a strong correlation between the positions of the zeros and the values of the coupling constants. But even in the extreme case in which no assumptions are made about the number and positions of the zeros, absolute (albeit rather weak) bounds on the admissible values of $g^2_{KN\Lambda}$ and $g^2_{KN\Sigma}$ are obtained. The results are insensitive to the specific assumptions made about the amplitude in the unphysical and asymptotic regions.

This method of analyzing data on the modulus of an amplitude has also been applied in Ref. (18) to the $K^o_L p \rightarrow K^o_S p$ regeneration amplitude, which contains the Σ pole. With no assuptions about the zeros, one finds the upper limit

$$g^2_{KN\Sigma} < 6.0. \quad (18)$$

A preliminary analysis of the regeneration amplitude by means of a logarithmic dispersion relation (24) has recently suggested that, apart from the zero at $\omega = 0$ required by the crossing properties of the amplitude $F_-(\omega) - F_+(\omega)$, this amplitude has a single pair of complex-conjugate zeros on the imaginary axis at $\omega = \pm ia$, with 0.3 GeV $\lesssim a \lesssim$ 0.4 GeV. If the Blaschke factors corresponding to such zeros are included explicitly in the analysis, it is found that

$$g^2_{KN\Sigma} = 2.5 \pm 0.3, \quad (19)$$

which represents a substantial refinement of (18). However, this result is in conflict with (13). It should be noted that the value (19) depends strongly on the zeros.

6. Analyses of discrepancy functions

All the methods considered so far require definite models for the evaluation of dispersion integrals over the unphysical and asymptotic energy regions, where no data are available. The amplitudes in these regions must be obtained by analytic continuation from the physical region. The early dispersion calculations were found to be very sensitive to the choice of models, particularly for the unphysical region, and much of the subsequent work involved efforts to suppress this dependence. For example, weighted dispersion relations were written (25) for functions obtained from the scattering amplitude by multiplying it by a kinematic factor which is small in the troublesome region. Other methods, such as that developed in Ref. (23), are also insensitive to these regions.

Even if the "unknown" regions are strongly suppressed, their presence in the dispersion integrals is a serious defect in principle. It is well known that analytic extrapolation is an unstable process: if an analytic function is known over a finite interval with some definite error corridor, one can always find different analytic approximations which pass strictly inside this error corridor but which differ by arbitrarily large amounts outside the original interval.

Of course, in practice one can compare the extrapolations obtained by different parametrization (such as K-matrix parametrizations for the unphysical region, or Regge parametrizations for the high-energy region) to obtain some idea of the size of the uncertainty. However, a serious danger here is that this uncertainty may be underestimated if, as is the case in practice, most of the parametrizations are based on the same model for the amplitude. It is therefore desirable to have recourse to methods which require no specific assumptions about the form of the amplitude in the experimentally inaccessible energy regions.

A technique which has become increasingly popular in this connection is the utilization of a discrepancy function. A discrepancy function is defined as the difference between the measured real part of a scattering amplitude and the sum of all the experimentally known contributions to it in a given dispersion relation. For example, in the case of the once-subtracted dispersion relation (3) for the K^-p scattering amplitude, the discrepancy function is

$$\Delta(\omega) = D_-(\omega) - \frac{\omega}{4\pi^2} P \int_{m_K}^{W} \frac{k'}{\omega} \left[\frac{\sigma_-(\omega')}{\omega' - \omega} - \frac{\sigma_+(\omega')}{\omega' + \omega} \right] d\omega', \quad (20)$$

where W is the highest energy at which $\sigma_\pm(\omega)$ are known experimentally.

By virtue of the dispersion relation from which it is derived, the discrepancy function is equal to the sum of all the unknown contributions to the dispersion relation; therefore the discrepancy function has only the singularities associated with those contributions, namely the hyperon poles, the unphysical cut, and cuts in the asymptotic energy regions. Since the discrepancy function is equal to a sum of contributions from distant singularities (for values of ω which are not too close to any of those singularities), it is expected to have relatively little structure over most of the accessible energy region. This means that the discrepancy function, unlike the amplitude itself, can be parametrized in a relatively simple form.

One possibility is to parametrize the discrepancy function directly as a sum of terms which represent the various unknown contributions; for example, one can include an effective pole term to represent the two hyperon poles, and one or more effective poles to represent the unphysical cut (26).

An alternative procedure (27) is to conformally map the ω-plane into a new complex ξ-plane in such a way that the cuts of the discrepancy function go onto the unit circle $|\xi| = 1$. If the discrepancy function has only the hyperon poles and the unphysical cut (for example, if a particular Regge model is assumed for the asymptotic region) a conformal mapping which achieves this is

$$\xi = \frac{\sqrt{\omega - m_k} - \sqrt{\omega - \omega_{\pi\Lambda}}}{\sqrt{\omega - m_k} + \sqrt{\omega - \omega_{\pi\Lambda}}} \qquad (21)$$

The discrepancy function $\Delta(\xi)$ is then analytic inside the unit circle in the ξ-plane, apart from the Λ and Σ poles. If these two poles are replaced by a single effective hyperon pole at $\omega = \omega_Y$, the discrepancy function can be expanded in the form

$$\Delta(\omega) = \frac{1}{\xi - \xi_Y} \sum_{n=0}^{\infty} a_n \xi^n , \qquad (22)$$

which is valid throughout the entire ω-plane.

More generally, if the discrepancy function $\Delta(\omega)$ also contains cuts in the region $|\omega| > W$ (which are disconnected from the unphysical cut), it can be represented as the sum of the two parts, $\Delta(\omega) = \Delta_1(\omega) + \Delta_2(\omega)$, where $\Delta_1(\omega)$, containing only the unphysical cut, is expanded in the form (22) and $\Delta_2(\omega)$, containing the remaining cuts, is expanded in terms of another appropriate variable which maps those cuts onto the unit circle; thus the complete expansion is valid throughout the whole of the original energy plane. The method of analysis is to truncate these expansions and obtain the best fit to the experimental data points for the discrepancy function (which lie inside the unit circle in each variable), using the first few expansion coefficients as adjustable parameters.

A modification of this method was employed in Ref. (28), where the conformal mapping into a circle was replaced by a mapping (involving elliptic integrals) into an "optimum" ellipse, which in principle leads to an optimally convergent expansion. In contrast with Ref. (27), where the analytic continuation of the discrepancy function to the hyperon pole was found to be highly unstable, a reasonably accurate value for G^2 was given in Ref. (28).

However, the accuracy of such discrepancy-function calculations was seriously questioned in Ref. (29), where the techniques that had been used to analyze the experimental data were applied to simulated data constructed from model functions which have the correct analytic properties and which qualitatively resemble the actual amplitudes. It was found that the previously employed methods were not capable of reproducing the input values of the pole strengths in these model functions with the accuracy that had been claimed in some previous analyses. In addition, it was shown that the use of the "optimum" ellipse offers no practical advantage over the much simpler mapping into a circle, since the gain in accuracy is far below the level of the statistical uncertainties inherent in the data.

Recent studies of analytic continuation techniques for various problems (30, 31) have shown that extrapolations by means of rational functions (Padé approximants of type III) are often much more accurate and stable than those based on polynomial fits. This idea was recently exploited in Ref. (14), where all available data on forward $K^{\pm}p$ scattering were reanalyzed by parametrizing the discrepancy function in terms of rational functions of the appropriate variable. This analysis led to the value $G^2 \approx 14 \pm 2$ for the effective KNY coupling, corresponding to a fit with $\chi^2/NDF = 215/170$.

The uncertainties in such a calculation are expected to be larger than in analyses of conventional dispersion relations, since the discrepancy-function method involves no specific assumptions about the structure in the unphysical region and hence allows a much larger class of admissible functions in the representation of the amplitude.

Since the asymptotic energy region ($\omega > W = 240$ GeV) was treated as an unknown contribution in the discrepancy-function analysis of Ref. (14), the fit to the data obtained there provides information about the asymptotic behaviour of the amplitude. It was found that only a single parameter, b_o, was required in the expansion of the high-energy part $\Delta_2(\omega)$. This parameter can be shown to have the value

$$b_o = \frac{1}{4\pi} \int_W^{\infty} \frac{\left[\sigma_-(\omega') - \sigma_+(\omega')\right] d\omega'}{\omega'} \qquad (23)$$

in the case in which the Pomeranchuk theorem is satisfied (in the more general case, b_o becomes a finite-contour integral in the complex plane). The best fit was obtained with the parameter value $b_o = (0.027 \pm 0.002)$ fm^2.

A variant of the discrepancy-function method in which the zeros of the amplitude play a prominent role (32) recently gave the value

$$G^2 = 10.4 \pm 3.0. \qquad (24)$$

This analysis will be described in greater detail at this Workshop in the talk by G.K. Atkin.

7. Conclusions

Owing to the abundance of accurate experimental data and new methods for their analysis, significant progress has recently been achieved in understanding the structure of the kaon-nucleon scattering amplitudes.

The couplings $g^2_{KN\Lambda}$ and $g^2_{KN\Sigma}$ now seem to be determined with reasonable accuracy and consistency by a variety of methods. We have quoted a number of different values obtained for these coupling constants, and we shall not attempt here to present any single set of "best" or "average" values.

It is interesting to note that the current predictions for these coupling constants from dispersion theory are consistent with the results of other analyses. For example, the value $G^2 = 15.2 \pm 2.3$ was obtained in Ref. (37) from an energy-independent phase-shift analysis of K^+p scattering data in which the higher partial waves were included through a conformal-mapping method. The value $G^2 = 15.2 \pm 0.8$ was found in Ref. (38) from a modified phase-shift analysis of K^+p scattering data, in which the coupling constants of the exchanged particles parametrize the high partial waves.

Some of the dispersion relations which have been evaluated provide fairly accurate predictions for the real parts of the scattering amplitudes. Tabulations and graphs of the calculated real parts can be found in the literature: $K^\pm p$ real parts in Refs. (14, 21, 33-36), $K^\pm n$ real parts in Ref. (12), and K^-p and K^-n real parts in Ref. (16). Comparison of these results shows that there are still some significant discrepancies between the various calculations in certain energy regions.

In this review we have not touched upon non-forward or partial-wave dispersion relations, on which relatively little work has been done. In view of the abundance of good experimental data, it would be interesting to analyze the non-forward differential cross section $d\sigma/d\Omega$ by dispersion methods. However, the description of non-forward scattering is complicated by the presence of spin, and it cannot be assumed that $d\sigma/d\Omega$ measures the squared modulus of any single amplitude. Perhaps a way of overcoming this difficulty is the observation that the familiar relations

$$\frac{d\sigma}{d\Omega} = |f|^2 + |g|^2, \qquad P\frac{d\sigma}{d\Omega} = -2\,\mathrm{Im}\,(f^* g) \tag{25}$$

(where P is the polarization) imply that

$$(1 \pm P)\frac{d\sigma}{d\Omega} = |f \pm ig|^2. \tag{26}$$

Good polarization data are available over a wide range of energies. Thus it appears that the methods of analyzing the modulus of an analytic function which we discussed earlier can be applied to this function. This possibility is being investigated.

Acknowledgments

I am greatly indebted to my collaborators G.K. Atkin, J.E. Bowcock, O. Dumbrajs and G. Violini for innumerable helpful communications and discussions.

References

1. Queen, N.M., Restignoli, M., and Violini, G.: 1969, Fortschr. Phys. 17, 467.
2. Queen, N.M., Restignoli, M., and Violini, G.: 1973, Fortschr. Phys. 21, 569.
3. Martin, B.R.: 1970, Springer Tracts in Modern Physics 55, 73.
4. Queen, N.M., and Violini, G.: 1974, Dispersion Theory in High-Energy Physics, Macmillan, London.
5. Samaranayake, V.K.: 1979, J. Phys. G: Nucl. Phys. 5, 657.
6. de Swart, J.J.: 1963, Rev. Mod. Phys. 35, 916.
7. Gürsey, F., and Radicati, L.A.: 1964, Phys. Rev. Lett. 13, 173.
8. Carroll, A.S. et al.: 1976, Phys. Lett. 61B, 303.
9. Di Claudio, B., and Violini, G.: 1978, Revista Brasileira de Fisica 8 , 189.
10. Dumbrais, O.V., Dumbrais, T.Yu., and Queen, N.M.: 1971, Fortschr. Phys. 19, 491.
11. Queen, N.M.: 1978, Report UB-KP-1-78, University of Birmingham.
12. Baillon, P. et al.: 1978, Nucl. Phys. B134, 31.
13. Perrin, R., and Woolcock, W.S.: 1968, Nucl. Phys. B4, 671; 1969, B12, 26.
14. Di Claudio, B., Violini, G., and Queen, N.M.: 1979, Nucl.Phys. B161, 238.
15. Martin, B.R., and Sakitt, M.: 1969, Phys. Rev. 183, 1345.
16. Martin, A.D.: 1976, Phys. Lett. 65B, 346.
17. Haber-Schaim, U.: 1956, Phys. Rev. 104, 1113.
18. Atkin, G.K., Di Claudio, B., Queen, N.M., and Violini, G.: 1980, Preprint, University of Birmingham.
19. Phillips, R.J.N.: 1974, Nucl. Phys. B72, 481.
20. Dumbrajs, O.: 1977, Nucl. Phys. B126, 542.
21. Dumbrajs, O., and Staszel, M.: 1975, J. Phys. G: Nucl. Phys. 1, 172.
22. Dumbrajs, O.V., and Staszel, M.: 1975, Nucl. Phys. B88, 52.
23. Atkin, G.K., Bowcock, J.E., and Queen, N.M.: 1980, J. Phys. G: Nucl. Phys. 6, 163.
24. Dumbrajs, O., private communication.
25. Restignoli, M., and Violini, G.: 1970, Nuovo Cimento 69A, 691.

26. López, C., and Yndurâin, F.J.: 1972, Phys. Lett. 41B, 183; 1973, Nucl. Phys. B64, 415.

27. Dumbrais, O.V., Dumbrais, T.Yu.,and Queen, N.M.: 1971, Nucl. Phys. B26, 497.

28. Chao, Y.-A., and Pietarinen, E.: 1971, Phys. Rev. Lett. 26, 1060.

29. Conti, A., and Queen, N.M.: 1975, Nuovo Cimento 29A, 452.

30. Dacunha, N.M.C.: 1978, M.Sc. thesis, University of Birmingham.

31. King, M.B., and Queen, N.M.: 1979, J. Chem. Eng. Data 24, 178.

32. Atkin, G.K., Bowcock, J.E., and Queen, N.M.: 1980, Preprint, University of Birmingham.

33. Baillon, P. et al.: 1976, Nucl. Phys. B105, 365.

34. Baillon, P. et al.: 1976, Nucl. Phys. B107, 189.

45. Dumbrajs, O., and Staszel, M.: 1976, Czech. J. Phys. B26, 68.

36. Dumbrajs, O.: 1978, Preprint HU-TFT-78-15, University of Helsinki.

37. Miller, R.C. et al.: 1972, Nucl. Phys. B37, 401.

38. Alcock, J.W., Cottingham, W.N., and Davis, A.C.: 1976, Nucl. Phys. B102, 173.

DISCUSSION

- P. Gensini:

One has to be very cautions, in using the discrepancy method, about the possible presence of systematic errors, particularly in the total cross-section measurements. Such errors introduce non-analytic functions exactly in that region where we assume the discrepancy function to be holomorphic. These of course would in turn destroy the smoothness of the extrapolation, and explain as well why the early attempts did fail to yield a definite answer, despite the mapping used.

- N.M. Queen :

It is true that the existence of systematic errors in the total cross sections means that the physical cut is not being correctly subtracted off, so that extra singularities remain in the discrepancy function. Nevertheless, one may hope that in practice the values constructed from the data provide a reasonable numerical approximation to the true discrepancy function. In any case, the problem of systematic errors is no worse than in conventional dispersion methods.

- M. Svec :

I would like to máke a comment about a very useful application of fixed-t dispersion relations (FTDR) in two - body reactions at high energies. When a complete set of observables is measured at a given energy, the amplitudes in a two - body reaction can be constructed directly from the data (up to an overall phase).

Experimentally, however, differential cross sections and polarizations are known at various energies. Using information from the low energy phase shift analysis and an effective Regge pole parametrization of the imaginary parts of amplitudes in the high energy region, FTDR have been used to construct amplitudes from incomplete data. In particular, A.P. Contogouris (1) and more recently D.W. Joynson and B.R. Martin (2) studied the structure of exchange amplitudes in KN and $\bar{K}N$ scattering using FTDR. Using the latest KN and $\bar{K}N$ phase shift analyses and all data at high energies, Joynson and Martin obtained the t- structure of A_2 - exchange amplitudes similar to that obtained directly from KN and $\bar{K}N$ charge exchange data (3). This method, when applied to πN charge exchange scattering (4), reproduced the ρ-exchange amplitudes known from the coupled πN data at 6 GeV/c. Exchange amplitudes in other two - body reactions (5) have also been studied using FTDR.

1. Contogouris, A.P., et al.: 1974, Phys. Rev., D10, 2095.
2. Joynson, D.W., and Martin, B.R. : 1978, Nucl. Phys. B134, 83.
3. Fusjisaki, M., et al. :1979, Nucl. Phys. B152, 232.
4. Argyres, E., et al. : 1974, Ann. Phys. 85, 283.
Svec, M. : 1977, Lett. Nuovo Cimento 18, 45.
5. Contgouris, A.P., and Svec, M. : 1978, Phys. Rev. D17, 806.
Svec, M. : 1980, Phys. Rev. D21, to appear.

THE DETERMINATION OF $g^2_{N\Sigma K}$ FROM K° REGENERATION DATA

G.K. Atkin[+], B. Di Claudio[+], N.M. Queen[*], and G. Violini[+]
+ Istituto di Fisica - Università di Roma
Istituto Nazionale di Fisica Nucleare - Sezione di Roma
* Department of Mathematical Physics - University of Birmingham

Presented by G. Violini

Although the symmetry of SU (3) gives a prediction for the $p\Sigma^oK^-$ coupling constant as definite as that for the $p\Lambda^oK^-$ one, namely

$$g^2_{p\Sigma^oK^-} = (1 - 2\alpha)^2 \, g^2_{\pi NN}$$

where $\alpha = f/(f+d)$ is the fraction of F type coupling, very few determinations of this quantity can be found in the literature.

The reason for this is that most of the works which attempt to determine these coupling constants using dispersion relations deal with K^-p scattering. This is rather obvious, since data for this process are much more abundant than for Kn scattering, but unfortunately this makes the $p\Sigma^oK^-$ coupling appear together with the $p\Lambda K^-$ one. In general it is not believed that the data allow a good determination of both individual couplings, and it has become customary in this type of analysis to assume that the Λ and Σ poles collapse into one effective pole with coupling

$$G^2_{pYK^-} = g^2_{p\Lambda K^-} + 0.84 \, g^2_{p\Sigma^oK^-}$$

Of course there exists a number of analyses of Kn scattering; the oldest ones are not very reliable because they make use of very crude and out-dated parametrizations for the $\bar{K}N$ unphysical and low-energy region. Moreover at that time the total absence of measured real parts forced one to make use of the so-called threshold-subtracted dispersion relations, which again gave still more weight to those crude parametrizations. To this one must also add, and this is true also for the most recent analyses, that Kn are intrinsically less good than Kp data, because their extraction from Kd experiments is not model-indipendent due to the necessity of taking into account the fact that in the deuteron the

E. Ferrari and G. Violini (eds.), Low and Intermediate Energy Kaon-Nucleon Physics, 131–138.

neutron is not free. In any case, in more recent analyses it has been possible to determine (A.D. Martin, P. Baillon et al.) a value of $g^2_{p\Sigma^o K^-}$ by using five Kn real part data (four K^-n and one K^+n). The K^+n point as well as one K^-n point is at 10 GeV; the remaining K^-n points are in the region just above 1 GeV/c. There is, however, an alternative way of determining $g^2_{p\Sigma^o K^-}$; this is provided by the K^o_L regeneration on hydrogen, since the amplitude for this process is given by

$$\frac{1}{2}\,(K^o p - \overline{K^o} p) = \frac{1}{2}\,(K^+ n - K^- n)$$

Thus in particular the real part of the regeneration amplitude is proportional to the difference of the $K^\pm n$ real parts. Experimentally the real part of the regeneration amplitude will be determined from the value of

$$\left(\frac{d\sigma}{dt}\right)_{reg}$$

taken in the forward direction, using for the imaginary part the value provided by the optical theorem applied to $K^\pm n$ reactions.

This procedure will obviously lead to the modulus of the real part of the regeneration amplitude. However, the sign can easily be fixed by taking advantage of the indications of dispersion relation calculations for Kn. It has to be pointed out that the use of different experimental data for $(d\sigma/dt)_{reg}$ and σ_{TOT} $(K^\pm n)$ may easily produce a violation of the optical limit. We shall return to this point later. For the moment we shall assume that the effect is not especially important, even if a more refined analysis of this problem may be desirable.

Before presenting our calculations we want to mention that the use of regeneration data for the application of $K^\pm n$ dispersion relations was first considered by Lusignoli, Restignoli and Violini. However, these authors were working in a rather different situation, since at that time (twelve years ago) there were almost no data (just one point) and they were suggesting that an accurate regeneration experiment giving a good point-to-point set of real parts could provide a very nice test of dispersion relations, highly independent of the details of the unphysical region or of the pole strength. In this connection there is another point to discuss. In the unphysical region, $\bar{K}N$ exhibits a resonance in the $I = 1$ channel, the Y^*_1 (1385). This is a p-wave resonance; however, since most of the parametrizations which are used in that region are purely s-wave, the contribution of the Y^*_1 will not be included in dispersion calculations using such parametrizations. Therefore any value of the $N\Sigma K$ coupling has to be considered as representing the effective strength of the Σ^o pole and the Y^*_1 resonance. It is to be recalled that conventional dispersion relations will receive an equal-sign contribution from both of these singularities.

We now proceed to a discussion of our calculations. The present experimental situation on regeneration is that there are 78 data points reported in the literature, of which 11 correspond to an extension of statistics of a previous experiment. Therefore we are left with 67 points, a few of which violate the optical limit. For these latter points we considered two possibilities: either excluding them from the fit, or including them as points where the real part vanishes with a suitable error depending on the degree of violation and the size of the experimental errors.

We analyzed the data by two different methods: a method of linear extrapolation in the variable ω^2 (Haber-Schaim method), and a method which makes use of a factorized reperesentation of a function constructed in a suitable way from the regeneration amplitude.

One might wonder why it is desirable to use different methods. If the data were free of errors, it is obvious that the analyticity property of the function would lead to the same results whatever method of analysis is used. However, it must be borne in mind that we are facing a situation where data are affected by experimental (and theoretical) uncertainties and the use of different methods may shed light on the stability of the results for variations on the input.

The Haber-Schaim method consists in evaluating the function

$$L(\omega^2) = (\omega_\Sigma^2 - \omega^2)\left[\frac{D_{reg}(\omega)}{\omega} - \frac{1}{\pi}\int_{\omega_{\pi\Lambda}}^{m_K}\frac{A_-(\omega')\,d\omega'}{\omega'^2 - \omega^2} - \frac{\omega^2}{4\pi^2}\int_{m_K}^{\bar{\omega}}\frac{k'(\sigma_- - \sigma_+)\,d\omega'}{\omega'^2(\omega'^2 - \omega^2)}\right]$$

where ω and K are the kaon energy and momentum in the lab, D_{reg} is the real part of the regeneration amplitude, A_- is the imaginary part of the K^-n amplitude, $\sigma_\pm$ are the $K^\mp n$ total cross-section, $\bar{\omega}$ is the highest energy at which $\sigma_\pm$ are available and ω_Σ is the position of the Σ pole.

The function $L(\omega^2)$ is exactly linear in ω^2, as one can show from dispersion relations that

$$L(\omega^2) = A + B\,\omega^2$$

Moreover, the coefficient B can be expressed as a sum rule

$$B = -\frac{1}{4\pi}\int_{m_K}^{\infty}\frac{k'\,(\sigma_- - \sigma_+)\,d\omega'}{\omega'^2}$$

Thus in the fitting procedure one can, by imposing the sum rule on B, perform a one parameter fit to the values of $L(\omega^2)$ as calculated from the experimental values of $D_{reg}(\omega)$.

From a general point of view this method presents the disadvantage that it requires the knowledge of both D_- and D_+ at the same energies; this disadvantage is compensated by the fact that in the fit one does not need to determine D(0). However, in the case of regeneration this disadvantage is not present, since the real part of the regeneration amplitude automatically corresponds to the real part of the crossing-antisymmetric amplitude. Therefore the Haber-Schaim method is most natural for the analysis of the regeneration amplitude.

In the calculation of the dispersion integral we made two choices for the parametrization of the amplitude in the $\bar{K}N$ unphysical region. We used the most recent parametrization (A.D. Martin, 1976) and that by B. Martin and Sakitt. The difference in the value of $g^2_{p\Sigma^oK^-}$ due to this choice corresponds to approximately 0.5, the larger value being obtained with Martin and Sakitt's parametrization.

For what concerns the sum rule, we found that unless it is imposed as a constraint, it is not satisfied. Nevertheless if one includes it in the fit, with the appropriate statistical error, this produces an increase in $g^2_{p\Sigma^oK^-}$ of 0.6. Finally, the inclusion of all points leads to a value of the coupling smaller by about 0.6 than the value obtained when the points violating the optical theorem are omitted.

In conclusion, the upper value we obtain by this type of analysis, corresponding to the use of Martin and Sakitt's parametrization, imposing the sum rule and omitting the points violating the optical theorem, is

$$g^2_{p\Sigma^oK^-} = 1.0 \pm 0.5$$

Obviously our error is purely statistical, and we have already discussed the effect on the coupling of different possible ways of fitting the data. If we compare our results with the evaluation of $g^2_{p\Sigma^oK^-}$ by Kn dispersion relations, we see that we find a smaller value with much better statistical error. It has to be considered that, as we have already pointed out, we neglected the $Y_1^*(1385)$. This implies that our results must be regarded as an upper limit for $g^2_{p\Sigma^oK^-}$ and that, whatever the size of the $Y_1^*(1385)$, its effect would be to reduce the value of the coupling.

The other method of calculation we shall discuss is based on a property of analytic functions. We do not give much detail here, since this part will overlap Atkin's talk. Briefly, if g(z) is analytic inside and on the unit circle, it can be written in the form

$$g(z) = C(z)\, X(z)$$

where C(z) has no zeros and $X(z) = \prod_{i=1}^{n} \frac{z-z_i}{1-zz_i^*}$, z_i being the n zeros of g(z) inside the unit circle.

Under certain mathematical conditions C(z) is uniquely determined as

$$C(z) = \exp\left\{ \frac{1}{2\pi i} \oint_{|z'|=1} \left[\ln|g(z')| \right] \frac{z'+z}{z'-z} \frac{dz'}{z'} \right\}$$

In order to apply this formalism to our case, we shall first map the ω plane inside the unit circle in the z plane, z being the new variable defined as

$$z = \frac{\sqrt{\omega_{\pi\Lambda}+\omega} - \sqrt{\omega_{\pi\Lambda}-\omega}}{\sqrt{\omega_{\pi\Lambda}+\omega} + \sqrt{\omega_{\pi\Lambda}-\omega}}$$

Then g(z) will be defined as

$$g(z) = [f_-(z) - f_+(z)] \frac{z^2 - z_\Sigma^2}{1 - (zz_\Sigma)^2}$$

$$z_\Sigma = z(\omega_\Sigma)$$

As one can check, on the unit circle $|g| = |f_- - f_+|$. This makes it possible to calculate C(z) from experimental data. Moreover, since on the unit circle one has $|X(z)| = 1$, this implies that the coupling constants $g^2_{p\Sigma^oK^-}$ should satisfy an inequality. Before proceeding to the numerical aspects of our calculations, without giving more details, we shall limit ourselves to just two remarks. First, the presence of the logarithm in the integral of C(Z) is extremely important since this makes the results very stable against any variation or error in the input. Second, in this type of calculation it is possible to use directly the experimentally known differential cross-section without having to extract from it the real part of the amplitude with the related ambiguities.

Thus we obtain the following as an upper limit for $g^2_{p\Sigma^oK^-}$

$$g^2_{p\Sigma^oK^-} \lesssim 6.0$$

This limit can be improved if the function's zeros are known. In fact, if its only zero is that at $\omega = 0$, then the coupling would be equal to the previously found upper limit. Thus in view of this argument and of the previously found value of $g^2_{p\Sigma^oK^-}$, one can infer the existence of further zeros, without having to use any general theorems about the number of the zeros of a function. A preliminary analysis by Dumbrajs and Oades, which will be the subject of one of the next talks this afternoon, suggests from logarithmic dispersion relations the presence of two zeros at $\omega_o = \pm ia$, with 0.3 GeV $\lesssim a \lesssim$ 0.4 GeV. (In the actual talk these preliminary values appear to have changed, but this does not affect the following discussion). If we assume, besides that at $\omega = 0$, two further zeros located at $\omega_o = \pm ia$, then we can calculate $g^2_{p\Sigma^oK^-}$ as a function of a. This result is shown in figure 1.

For the previously mentioned values of the zeros one finds $g^2_{p\Sigma^oK^-} = 2.5 \pm 0.3$. This value can be compared with those presented yesterday in Oades' talks. We recall that he found, assuming for the Y^*_1 coupling

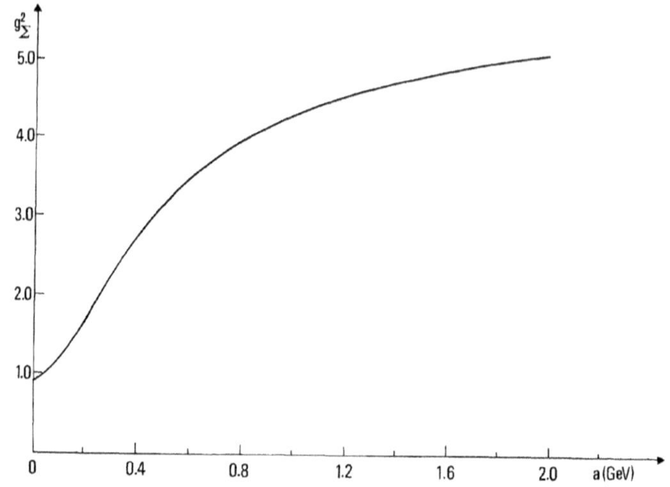

Figure 1 - Plot of the $g^2_{p\Sigma^oK^-}$ as a function of the zero position on the imaginary axis

the SU(3) values, the values shown in Table 1. The agreement with the second column reflects the assumption about the zero position.

Table 1 - Results of Oades' analysis, using two methods, corresponding to the use of different $\bar{K}N$ phase shifts. The error corresponds to the effect of a violation of 25% in the SU(3) value of the Y_1^* coupling

	RLIC	LBL
Fixed t analysis	$4.4^{+0.2}_{-0.7}$	$1.9^{+0.3}_{-0.6}$
Logarithmic dispersion relations	$4.5^{+0.35}_{-0.25}$	2.65 ± 0.25

Comparison with the Haber-Schaim calculation suggests that either the two zeros should lie close to the origin, or a more complicated zero structure is present. (In this connection, however, it has to be noticed that if there is a zero at say $\omega = \omega_1 + i\omega_2$ then crossing implies that there are four zeros).

For what concerns the effect of the Y_1^* on this type of analysis we can mention that we also performed a calculation using the parametrization of Berley et al. If in this calculation we turn off the $p_{1/2}$ and $p_{3/2}$ waves, then the coupling is reduced by 10%.

Finally we want to compare our results with SU(3). Obviously, it has to be borne in mind that in view of the various uncertainties mentioned in this talk, this has only the value of an indication. First of all, if one uses the upper limit found $g^2_{p\Sigma^o K^-}$ together with the values of the effective KNY coupling found in the most recent analysis, then one would obtain $g^2_{p\Lambda K^-}$ a value of the same order of magnitude as $g^2_{p\Sigma^o K^-}$.

Remembering that SU(3) establishes the relation

$$g^2_{p\Lambda^o K^-} = \frac{(1 + 2\alpha)^2}{3} \quad g^2_{\pi NN}$$

and taking this together with the similar relation for $g^2_{p\Sigma^o K^-}$, then one finds that $\alpha \approx 0$ which is not realistic.

Let us now use the value of the Haber-Schaim analysis together with the value of $g^2_{\pi NN}$, which is fairly well established (for the sake of definiteness we shall take that recently determined by Langbein), and also the values of $g^2_{p\Lambda^o K^-}$ as obtained from the previously mentioned values of G^2_{KNY}. In this way we find that the various values of the couplings are in fair agreement with SU(3), the agreement being slightly better in the case of the parametrization of Martin and Sakitt. In this case the value of α we find is $\alpha = 0.37 \pm 0.02$.

To conclude, we want to mention briefly the future developments of this work. We are going to include in the analysis the five $K^{\pm}n$ real parts. This will lead to the determination of $D^{\pm}(0)$ and this should allow us to calculate Kn real parts as a function of the energy and compare the calculated with the experimental values of charge exchange differential cross section. The natural development of this program will be a simultaneous analysis of elastic, charge-exchange and regeneration data. A final problem which should deserve attention is the determination of the positions of the zeros of the scattering amplitude.

Hopefully, the result one can obtain from this program will give some indication of how to resolve some of the inconsistencies which, as it appears from this as well as from Oades talk yesterday, are present in the data.

References

- Atkin, G.K., Di Claudio, B., Queen, N.M., and Violini G.: To be published as a letter and references quoted therein.

- Proceedings of this Workshop.

DISCUSSION

- J. Bowcock :

I wanted to voice the same concern previously expressed by Zovko over the relatively small errors being attached to values of the coupling constants. Although the data is used in somewhat different ways, it should be possible to get a more accurate estimate of the errors arising in each of the methods simply by applying the different methods to a mathematical function which has the appropriate analytic structure and with a known value of the coupling constants; errors may be applied to simulate the experimental situation.

- G. Violini :

There are two different points in your comment. About the first, that of attaching a small error to the coupling despite much larger uncertainties, I confirm that my point of view is that this is the proper way of doing it. It is important that in phenomenological analyses one distinguishes the error whose source is statistical from other uncertainties depending on models, data used, etc. Thus one will know how any change in these will affect the calculated quantities. I must add that this is true also for experimental work, since often we are facing data which include what is the very subjective estimate of possible systematic effects into the statistical errors. For what concerns the second point, the possibility of using a model function approach to the problem under consideration, I am afraid that it will be of little help, since here we know the exact form of the function which should fit the experimental data, whereas this is not true in the case where model functions have been fruitfully used.

- J. Bowcock :

In fact by having control over the errors one can calculate the accuracy to which experiments must be done in order to achieve a given limit of error on the coupling constants.

- G. Violini :

Yes, I agree that this type of analysis can be done and give an important indication to the experimentalists about the quality of data we need if we want to determine from this sort of data $g^2_{N\Sigma K}$.

THE ZEROLOGY OF KAON-NUCLEON FORWARD SCATTERING AMPLITUDES*

O. Dumbrajs
Institute of Physics - Aarhus University

1. Introduction

It has been realized for a long time (1-4) that zeros of the forward kaon-nucleon scattering amplitudes are useful in correlating different low and high-energy scattering parameters and in providing a consistency test of available data. The simplest possibility of exploring zeros is to evaluate the ordinary dispersion relations in the complex energy plane (4). The more natural way of bringing zeros of amplitudes into play is to consider either one of the more sophisticated forms of dispersion relations:

i) phase dispersion relations (2, 3),
ii) inverse-amplitude dispersion relations (1, 3),
iii) logarithmic dispersion relations (5),

or to apply the maximum modulus theorem and a factorization theorem (6, 7).

In this talk I concentrate on the use of logarithmic dispersion relations because just this approach seems to be the most convenient one for future extensions to nonforward scattering data analyses based on the zeros of the amplitude.

2. Number of zeros of forward kaon-nucleon scattering amplitudes

We consider the ordinary $K^{\pm}$ N forward scattering amplitudes $F_{\pm}(\omega)$ and their symmetric and antisymmetric combinations

$$F^{+}(\omega) = F_{+}(\omega) + F_{-}(\omega), \quad (1)$$

$$F^{-}(\omega) = F_{-}(\omega) - F_{+}(\omega), \quad (2)$$

respectively.

E. Ferrari and G. Violini (eds.), Low and Intermediate Energy Kaon-Nucleon Physics, 139–144.

In general, the number of zeros of the scattering amplitude depends on the number of poles, on the number of times the imaginary part of the amplitude changes the sign, on the growth order of the amplitude with energy and on its phase at infinity and at the threshold (8). It was shown later (9) that a knowledge of the actual growth order of the amplitude and of the values of its phase at infinity is not really needed for deduction of the number of zeros.

In my talk I will not discuss the general formulae but will proceed to specific amplitudes. I would like to stress that in what follows I assume that the $K^{\mp}p$ forward scattering amplitude has only one pole, the combined Λ/Σ^o-pole, placed at the position of the Λ^o-pole. This remark is important because in (1-7) where the $K^{\pm}p$ scattering was discussed the two poles (the Λ^o-pole and the Σ^o-pole) were considered separately thus increasing the number of zeros by one real zero for $F_{\pm}(\omega)$ and by two real zeros for $F^{+}(\omega)$ and for $F^{-}(\omega)$. Taking this into account one can conclude that in (4) it was shown that $F_{+}(\omega)$ has two complex conjugate zeros, in (2) it was shown that $F^{+}(\omega)$ has either four complex conjugate zeros or four imaginary zeros (10), and in (3) it was shown that $F^{-}(\omega)$ has two imaginary zeros in addition to one real zero at $\omega=0$.

3. Logarithmic dispersion relations

Using the method outlined in (5) one can write down the following logarithmic dispersion relations

$$\phi_{\pm}(\omega) = -\frac{\sqrt{(\omega\mp m_K)(\omega\pm\omega_{\pi Y})}}{\pi}\left[\int_{m_K}^{\infty}\frac{\ln|F_{+}(\omega')|\,d\omega'}{\sqrt{(\omega'-m_K)(\omega'+\omega_{\pi Y})}(\omega'\mp\omega)} + \right.$$

$$\left. + \int_{\omega_{\pi Y}}^{\infty}\frac{\ln|F_{-}(\omega')|\,d\omega'}{\sqrt{(\omega'+m_K)(\omega'-\omega_{\pi Y})}(\omega'\pm\omega)}\right]\pm\frac{\pi}{2} \tag{3}$$

$$\mp \arcsin\left[\frac{2m_K\omega_{\pi Y} \mp 2\omega\omega_Y \mp(\omega_{\pi Y}-m_K)(\omega\pm\omega_Y)}{(m_K+\omega_{\pi Y})(\omega_Y\mp\omega)}\right]$$

$$\mp i\ln\left\{\frac{\left[2(\omega\mp m_K)(\omega\pm\omega_{\pi Y})+(\omega_{\pi Y}-m_K\pm 2\omega)(z^{*}\mp\omega)\pm 2\sqrt{(\omega\mp m_K)(\omega\pm\omega_{\pi Y})}\sqrt{(z^{*}-m_K)(z^{*}+\omega_{\pi Y})}\right](z\mp\omega)}{\left[2(\omega\mp m_K)(\omega\pm\omega_{\pi Y})+(\omega_{\pi Y}-m_K\pm 2\omega)(z\mp\omega)\pm 2\sqrt{(\omega\mp m_K)(\omega\pm\omega_{\pi Y})}\sqrt{(z-m_K)(z+\omega_{\pi Y})}\right](z^{*}\mp\omega)}\right\}$$

where $\omega_{\pi Y}$ is the beginning of the unphysical cut, ω_Y is the position of the hyperon pole, z is the position of the complex zero and the phase ϕ is defined as usual:

$$\phi_\pm(\omega) \equiv \arctan\left[\frac{\mathrm{Im}F\pm(\omega)}{\mathrm{Re}F\pm(\omega)}\right],$$

$$\phi^+(\omega) = -\frac{\sqrt{\omega^2-\omega^2_{\pi Y}}}{\pi}\int_{\omega_{\pi Y}}^{\infty}\frac{\ln|F^+(\omega')|\,d\omega'^2}{\sqrt{\omega'^2-\omega^2_{\pi Y}}\,(\omega'^2-\omega^2)} - \pi$$

$$+ \arcsin\left[\frac{\omega^2_{\pi Y}+\omega\omega_Y}{\omega_{\pi Y}(\omega_Y+\omega)}\right] + \arcsin\left[\frac{\omega^2_{\pi Y}-\omega\omega_Y}{\omega_{\pi Y}(\omega-\omega_Y)}\right]$$

$$+ i\ln\left\{\frac{\left[2\omega^2_{\pi Y}-\omega^2-z^{*2}+2\sqrt{\omega^2-\omega^2_{\pi Y}}\sqrt{z^{*2}-\omega^2_{\pi Y}}\right](z^2-\omega^2)}{\left[2\omega^2_{\pi Y}-\omega^2-z^2+2\sqrt{\omega^2-\omega^2_{\pi Y}}\sqrt{z^2-\omega^2_{\pi Y}}\right](z^{*2}-\omega^2)}\right\}, \quad (4)$$

$$\phi^-(\omega) = -\frac{\sqrt{\omega^2-\omega^2_{\pi Y}}}{\pi}\int_{\omega_{\pi Y}}^{\infty}\frac{\ln|F^-(\omega')|\,d\omega'^2}{\sqrt{\omega'^2-\omega^2_{\pi Y}}\,(\omega'^2-\omega^2)} - \frac{\pi}{2} - \arcsin\left(\frac{\omega_{\pi Y}}{\omega}\right)$$

$$+ \arcsin\left[\frac{\omega^2_{\pi Y}+\omega\omega_Y}{\omega_{\pi Y}(\omega_Y+\omega)}\right] + \arcsin\left[\frac{\omega^2_{\pi Y}-\omega\omega_Y}{\omega_{\pi Y}(\omega-\omega_Y)}\right]$$

$$+ i\ln\left\{\frac{\left[\omega^2_{\pi Y}-i\omega y+\sqrt{\omega^2-\omega^2_{\pi Y}}\sqrt{-y^2-\omega^2_{\pi Y}}\right](\omega+iy)}{\left[\omega^2_{\pi Y}+i\omega y+\sqrt{\omega^2-\omega^2_{\pi Y}}\sqrt{-y^2-\omega^2_{\pi Y}}\right](\omega-iy)}\right\}, \quad (5)$$

where y is the coordinate of the imaginary zero (11).

4. Calculations

We have fitted experimental data on both the $K^\pm p$ and the $K^\pm n$ scattering amplitudes to eqs. (3), (4), (5) leaving the position of zeros as free parameters. In the case of the antisymmetric amplitude also data on the $K^o_L p \to K^o_S p$ and on the $K^o_L n \to K^o_S n$ regeneration were included. Below $\omega \approx 1.5$ GeV instead of directly using the existing experimental data and the existing parametrizations in the unphysical region we used as an input for the phase and the modulus values reconstructed from the fixed-t KN and $\bar{K}$N amplitudes as given by G. Oades (12).

5. Results

The positions of zeros found from these fits and from earlier investigations are shown in the tables. Errors ascribed to the positions of zeros found in the present work correspond to the use of different "acceptable" sets of amplitudes from (12).

$K^{\pm}p$ scattering					
$F_{\pm}(\omega)$		$F^{+}(\omega)$		$F^{-}(\omega)$	
Positions of zeros in GeV	Reference	Positions of zeros in GeV	Reference	Positions of zeros in GeV	Reference
(-0.05;±0.50)	(4)	±i 0.40 ±i 0.32	(3)	0. ±i 0.59	(3)
(-0.11;±0.32)	(13)	(±0.23±0.01; ±0.16±0.01)	present work	0. ±i(0.58±0.11)	present work
(-0.18;±0.34)	(14)				
(+0.12;±0.60)	(15)				
(-0.21±0.01; ±0.38±0.03)	present work				

$K^{\pm}n$ scattering					
$F_{\pm}(\omega)$		$F^{+}(\omega)$		$F^{-}(\omega)$	
Positions of zeros in GeV	Reference	Positions of zeros in GeV	Reference	Positions of zeros in GeV	Reference
(-0.33±0.03; ±0.18±0.09)	present work	±0.21 (±0.42;±0.04)	present work with LBL (12)	0. ±i(0.44±0.23)	present work
		(±0.40;±0.06)	present work with RLIC (12)		

6. Conclusions

The positions of the complex zeros of the ordinary $K^{\pm}p$ scattering amplitude found by different methods are in reasonable mutual agreement with the exception of the value of ref. (15) where the real part of the zero locations was found to be positive. We have discovered that the zeros of the symmetric $K^{\pm}p$ scattering amplitude are complex contrary to expectations of ref. (3) where these zeros were found to be imaginary. There is a good agreement between locations of the zeros of the antisymmetric $K^{\pm}p$ scattering amplitude calculated in present work and those determined in (3). I am not aware of earlier determinations of zero positions of the $K^{\pm}n$ scattering amplitudes. There seems to be no simple way of comparing the locations of zeros of $K^{\pm}p$ and $K^{\pm}n$ forward scattering amplitudes.

The knowledge of locations of zeros may be useful in stabilizing the analytic continuation of scattering data (16) and in reducing ambiguities of phase shift analysis. For the latter purpose one would need to trace the zero trajectories both in the ω- and in the t-planes (17). The logarithmic dispersion relations offer a convenient possibility of extending the present analysis to include the nonforward scattering data.

References and notes

*. Work supported by a grant from the Danish Natural Science Research Council.

1. Zovko, N.: 1966, Phys. Lett. 23, 143.
2. George, D.J., and Tubis, A.: 1968, Phys. Rev. 174, 1871.
3. Zovko, N.: 1971, Nuovo Cimento 6A, 304.
4. Dumbrajs, O.: 1972, Nucl.Phys. B38, 600.
5. Dumbrajs, O., and Staszel, M.: 1975, J. Phys. G2, 172.
6. Dumbrajs, O., and Staszel, M.: 1975, Nucl. Phys. B88, 52.
7. Atkin, G.K., Bowcock, J.E., and Queen, N.M.: 1980, J. Phys. G6, 163.
8. Sugawara, M., and Tubis, A.: 1963, Phys. Rev. 130, 2127.
9. Kurbatov, A.F.: 1977, Theor. Mat. Fiz. 33, 354.
10. In the case of the $K^{\pm}n$ scattering some models predict $\mathrm{Re}F^{+}(\omega_{\pi\Sigma})<0$. Therefore $F^{+}(Kn)$ has sometimes two extra real zeros.
11. As noted in note (10) in the case of $K^{\pm}n$ scattering the amplitude F^{+} has sometimes two extra real zeros. In such a case two extra terms have to be added to the right hand side of eq. (4):

$$-\left[\frac{\omega_{\pi Y}^2+\omega\omega_o}{\omega_{\pi Y}(\omega_o+\omega)}\right] \text{ and } -\left[\frac{\omega_{\pi Y}^2-\omega\omega_o}{\omega_{\pi Y}(\omega-\omega_o)}\right],$$

where ω_o is the position of the real zero.

12. Oades, G.: Contributions to this Workshop.

13. López, C., and Ynduráin, F.J.: 1973, Nucl. Phys. B64, 315.

14. Dumbrajs, O.: 1978, University of Helsinki Preprint HU-TFT-78-15.

15. Atkin, G.K.: 1979, Thesis, University of Birmingham.

16. Atkin, G.K.: Contributions to this Workshop.

17. Höhler, G., and Sabba Stefanescu, I.: 1979, Z. Phys. C2, 253.

DISCUSSION

- N. M. Queen :

Mathematically, there is nothing special about the zeros of an amplitude, since, if a constant is added to a given function, one obtains another fuction with exactly the same analytic structure but with zeros at different positions. The zeros certainly provide a useful characterization of an amplitude, but would it not be equally good in principle to consider the points at which the amplitude takes some other particular value?

- O. Dumbrais :

From the formal point of view you are, of course, right that there is nothing special about the zeros of a given analytic function. However, when talking about the scattering amplitude, the locations of zeros are closely related to experimentally measured quantities (e.g. the dips in the differential cross-sections, the vanishing of the difference of particle and antiparticle total cross-sections) and any redefinition of the function would be rather artificial.

THE USE OF ZEROS IN STUDYING $\bar{K}N$ PHYSICS

G. Atkin
Istituto di Fisica - Università di Roma
Istituto Nazionale di Fisica Nucleare - Sezione di Roma

1. Introduction

The determination of the KΣN coupling constant g_Σ^2 has received much less attention than its companion g_Λ^2 (1, 2). Early applications of Kaon-nucleon dispersion relations (3) provided values of g_Σ^2 based on $K^\pm n$ analysis. However, these calculations are now considered unreliable as they are based on crude parametrizations of the low energy $\bar{K}N$ scattering amplitude. The lack of real part data in these dispersion relations made the use of a threshold subtraction necessary so that the results were highly model dependent. One must also note that $K^\pm n$ data are in general not as good as $K^\pm p$ data because of the problems related to the deuteron structure. Thus for many years it has become usual to determine the value of an effective coupling $G^2 = g_\Lambda^2 + 0.84\, g_\Sigma^2$ making use of $K^\pm p$ analyses.

Gradually a large collection of experimental data has been accumulated on the regeneration amplitude for the reaction $K_L^o p \rightarrow K_S^o p$. Because of isospin invariance this amplitude is proportional to the difference of the $K^\pm n$ scattering amplitudes $f^\pm$. The $K^- n$ amplitude corresponds to a pure isospin 1 state so that the crossing antisymmetric amplitude in the forward direction has the bound state pole corresponding to the sigma particle. In the first analysis we show how the factorization theorem can be used together with the regeneration amplitude data to calculate an upper bound for the KΣN coupling constant. Imposing constraints on the zeros of the regeneration amplitude gives a definite value for g_Σ^2.

The real parts of the $K^\pm p$ forward scattering amplitudes have been predicted (4) on the basis of several types of dispersion relations. The various predictions for the $K^+ p$ real part are generally consistent with one another and with the available experimental data. However, the situation is much worse for $K^- p$ scattering amplitude; here the predictions are in conflict with each other, especially at Kaon laboratory momenta in the region of 1.0 GeV/c, where the sign of the $K^- p$ real part is still not determined with certainty. These disagreements are due to the use of

E. Ferrari and G. Violini (eds.), Low and Intermediate Energy Kaon-Nucleon Physics, 145–163.

different sets of data on the total cross-sections, different models for the imaginary part of the scattering amplitude on the unphysical cut below the K^-p threshold and at ultra high energies, and different values for the $KN\Lambda$ and $KN\Sigma$ coupling constants in the case of calculations of the real parts. The residues of the Λ and Σ poles of the scattering amplitude have also been calculated with a variety of methods, but the values determined again show considerable differences (1, 2).

In the second analysis to be described here we use the factorization theorem to parametrise a discrepancy function, values of which can be constructed by using only experimental data on the real and imaginary parts of the K^-p amplitude in a finite interval of physical energies, together with a model for the amplitudes in the high energy region. We show that our method stabilizes analytic continuation by imposing constraints on the number and positions of the zeros of the discrepancy function.

2. Mathematical preliminaries

2.1 Carleman function

The Carleman function C(z) is defined to be (5, 6) that function which has a given modulus on the unit circle Γ in the complex plane, has no zeros inside or on the unit circle and is analytic inside Γ. The Poisson Integral formula (7) can be used to construct this function from its modulus on the unit circle

$$C(z) = \exp\left\{ \frac{1}{2\pi i} \oint_{\Gamma} \left[\ln|C(z')| \right] \frac{z'+z}{z'-z} \frac{dz'}{iz'} \right\} \qquad (2.1.1)$$

where the integration is over the unit circle $|z| = 1$. If we constrain the function C(z) to satisfy the Schwarz reflection property then we can further represent it with the form

$$C(z) = \exp\left(\sum_{k=0}^{\infty} C_k z^k \right) \qquad (2.1.2)$$

where the coefficients are real and given by

$$C_o = \frac{1}{\pi} \int_0^{\pi} d\theta \;\; \ln|C(e^{i\theta})| \qquad (2.1.3)$$

$$C_k = \frac{2}{\pi} \int_0^{\pi} d\theta \;\; \ln|C(e^{i\theta})| \cos k\theta \qquad k > 0. \qquad (2.1.4)$$

Expression (2.1.2) is the easiest form in which to numerically calculate the Carleman function as the coefficients C_k are easily found. Furthermore the Carleman function has some interesting mathematical properties.

Firstly it is insensitive to the effects of errors on the modulus data which we use to calculate the C_k. This is because we use $\ell n|C(e^{i\theta})|$ in the integrand. Secondly the Carleman function can be analytically continued inside the unit circle and the result will be stable against errors on the modulus data. This is because the maximum modulus theorem requires that any errors added to the modulus data on Γ will not become larger inside the unit circle.

2.2. Blaschke factors

We define the Blaschke factor to be the function

$$\frac{z-z_o}{1-zz_o^*} \qquad |z_o| < 1 \tag{2.2.1}$$

which has a zero at z_o and a simple pole outside the unit circle at $1/z_o^*$. Furthermore it has the useful property of having unit modulus on the unit circle.

2.3 Factorization theorem

Our analyses are based on the following factorization theorem (8). If the function $g(z)$ is analytic inside the unit circle in the complex z plane and has N zeros z_i $(i = 1, 2, \ldots N)$ in this region, then it can be factorized inside and on the unit circle as a product of Blaschke factors

$$\chi(z) = \prod_{i=1}^{N} \frac{z-z_i}{1-zz_i^*} \tag{2.3.1}$$

and an analytic function $C(z)$ having no zeros

$$g(z) = \chi(z)\, C(z). \tag{2.3.2}$$

If $g(z)$ is polynomially bounded on the unit circle and has the Schwarz reflection property then (5) $C(z)$ is just the Carleman function and can be determined in terms of the values of $|g(z)|$ on the unit circle by means of (2.1.1)

$$C(z) = \exp\left\{ \frac{1}{2\pi} \oint \left[\ell n|g(z')|\right] \frac{z'+z}{z'-z} \frac{dz'}{iz'} \right\}. \tag{2.3.3}$$

3. The K^o regeneration amplitude

3.1 Application of the factorization theorem

By isospin invariance it can be shown that the amplitude for the reaction $K_L^o p \to K_S^o p$ is equal to $\frac{1}{2}(f_- - f_+)$ where $f_\pm$ are the elastic amplitudes for the $K^{\pm}n$ reactions. In the forward direction the combination of amplitude $(f_- - f_+)$ has the crossing antisymmetric property so it has a zero at $\omega = 0$, where ω is the Kaon laboratory energy. The K^-n forward scattering amplitude is pure isospin 1 and hence has an unphysical region extending from the $\pi\Lambda$ threshold energy $\omega_{\pi\Lambda}$ to the K^-n threshold energy m_K (the Kaon mass) together with a bound state Σ pole at energy ω_Σ. The K^-n amplitude also has a physical cut for $m_K \leq \omega < \infty$ and a crossed (K^+n) cut for $-\infty < \omega \leq -m_K$.

We assume that the modulus $|f_-(\omega) - f_+(\omega)|$ is known everywhere along its cuts. This information can be obtained from existing K matrix parametrizations for the K^-n interactions in the low energy and unphysical regions, experimental data on the regeneration amplitude $K^o_Lp \to K^o_Sp$ over a wide range of energies, Regge parametrizations for the asymptotic energies and C.S.L. parametrizations near the K^+n threshold.

To apply the factorization theorem (2.3.2) to the amplitude $[f_-(\omega) - f_+(\omega)]$ we map the cut ω plane into the interior of the unit circle in the z plane by means of the conformal mapping

$$z = \frac{\sqrt{\omega_{\pi\Lambda} + \omega} - \sqrt{\omega_{\pi\Lambda} - \omega}}{\sqrt{\omega_{\pi\Lambda} + \omega} + \sqrt{\omega_{\pi\Lambda} - \omega}} \, . \qquad (3.1.1)$$

It is easy to see that this transformation $z = z(\omega)$ has the properties $z(\pm\omega_{\pi\Lambda}) = \pm 1$ and $z(\omega = 0) = 0$. The cuts in the ω plane are mapped onto the unit circle $|z| = 1$ so that $z(\omega = \pm\infty + i\varepsilon) = +i$ and the cut free interval $-\omega_{\pi\Lambda} < \omega < \omega_{\pi\Lambda}$ is mapped inside the circle onto the interval $-1 < z < 1$ which in particular includes the points $z(\pm\omega_\Sigma) = \pm z_\Sigma$.

3.2 Construction of the limit on the KNΣ coupling constant

We now define the function

$$g(z) = [f_-(z) - f_+(z)] \left[\frac{z - z_\Sigma}{1 - z z_\Sigma}\right] \left[\frac{z + z_\Sigma}{1 + z z_\Sigma}\right] \qquad (3.2.1)$$

which is analytic inside the unit circle $|z| < 1$ and can thus be factorized in the form (2.3.2). However, we know that $[f_-(z) - f_+(z)]$ has a zero at $z = 0$ so that this additional information gives us

$$g(z) = C(z)\ \chi(z)\ z \qquad (3.2.2)$$

where $\chi(z)$ represents the Blaschke product for the other unknown zeros of the crossing antisymmetric amplitude $[f_-(z) - f_+(z)]$ and $C(z)$ has the representation (2.3.3). Since the last two factors of (3.2.1) have unit modulus on the unit circle the quantity $|g(z')|$ in the integrand of (2.3.3) can be replaced by $|f_-(z') - f_+(z')|$. Thus equation (2.3.3) enables us to construct $C(z)$ if $|f_-(z) - f_+(z)|$ is known on the unit circle.

Near the Σ pole the amplitude $[f_-(\omega) - f_+(\omega)]$ will take the form of a simple pole with residue $2\beta_\Sigma$ where β_Σ corresponds to the $\bar{K}N\Sigma$ coupling constant and the factor 2 arises because the K^-n amplitude is pure isospin 1. It then follows from (3.2.1) and (3.2.2) that

$$\chi(z_\Sigma) = \frac{4\beta_\Sigma}{C(z_\Sigma)\,(1 - z_\Sigma{}^4)} \tag{3.2.3}$$

where $\chi(z_\Sigma)$ is the value of the Blaschke product, for the remaining unknown zeros, at $z = z_\Sigma$. Now $|\chi(z)| = 1$ on the unit circle, hence by the maximum modulus theorem

$$-\left|\frac{C(z_\Sigma)\,(1 - z_\Sigma{}^4)}{4}\right| < \beta_\Sigma < \left|\frac{C(z_\Sigma)\,(1 - z_\Sigma{}^4)}{4}\right| . \tag{3.2.4}$$

The residue β_Σ is related to the KNΣ coupling constant g_Σ^2 by (1)

$$g_\Sigma^2 = \frac{-4m_p^2\ \beta_\Sigma \left(\frac{d\omega}{dz}\right)_{z = z_\Sigma}}{(m_\Sigma - m_p)^2 - m_K^2} \tag{3.2.5}$$

where m_K, m_p and m_Σ are the masses of the kaon, proton and hyperon respectively. Since we are only interested in positive values of the coupling constant, expression (3.2.4) gives us an upper bound on g_Σ^2.

3.3 Results and discussion

The method described above was used to construct an upper limit on the value of the KNΣ coupling constant. The method requires a knowledge of $|f_-(\omega) - f_+(\omega)|$ everywhere along the cuts in order to evaluate $C(z_\Sigma)$ by means of equation (2.3.3). For this purpose we made use the isospin 1 K-matrix parametrization of Martin (9) for the K^-n scattering amplitude in the unphysical and low energy regions and his constant scattering length parameters for the low energy K^+n scattering amplitude. The regeneration amplitude data presented by G. Violini at this conference was used along the physical cut. Using the conformal mapping (3.1.1) places the energy region $|\omega| > 125$ GeV onto a very small arc on the unit circle in the z plane and the integral over this region in (2.3.3) was ignored. No account was taken of the errors in the experimental values of $|f_-(\omega) - f_+(\omega)|$ used to evaluate the integral (2.3.3); owing to the logarithmic terms, the effect of these errors is quite small (5).

The value of the upper limit obtained in this way was

$$g_\Sigma^2 < 6.0. \tag{3.3.1}$$

Most of the previous values of the KNΣ coupling constant (1, 2) lie in this range.

Finally we can investigate what additional information can be deduced from a knowledge of the zeros of the scattering amplitude $[f_-(\omega) - f_+(\omega)]$. Dumbrajs and Oades (10) have recently carried out an analysis of the crossing-antisymmetric $K^{\pm}n$ amplitude by means of logarithmic dispersion relations, using the regeneration data. They have found that this amplitude has only a single pair of complex conjugate zeros at $\omega_o = \pm ia$

$$0.3 \text{ GeV} \leq a \leq 0.4 \text{ GeV} \tag{3.3.2}$$

If the regeneration amplitude has only three zeros it is therefore possible by our analysis to calculate g^2_Σ as a function of a. In expression (3.2.2.) we can replace $\chi(z)$ by the corresponding Blaschke factors to obtain

$$\beta_\Sigma = \frac{C(z_\Sigma)\,(1 - z_\Sigma{}^4)}{4} \left[\frac{z_\Sigma - z_o}{1 - z_\Sigma z_o^*}\right]\left[\frac{z_\Sigma - z_o^*}{1 - z_\Sigma z_o}\right] \tag{3.3.3}$$

where z_o corresponds to the position of the complex conjugate zeros in the z plane. Using our modulus data figure 1 shows g^2_Σ as a function of the zero position ia in lab energy plane. As can be seen for Dumbrajs' value of the complex conjugate zeros (3.3.2) we obtain

$$g^2_\Sigma = 2.5 \pm 0.25 \tag{3.3.4}$$

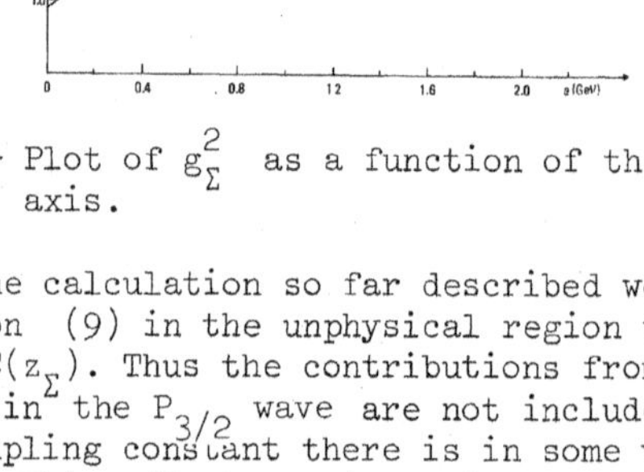

Figure 1 - Plot of g^2_Σ as a function of the zero position on the imaginary axis.

In the calculation so far described we have used an S wave $\bar{K}N$ parametrization (9) in the unphysical region to calculate the Carleman function $C(z_\Sigma)$. Thus the contributions from the $P_{1/2}$ wave and the $Y_1^*(1385)$ resonance in the $P_{3/2}$ wave are not included. This means that in our value of the coupling constant there is in some way some contribution from this resonance. This effect was investigated using instead the S, $P_{1/2}$ and $P_{3/2}$ parametrization of Berkley et al. (16). Using only their S wave parametrization to calculate $C(z_\Sigma)$ gave an upper bound of $g^2_\Sigma < 5.1$. Using all three partial wave parametrizations to calculate $C(z_\Sigma)$ reduced this bound by 10% to $g^2_\Sigma < 4.6$. If the additional information of the two imaginary zeros (3.3.2) are incorporated we obtain $g_\Sigma^2 = 2.1 \pm 0.2$ and $g_\Sigma^2 = 1.9 \pm 0.2$ respectively for the two cases described above. Hence the effect of the higher partial waves in this calculation appears to be small.

4. Discrepancy function analysis

4.1 Introduction

We mentioned earlier that typical K^-p dispersion relations made conflicting predictions about the sign of the K^-p real part in the momentum region of 1.0 GeV/c. This problem was associated with the unphysical region of the amplitude where some model was required to describe the amplitude there. To reduce the model dependence of these results Dumbrajs et al. (11) introduced a new method based on the analysis of a discrepancy function, values of which could be constructed using only experimental data on the real and imaginary parts of the $K^{\pm}p$ scattering amplitudes in a finite interval of the physical region. In this way they obtained reliable fits to the real parts of the amplitudes at moderate energies without invoking any assumptions about the strengths of the Λ and Σ poles or the values of the amplitudes on the unphysical cut and in the asymptotic region. Their method also allowed them to determine a parameter which describes the asymptotic behaviour of the difference between the $K^{\pm}p$ scattering amplitudes. The value found for this parameter favoured a zero asymptotic limit for the $K^{\pm}p$ cross-section difference but required a slower approach to zero than that predicted by Regge pole models at that time.

The parametrization of the discrepancy function was also used in (11) to analytically continue the scattering amplitude from the physical region to the Λ pole. However, it was found that an accurate determination of the $KN\Lambda$ coupling constant could not be made in this way because of the inherent instabilities in the analytic extrapolation. This conclusion was subsequently confirmed by tests of the method using model functions (12).

Several methods of stabilizing such analytic continuations have been proposed (6, 13). One such method (14) is to impose a specific bound on the modulus of the admissable functions on the boundary of their domain of analyticity.

In the method to be described here we use the factorization theorem to parametrise a discrepancy function, values of which can be constructed using only experimental data on the real and imaginary parts of the amplitude in a finite interval of physical energies, together with a model for the amplitudes in the high energy region. As in the previous discrepancy function analysis (11), no assumptions are required about the Λ and Σ poles or the amplitude on the subthreshold unphysical cut. Our method differs from the previous work in that we stabilize the analytic continuation by imposing constraints on the number and positions of the zeros of the discrepancy function. Further we show how a bound on the modulus of the fit can be imposed.

4.2 The discrepancy function

Let $f_{\pm}(\omega) = D_{\pm}(\omega) + iA_{\pm}(\omega)$ be the forward $K^{\pm}p$ scattering amplitudes in the laboratory system normalized so that the optical theorem takes the form

$$\sigma_{\pm}(\omega) = \frac{4\pi A_{\pm}(\omega)}{k} \qquad (4.2.1)$$

where k and ω are the laboratory momentum and energy of the incident Kaon, respectively. The discrepancy function $\Delta(\omega)$ is defined for arbitrary couplex ω to be

$$\Delta(\omega) = f_-(\omega) + \left(\frac{\omega+m_K}{4\pi^2}\right) \int_{m_K}^{\infty} \left[\frac{\sigma_+(\omega')}{(\omega'-m_K)(\omega'+\omega)} - \frac{\sigma_-(\omega')}{(\omega'+m_K)(\omega'-\omega)}\right] k'd\omega' \qquad (4.2.2)$$

where m_K is the Kaon mass. Note that the imaginary parts of the two terms on the right hand side of (4.2.2) cancel identically for values of ω along the cuts in the region $|\omega| > m_K$ so that $\Delta(\omega)$ is real in this region.

The definition of the discrepancy function (4.2.2) presupposes the validity of the Pomeranchuk theorem.

Using a conventional dispersion relation for $f_-(\omega)$ with one subtraction at the K^+p threshold, it can be shown that (1, 2, 11) the discrepancy function can be written as

$$\Delta(\omega) = \left(\frac{\omega+m_K}{\pi}\right)\left[\sum_Y \frac{\pi X(Y)}{(\omega_Y+m_K)(\omega_Y-\omega)} + \int_{\omega_{\pi\Lambda}}^{m_K} \frac{A_-(\omega')\,d\omega'}{(\omega'+m_K)(\omega'-\omega)}\right] + D_+(m_K) \qquad (4.2.3)$$

where Y denotes either of the hyperons Λ or Σ^o; $\omega_{\pi\Lambda}$ is the πΛ threshold energy,

$$\omega_Y = \frac{m_Y^2 - m_p^2 - m_K^2}{2m_p} \qquad (4.2.4)$$

are the positions of the hyperon poles with residues

$$X(Y) = \frac{g_Y^2\left[(m_Y - m_p)^2 - m_K^2\right]}{4m_p^2} \qquad (4.2.5)$$

where g_Y^2 is the K^-pY coupling constant.

If we suppose that the total cross-sections are known over the range $m_K \le \omega \le W$ and that above the energy W we have a Regge parametrization, then the integral in (4.2.2) can be evaluated for all values of ω. Thus, for each energy ω for which $D_\pm(\omega)$ is known, the quantity $\Delta(\mp\omega)$ is determined by (4.2.2) in terms of experimental information, applying the crossing relation $D_-(-\omega) = D_+(\omega)$ in the case of K^+p scattering. Equation (4.2.3), on the other hand, expresses $\Delta(\omega)$ as a sum of unknown contributions from the unphysical cut ($\omega_{\pi\Lambda} \le \omega \le m_K$) and from the Λ and Σ poles, and it can be seen from this expression that these are in fact the only singularities of the function $\Delta(\omega)$.

To obtain a rapidly convergent global expansion for the discrepancy function we conformally map this cut ω-plane into the interior of the unit circle in the complex ζ plane by means of the relation

$$\omega' = \frac{\left[m_K + \omega_{\pi\Lambda} - 2\omega_o\right]\omega + \omega_o\,(m_K + \omega_{\pi\Lambda}) - 2m_K\,\omega_{\pi\Lambda}}{(\omega - \omega_o)\,(\omega_{\pi\Lambda} - m_K)} \qquad (4.2.6)$$

$$\zeta = \frac{\sqrt{1+\omega'} - \sqrt{1-\omega'}}{\sqrt{1+\omega'} + \sqrt{1-\omega'}} \qquad (4.2.7)$$

where ω_o is some point in the interval $\omega_{\pi\Lambda} < \omega < m_K$, whose position can be varied to test the stability of the method with respect to variations of the mapping. The mapping $\zeta(\omega)$ defined by (4.2.6) and (4.2.7) has the properties $\zeta(\omega_{\pi\Lambda}) = +1$, $\zeta(\omega_o \pm i\varepsilon) = \pm i$ and $\zeta(m_K) = -1$. The Λ and Σ poles of $\Delta(\omega)$ are mapped onto certain real points inside the unit circle ζ_Λ and ζ_Σ respectively. The energy region in which $\Delta(\omega)$ can be evaluated using experimental data and a Regge parametrization is mapped onto a certain interval of the real axis to the left of ζ_Λ.

4.3 Zeros of the discrepancy function and a bound on its modulus

If we can specify the number of zeros of the discrepancy function in the complex ω plane, we can greatly restrict the class of allowed functions and hence to some extent stabilize any analytic extrapolation. To get information on the number of zeros we can apply the Argument theorem in the case of a closed contour C_1 surrounding the cut of $\Delta(\omega)$ in the clockwise sense and a counter-clockwise circular contour C_2 at large $|\omega|$. Thus we can write

$$\oint_{C_1} \frac{\Delta'(\omega)}{\Delta(\omega)}\, d\omega + \oint_{C_2} \frac{\Delta'(\omega)}{\Delta(\omega)}\, d\omega = 2\pi i\,(N - P) \qquad (4.3.1)$$

where N and P are the number of zeros and poles respectively within C_2. Equation (4.2.3) shows that $\Delta(\omega) \to$ const as $|\omega| \to \infty$, so that the second integral in (4.3.1) vanishes as the radius of C_2 becomes large. Consequently N - P with N = 2 equals the phase change of $\Delta(\omega)$ around the unphysical cut.

We evaluated this phase change by introducing trial values of the necessary parameters in equation (4.2.3). We chose parameters typical of those given in the literature (1, 2) e.g.

$$g_\Lambda^2 = 10$$
$$g_\Sigma^2 = 2 \qquad (4.3.2)$$
$$D_+(m_K) = -0.30 \text{ fm}$$

together with one of the existing zero range K matrix parametrizations of the K^-p scattering amplitude (15, 16). Any of these sets of parameters led to a phase change of 2π around the unphysical cut. This result is due to the fact that the real part of the integral in (4.2.3) changes sign along the unphysical cut. Thus we conclude that $\Delta(\omega)$ has three zeros.

As the residues X(Y) necessarily have the same sign, one of the three zeros of $\Delta(\omega)$ must be between the Λ and Σ poles and we use this condition as a constraint. It is assumed that the remaining two zeros form a complex conjugate pair.

Finally, a plausible upper hound $M(\omega)$ on $|\Delta(\omega)|$ along the unphysical cut was determined from equation (4.2.3) by using the coupling constants and the K^+p scattering length (4.3.2) and chosing in turn each of several alternative K matrix parametrizations of low-energy K^-p scattering (15, 16) and regarding it as a possible candidate for the amplitude along the unphysical cut.

4.4 Parametrization of the discrepancy function

Because of their small separation, it is not possible to distinguish the Λ and Σ poles in the K^-p scattering amplitude. We therefore replace these two poles by a single effective hyperon pole at the position of the Λ (whose coupling constant is known to be much stronger than that of the Σ (1, 2)) and accordingly omit the zero between the original two poles, assuming instead the existence of only two complex conjugate zeros of the discrepancy function. The residue of the effective hyperon pole is $X(Y) = X(\Lambda) + X(\Sigma)$, which by virtue of the kinematic factor in (4.2.5), is proportional to the square of an effective coupling constant G defined by $G^2 = g_\Lambda^2 + 0.84\, g_\Sigma^2$.

Since the function

$$g(\zeta) = (\zeta - \zeta_\Lambda)\,\Delta(\zeta) \qquad (4.4.1)$$

is analytic inside the unit circle and has exactly two complex conjugate zeros ζ_1 and ζ_1^* in this region, it can be represented in the form (2.3.2) as

$$g(\zeta) = C(\zeta) \left[\frac{\zeta - \zeta_1}{1 - \zeta\zeta_1^*}\right] \left[\frac{\zeta - \zeta_1^*}{1 - \zeta\zeta_1}\right] \qquad (4.4.2)$$

where the second term is a product of two Blaschke factors, each with unit modulus on the unit circle, and $C(\zeta)$ is analytic inside the circle and has no zeros there. As we showed in section 2.1, when $C(\zeta)$ has the Schwarz reflection property it can be represented in the form

$$C(\zeta) = \exp\left[\sum_{k=0}^{\infty} C_k \zeta^k\right] \qquad (4.4.3)$$

where the coefficients C_k are real.

We assume that the series expansion in (4.4.3) converges sufficiently rapidly that it may be truncated after the first few terms. In other words, we assume that the function $g(\zeta)$ can be parametrized within the unit circle in the form

$$g(\zeta) = \exp\left[\sum_{k=0}^{N} C_k \zeta^k\right] \left[\frac{\zeta - \zeta_1}{1 - \zeta\zeta_1^*}\right] \left[\frac{\zeta - \zeta_1^*}{1 - \zeta\zeta_1}\right]. \qquad (4.4.4)$$

We use this parametrization to obtain a least-squares fit to the experimental data on the discrepancy function in the interior of the unit circle, taking the C_k and ζ_1 as fitting parameters. This procedure has certain important advantages. Firstly we have incorporated the fact that the fit must two complex conjugate zeros. Secondly on the unit circle $|\zeta| = 1$ we have

$$|g(e^{i\theta})| = |C(e^{i\theta})| = \exp\left[\sum_{k=0}^{N} C_k \cos k\theta\right]. \qquad (4.4.5)$$

As the coefficients C_k directly determine the modulus of our fit on the unit circle it is feasible to search for fits to the data which satisfy a bound on the modulus of the discrepancy function along its cuts.

4.5 Fit to the discrepancy function data

The discrepancy function can be evaluated at each energy for which the real part of the amplitude is known experimentally. For this purpose we made use of 112 experimental values of $|D_+|$ for kaon laboratory momenta between 0.125 and 150 GeV/c and 94 values of $|D_-|$ between 0.3 and 150 GeV/c, taken from a recent compilation (17) of these data. The signs of these real part were fixed according to a procedure described in (11).

The evaluation of the integral in the expression (4.2.2) for the discrepancy function requires a knowledge of the cross-sections at all energies. We evaluated this integral using experimental data up to ω = 240 GeV from the compilation of Di Claudio and Violini (18) together with their Regge parametrization at higher energies. The K^+p total cross-section was assumed to have the constant value 12.06 mb from threshold up to 0.515 GeV, thus allowing an analytic expression to be found for the integral over this region. The K matrix parametrization of Martin and Ross (15) was used to evaluate the integral over the K^-p cross-section from threshold to 0.551 GeV.

Since the total cross-sections are known with great accuracy, the uncertainties in the determination of the discrepancy function from (4.2.2) come almost entirely from the statistical errors in the data on the real parts $D_\pm$. We therefore assigned each value of $\Delta(\omega)$ an error equal to the statistical error on the corresponding real part. In accordance with equations (4.4.1) and (4.4.4) and our consideration regarding the zeros the values of the discrepancy funtion constructed in this way were fitted to the parametrization

$$\Delta(\zeta) = \frac{1}{\zeta-\zeta_\Lambda} \exp\left[\sum_{k=0}^{N} C_k \zeta^k\right] \left| \frac{\zeta - M_1 e^{i\phi_1}}{1 - \zeta M_1 e^{-i\phi_1}} \right|^2 \qquad (4.5.1)$$

subject to the constraints

$$0 \leq M_1 \leq 1$$
$$0 \leq \phi_1 \leq \pi$$

where C_o, C_1, ..., C_N, M_1, and ϕ_1 are real parameters. The variable ζ is defined in (4.2.6) and (4.2.7) and the adjustable parameter ω_o in (4.2.6) was chosen initially as ω_o = 0.44 GeV. With this choice the hyperon poles correspond to the points ζ_Λ = 0.513 and ζ_Σ = 0.609, and the available data points lie between -0.016 and 0.388 on the real axis in the ζ plane.

A least squares fit of the parametrization (4.5.1) was made to the 206 experimental values of $\Delta(\omega)$ assuming a fixed set of signs for $D_\pm(\omega)$. The iterative procedure of Dumbrajs et al. (11) was then applied until all the output signs of $D_\pm(\omega)$, found by solving equation (4.2.2) for the amplitude after obtaining a fit to $\Delta(\omega)$, agreed with the corresponding input signs of $D_\pm(\omega)$ in obtaining this fit. As was to be expected, some of the experimental data on $D_\pm(\omega)$ were very poor. Whenever any of the input data points of $D_\pm$ differed by more than three standard deviations from the output value given by a least squares fit, these points were rejected and a new fit was made without them.

4.6 Results

A good fit to the data was found with N=0 in equation (4.5.1). In this three parameter fit 15 D_+ and 6 D_- data points were rejected leaving a total of 185 fitted points, for which the fit gave $\chi^2 = 243$. With $N = 1$ in equation (4.5.1) the χ^2 remained at 243. In fact the inclusion of more than three fitting parameters did not give a substantial improvement and in fact leads to unreasonably large errors. Since expression (4.5.1) evaluated at ζ_Λ gives directly the residue of the effective pole of $\Delta(\zeta)$ in the ζ plane then the effective coupling constant is given by (1)

$$G^2 = \frac{-\left(\frac{d\omega}{d\zeta}\right)_{\zeta=\zeta_\Lambda} \exp\left(\sum_{k=0}^{N} C_k \; \zeta_\Lambda^k\right) \left|\frac{\zeta_\Lambda - M_1 e^{i\phi_1}}{1-\zeta_\Lambda M_1 e^{-i\phi_1}}\right|^2 4m_p^2}{(m_\Lambda - m_p)^2 - m_K^2} \qquad (4.6.1)$$

The three parameter best fit gave a value $G^2 = 10.4 \pm 3.0$.

The stability of the parametrization under different conformal mappings can be tested by varying the free parameter ω_o in (4.2.6). Table 1 gives values of G^2, for different values of ω_o, using (4.5.1) with three fitting parameters. Errors in G^2 are of the order of ± 3.0 and the method is seen to be stable under different conformal mappings.

Table 1 - Values of G^2 found using the three parameter fit and using different mapping parameters ω_o in (4.2.6)

ω_o	0.46	0.44	0.42	0.38	0.34	0.32
G^2	10.4	10.4	10.3	10.1	10.1	10.1

Our predictions for $D_\pm$ for the three parameter fit for kaon momenta between 0.5 and 150 GeV/c are presented in table 2.

Table 2

k(GeV/c)	D_+(GeV /c)$^{-1}$	D_-(GeV /c)$^{-1}$	k(GeV/c)	D_+(GeV /c)$^{-1}$	D_-(GeV /c)$^{-1}$
.50	-2.50	1.21	15.00	-10.90	1.41
.60	-2.54	1.91	20.00	-11.58	1.74
.70	-2.27	2.74	25.00	-11.88	3.60
.80	-2.18	3.27	30.00	-11.80	5.49
.90	-2.08	3.77	35.00	-11.34	7.40
1.00	-1.94	2.94	40.00	-10.49	9.32
1.10	-2.12	.13	45.00	-9.07	11.26
1.20	-2.28	-.10	50.00	-7.48	13.22
1.30	-2.60	.56	55.00	-6.20	15.20
1.40	-3.02	1.41	60.00	-5.04	17.19
1.50	-2.96	2.05	65.00	-3.84	19.21
1.60	-3.14	1.56	70.00	-2.59	21.24
1.70	-3.22	.98	75.00	-1.29	23.40
1.80	-3.38	.70	80.00	.05	25.54
1.90	-3.65	.81	85.00	1.42	27.65
2.00	-3.59	.89	90.00	2.04	29.73
2.10	-3.63	.93	95.00	4.30	31.79
2.20	-3.68	.96	100.00	5.80	33.81
2.30	-3.73	.94	105.00	7.22	35.65
2.40	-3.69	.90	110.00	8.75	37.53
2.50	-3.81	.88	115.00	10.37	39.48
2.60	-4.09	.89	120.00	12.09	41.48
2.70	-4.20	.81	125.00	13.91	43.53
2.80	-4.28	.67	130.00	15.84	45.65
2.90	-4.37	.67	135.00	17.86	47.82
3.00	-4.48	.82	140.00	19.99	50.04
5.00	-4.02	.63	145.00	22.22	52.32
10.00	-8.53	.07	150.00	24.54	54.66

The predictions for D_+ over the entire momentum range are in excellent agreement with those tabulated in (4). Our results for D_- however show a small negative region for laboratory momentum of about 1.2 GeV/c. Dumbrajs et al. (11) found a fit in which D_- remains positive for all $k > 0.55$ GeV/c. That D_- remains positive like this was also found by Di Claudio et al. (19) in a reanalysis of the discrepancy function using a larger data set and using Padé approximants in the fitting procedure.

It was shown how expression (4.2.3) could be used to determine a plausible upper bound $M(\omega)$ on $|\Delta(\omega)|$ along the unphysical cut which is mapped onto the unit circle $\zeta = e^{i\theta}$ in the ζ plane. Figure 2 compares this modulus bound $M(e^{i\theta})$ with the modulus of the three and four parameter fits on the unit circle in the ζ plane.

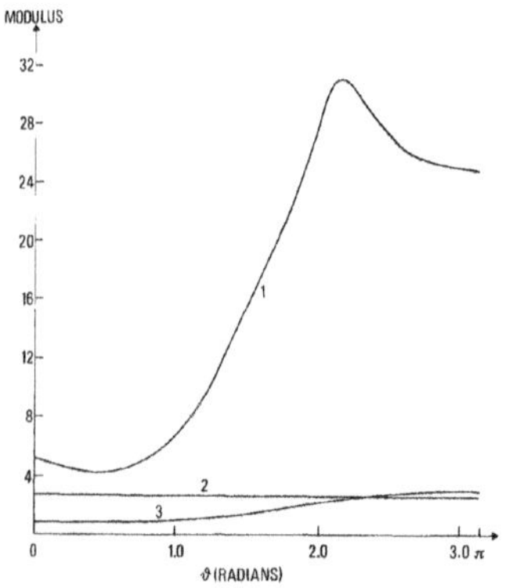

Figure 2 - Curve I shows the modulus bound $M(e^{i\theta})$. Curves 2 and 3 show the modulus of the three and four parameter fit respectively on the unit circle as a function of θ

As can be seen when analytically continued to the unphysical cut the fit is found to be compatible with typical K matrix parametrizations of the K^-p scattering amplitude.

The $K^{\pm}p$ scattering amplitude has been used to create logarithmic dispersion relations (20) and dispersion relations using the inverse of the scattering amplitude (21). When this is done the zeros of the scattering amplitude become the singularities of the logarithmic and of the inverse amplitude. Hence the location of the zeros is important for these dispersion relations. Here effectively the zeros play the same role as the coupling constants in usual dispersion relations. Dumbrajs (22) has found that the $K^{\pm}p$ amplitude has three zeros by using a theorem of Sugawara and Tubis (23). This is the same number as our discrepancy function and would be consistent with Rouché's theorem (24). From our three parameter fit to the discrepancy function we have returned to (4.2.2) to calculate $f_-(\omega)$ in the complex ω plane. In this way the couplex conjugate zeros were located at

$$-0.135 \text{ GeV} \quad \pm \quad i\ 0.59 \text{ GeV}$$

Contours of equal modulus of $f_-(\omega)$ are shown in figure 3.

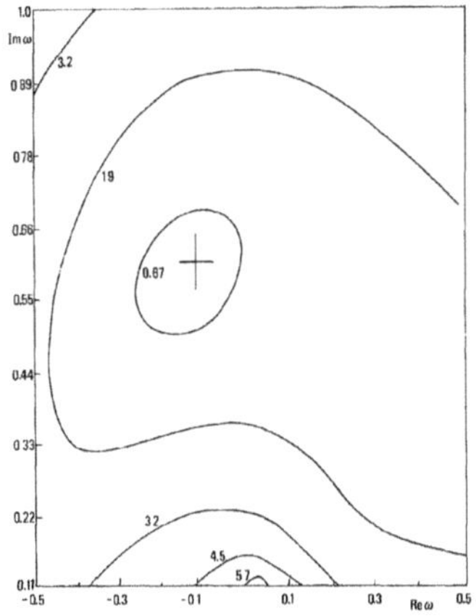

Figure 3 - Contours of equal modulus of $f_-(\omega)$ in the complex lab energy plane

Parametrizing the discrepancy function in this way gives a simple method of seeing if the bound state Λ and Σ poles can be resolved. With the two poles incorporated, the discrepancy function has three zeros so that our parametrization (4.5.1) must be replaced by

$$\Delta(\zeta) = \frac{1}{(\zeta-\zeta_\Lambda)\,(\zeta-\zeta_\Sigma)} \exp\left[\sum_{k=0}^{N} C_k\,\zeta^k\right]\left(\frac{\zeta-M_o}{1-\zeta M_o}\right)\left|\frac{\zeta-M_1 e^{i\phi_1}}{1-\zeta M_1 e^{-i\phi_1}}\right|^2 \qquad (4.6.2)$$

where M_o is the position of the real zero, constrained to remain between the two bound state poles. M_o can be fixed at different positions between ζ_Λ and ζ_Σ and this parametrization can be used to minimize χ^2 using the fitting parameters $C_k (k = 1, 2, \ldots, N)$, M_1 and ϕ_1. The effect of fixing M_o is to distribute different residue values between the two poles. For the case $N = 0$ the χ^2 value was seen to vary by only 2.5 as M_o was fixed at different positions between ζ_Λ and ζ_Σ. From this small variation we can conclude that this method cannot resolve the two poles.

5. Further applications

5.1 The $\pi^\pm p$ scattering amplitude

The factorization theorem (2.3.2) can be used to explain a recent result of Dumbrajs and Staszel (25). They found that the two complex zeros of the $\pi^\pm p$ scattering amplitude in the lab energy plane lie very close to the edge of a forbidden region.

In their method the cut lab energy plane of the $\pi^{\pm}p$ scattering amplitude was mapped inside the unit circle of the z plane so that the bound state pole of the amplitude was at the origin. Assuming that the position of the zeros of the $\pi^{\pm}p$ scattering amplitude f(z) are z_o and z_o^* they constructed the function

$$G(z) = \left[\frac{1-zz_o^*}{z-z_o}\right]\left[\frac{1-zz_o}{z-z_o^*}\right] z\, f(z) \qquad (5.1.1)$$

which is analytic inside the unit circle and further has <u>no</u> zeros. Thus G(z) is just the Carleman function and can be constructed from the modulus data $|f(z)|$ using (2.3.3). Using modulus data, on the cut, of the $\pi^{\pm}p$ scattering amplitude Dumbrajs and Staszel evaluated G(0). Setting z = 0 in (5.1.1) gives

$$a_{-1} = |z_o|^2\, G(0) \qquad (5.1.2)$$

where a_{-1} is the residue of the bound state pole of the $\pi^{\pm}p$ scattering amplitude in the z plane. In fact Dumbrajs and Staszel use an inequality sign in (5.1.2) as they misuse a certain theorem by Szego.

Using the known value of the coupling constant

$$g^2_{\pi^o pp} = 14.6 \qquad (5.1.3)$$

and G(0) then $|z_o|$ can be calculated using (5.1.2) and hence also a closed contour in the lab energy plane upon which the zeros must lie. They determined this contour and found that the known position of the $\pi^{\pm}p$ zeros calculated by an independent method (26) lie practically on the boundary. Their result is shown in figure 4. The equality sign in (5.1.2) is therefore explained by the factorization theorem (2.3.2) and hence also their result.

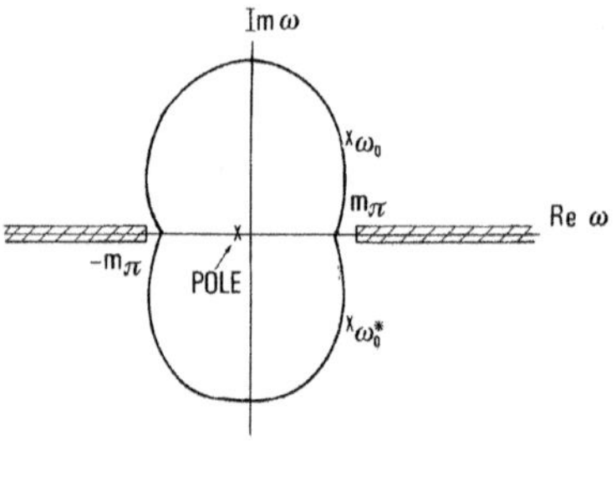

Figure 4 - Closed contour found by Dumbrajs and Staszel in the complex ω plane together with the zeros ω_o and ω_o^* found in reference (26)

6. Conclusions

In this paper we have shown that the Carleman function and Blaschke factor can be used to factorize functions inside the unit circle of the complex plane. The Blaschke factors contain the zeros of the function being studied and the maximum modulus theorem imposes a constraint on the Blaschke factor. Applied to the regeneration amplitude we saw that the number and position of the zeros determined directly the $K^-p\Sigma$ coupling constant. In the discrepancy function analysis the factorization theorem was used as a fitting parametrization and the zeros in this case became fitting parameters determining directly both the effective coupling constant and the fit to the real part. Very few parameters were needed in this case to obtain a good χ^2. Further the fit was specified to have the same number of zeros as the discrepancy function. Hence we have greatly restricted the class of allowed functions which can fit the data and hence to some extent stabilized any analytic continuation.

The correlation between the zeros of the K^-p amplitude and the strengths of its poles has also been demonstrated in a domain calculation (27) using the factorization theorem.

Acknowledgments

I am indebted to Dr. N.M. Queen and Dr. J.E. Bowcock for many informative and stimulating discussions. I also had many useful discussions with Professor S. Ciulli concerning analytic continuation. Professor G. Violini and Dr. B. Di Claudio provided me with the data for the regeneration amplitude and I had many useful discussions with them. I would like to thank the Science Research Council for financial support through a Nato fellowship during the course of this work.

References

1. Queen, N.M., Restignoli, M., and Violini, G.: 1969, Fortschr. Phys. 17, 467.
2. Queen, N.M., Restignoli, M., and Violini, G.: 1973, Fortschr. Phys. 21, 569.
3. Lusignoli, M., Restignoli, M., Snow, G.A., and Violini, G.: 1966, Physics Letters 21, 229.
4. Dumbrajs, O.V., and Staszel, M.: 1976, Czech. J. Phys. B26, 68.
5. Atkin, G.K.: 1979, Ph. d. Thesis, University of Birmingham.
6. Ciulli, S., Pomponiu, C., and Stefanescu, S.: 1975, Phys. Rep. 17C, 133.
7. Greenleaf, P.: 1972, Introduction to Complex Variables, W.B.Saunders, 457.

8. Rudin, W.: 1966, Real and Complex Analysis, McGraw-Hill, New York, 365.

9. Martin, A.D.: 1976, Phys. Lett. 65B, 346.

10. Dumbrajs, O.V., and Oades, G.C., University of Aarhus, Denmark, Private communication.

11. Dumbrajs, O.V., Dumbrajs, T.Yu., and Queen, N.M.: 1971, Nucl. Phys. B26, 497.

12. Conti, A., and Queen, N.M.: 1975, Nuovo Cimento 29A, 452.

13. Atkinson, D.: 1977, Analytic Extrapolations and Inverse Problems, Talk published in proceedings of 'Etudes Interdisciplinaires des Problémes Inverses', Montpellier.

14. Ciulli, S., and Nenciu, C.: 1973, J. Math. Phys. 14, 1675.

15. Martin, A.D., and Ross, G.G.: 1970, Nucl. Phys. B16, 479.

16. Thompson, J., Proc. Duke Conf. on Hyperon Resonances, 61.
Berley, D., et al.: 1970, Phys. Rev. D1, 1996.
Chao, Y.A., et al.: 1973, Nucl. Phys. B56, 46.

17. Queen, N.M.: 1978, University of Birmingham Report UB-KP-1-78,January.

18. Di Claudio, B., and Violini, G.: 1978, Revista Brasileire de Fisica 8, 189.

19. Di Claudio, B., Violini, G., and Queen, N.M.: 1979, Nucl. Phys. B161, 238.

20. Odorico, R.: 1968, Nuovo Cimento 54A, 96.
Dumbrajs, O.V., and Staszel, M.: 1975, Journal of Physics G2, 172.

21. Zovko, N.: 1971, Nuovo Cimento 6A, 304.

22. Dumbrajs, O.V.: 1972, Nucl. Phys. B38, 600.

23. Sugawara, M., and Tubis, A.: 1963, Phys. Rev. 110, 2127.

24. Spiegel, M.R.: 1964, Theory and Problems of Complex Variables, McGraw-Hill Co., 128.

25. Dumbrajs, O.V., and Staszel, M.: 1975, Nucl. Phys. B88, 52.
Dumbrajs, O.V., Private communication

26. Jorna, S., and McClure, J.A.: 1969, Nucl. Phys. B13, 68.

27. Atkin, G.K., Bowcock, J.E., and Queen, N.M. : 1980, Journal of Physics G6, 163.

DISCUSSION

- E. Ferrari :

It seems to me that all this procedure of looking for the amplitude zeros in order to go back to the coupling constants and amplitude determination might turn out to be an unnecessary complication. As a matter of fact, the location of the zeros (at least, of those nearest to the pole region) is strongly dependent on the pole residues, and cannot be known without knowing the latter beforehand. One must then make a fit to the zero positions and to the other parameters introduced by the mapping procedure (e.g. the coefficients of the expansion of the Carleman function); but at this point I do not see clearly what is the advantage with respect to a direct fit to the coupling constants and the unphysical-region parameters in the original dispersion relation. In particular, for pion-nucleon forward amplitudes, where there is no unphysical region and the coupling constant is well known, I do not see any reason for not evaluating directly the dispersion integrals for complex argument; as a matter of fact, already in 1973, I performed such a calculation for the crossing-odd amplitude and located its zeroes quite accurately (Nuovo Cimento 17, A505 (1973)). Subsequently, the same work was carried out by myself and Lusignoli (Nuovo Cimento 23, A51 (1974)) also for the crossing-even amplitude.

- G. K. Atkin :

Direct analytic continuation using experimental data is a highly unstable procedure (see for instance a proof by S. Ciulli et al. in Phys. Rep. 17 (1975) 133). Therefore, the discrepancy function method which eliminates the need to consider the unphysical cut of the K^-p amplitude is an obvious choice. However, the cross-section data and the real part data of the amplitude continue to have errors and are only measured at discrete points so we can expect that any analytic continuation of the discrepancy function will still be unstable. By using the factorization theorem as a parametrization inside the unit circle we incorporate the fact that the K^-p amplitude has three zeros (O. Dumbrais, Nucl. Phys. B38 (1972) 600). In this way we have reduced the class of functions which can fit the data since we only consider functions with three zeros (two zeros if we consider the $K^-p\Lambda$ and $K^-p\Sigma$ poles as combined into one effective pole). This is the reason for using the product of Blaschke factors and Carleman function as a parametrization inside the unit circle.

KAON-NUCLEUS INTERACTIONS*

C.B. Dover
Physics Department - Brookhaven National Laboratory, Upton,N.Y.

Most of the emphasis at this meeting has been on the elementary interaction of kaons ($K^{\pm}$, K^{o}_{L}) with nucleons. I would like to shift the emphasis for a moment to a discussion of possibilities for the study of nuclear and hypernuclear structure physics with kaon beams (1-4). With the availability of K^+ and K^- beams (3, 4) at CERN, Brookhaven, and KEK in Japan, there has been increased interest in the last few years in kaon-nucleus physics, particularly in hypernucleus formation. I propose to review briefly the recent progress in hypernuclear physics, in particular the studies of Λ and Σ states via the strangeness-exchange (K^-, π^-) reaction. Prospects for future investigations with (proposed) intense kaon beams are also evaluated, for instance the production of high spin hypernuclei via the (π^+, K^+) reaction, and the formation of strangeness $S = -2$ hypernuclei ($\Lambda\Lambda$ or Ξ^-) by means of the (K^-, K^+) process. We close with a very brief resumé of elastic, inelastic and charge exchange reactions induced by the interaction of K^+ mesons with nuclei (5).

To appreciate the striking difference between K^- and K^+ interactions with nuclei, it is worthwhile to review the main features of the elementary $K^{\pm}N$ interactions.

The average interactions of K^+ and K^- with nucleons are very different in character. The low energy K^-N cross sections (6) for isospin $I = 0, 1$ are shown in figure 1. The average K^- cross section is roughly 40mb, typical of hadronic processes. The existence of a number of $S = -1$ Y^* resonances is evident from the data. In figure 2, we show for comparison the K^+N cross sections (7) for $I = 0,1$. In contrast to K^-N, the K^+N cross sections are small ($\simeq$ 10mb) and almost energy independent for $p_{lab} \leq 700$ MeV/c, below the onset of inelastic processes.

The opposite strangeness $S = \pm 1$ for $K^{\pm}$ plays the dominant role in understanding the large cross section differences evident from figures 1 and 2.

E. Ferrari and G. Violini (eds.), Low and Intermediate Energy Kaon-Nucleon Physics, 165–186.

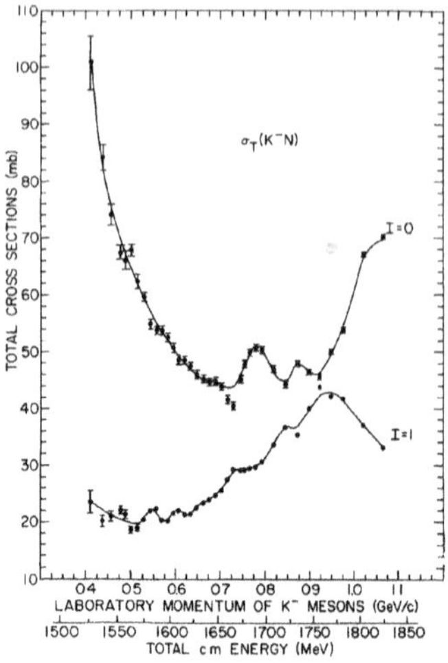

Figure 1

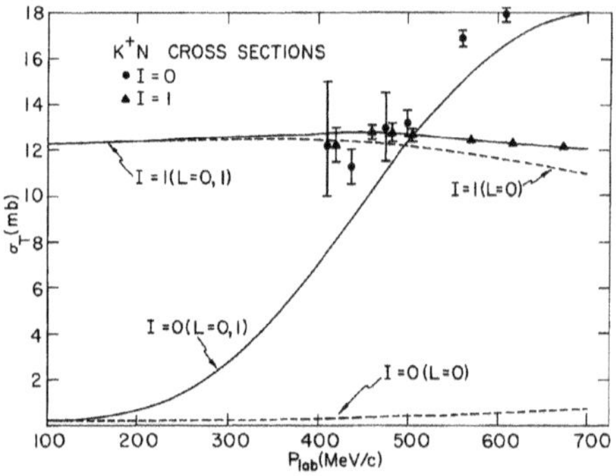

Figure 2

In figure 3, we illustrate (8) how S = -1 Y^* resonances are formed in a natural way in the quark model through the annihilation of $q\bar{q}$ pair. The Y^* resonances correspond to ordinary low-lying three quark states. In contrast, we see from figure 3 that there is no corresponding quark annihilation process for K^+N, so possible intermediate states (exotic Z^* resonances with S = +1) cannot be composed of three quarks, the minimum complexity being $\bar{s}qqqq$.

Due to the qualitative differences in their elementary interactions with nucleons, the K^+ and K^- are quite distinct probes of the nucleus.

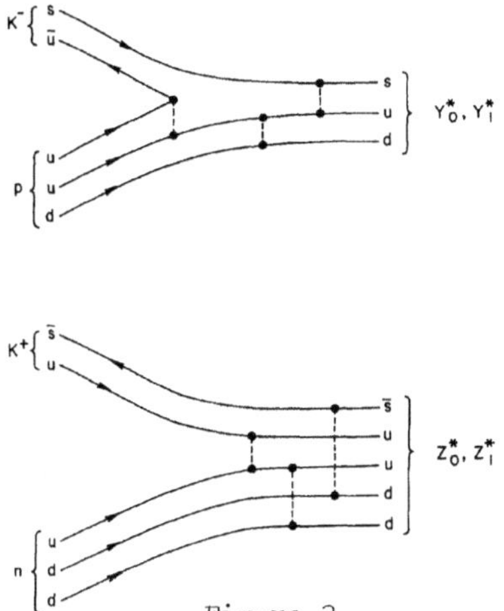

Figure 3

One way of expressing this difference is in terms of the mean free path $\lambda = (\rho\sigma)^{-1}$ in nuclear matter. Here $\rho = 0.16\text{fm}^{-3}$ is the nuclear matter density and σ is the average total cross section $(\sigma_{K^\pm p} + \sigma_{K^\pm n})/2$. For K^-, we have $1\text{fm} \leq \lambda_{K^-} \leq 2\text{fm}$ for $p_{lab} \leq 800$ MeV/c, while for K^+ we obtain $5\text{fm} \leq \lambda_{K^+} \leq 7\text{fm}$. The K^+ mean free path is predicted to be very long (5), comparable to the size of a large nucleus like ^{208}Pb. The K^-, on the other hand, is a strongly absorbed particle in nuclei; its interactions are thus restricted to the nuclear surface.

The K^-, unlike the K^+, can transfer one or two units of strangeness to the nucleus. Some of the strangeness exchange reactions which lead to Λ and Σ hypernuclear formation are the following

$$
\begin{aligned}
&K^- + n \rightarrow \pi^- + \Lambda \text{ or } \pi^- + \Sigma^0 \\
&K^- + p \rightarrow \pi^+ + \Sigma^-
\end{aligned}
\tag{1}
$$

Another reaction which could be used to produce hypernuclei, although under quite different kinematical conditions, is associated production

$$\pi^+ + n \rightarrow K^+ + \Lambda \tag{2}$$

Finally, Ξ^- hypernuclei (S = -2) can be produced via the reaction

$$K^- + p \rightarrow K^+ + \Xi^- \tag{3}$$

Systems containing a Σ or Ξ particle are unstable with respect to strong conversion processes of the type

$$
\begin{aligned}
&\Sigma^- + p \rightarrow \Lambda + n \\
&\Xi^- + p \rightarrow \Lambda + \Lambda
\end{aligned}
\tag{4}
$$

One of our tasks will be to estimate the widths of Σ and Ξ states arising from such reactions.

Reactions (1) and (2) are distinguished by the momentum transfer q, for a nuclear target at rest in the lab system. The value of q at $\theta_{LAB} = 0^\circ$ is plotted in figure 4 as a function of meson incident lab momentum. For the reaction (K,π), there exists a "magic momentum" such that the Λ or Σ is created at rest in the lab.

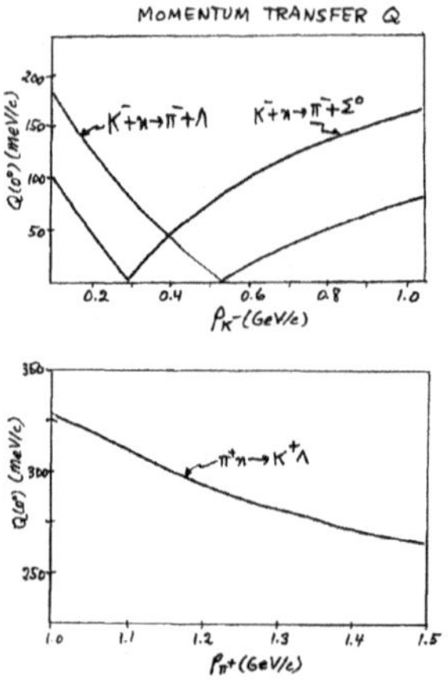

Figure 4

The importance of this fact for hypernuclear production was emphasized by Feshbach and Kerman (9). The Λ or Σ at rest has a sizable "sticking probability" for remaining bound to the nucleus. The excitation of 0^+ configurations where the Λ or Σ coherently replaces the last valence nucleon (i.e., occupies the same shell model orbit as vacated by the nucleon) is favored. The magic momenta are about 300 MeV/c for Σ° and 530 MeV/c for Λ production. Currently, kaon beams in this momentum region are not available, so experiments are done in the 700-900 MeV/c region. However, from figure 4, we see that q remains small ($\lesssim$ 50 MeV/c) compared to the nucleon Fermi momentum p_F = 270 MeV/c for $p_{K^-} \lesssim$ 800 MeV/c, so coherent substitutional processes are important at 0° in the whole low momentum region. The situation is quite different for the $\pi^+ n \rightarrow K^+\Lambda$ reaction, as seen in figure 4. Here $q \gtrsim p_F$ for all momenta of interest. The low spin states emphasized by the (K^-, π^-) process will be only very weakly excited in (π^+, K^+). Instead, the Λ has a measurable "sticking probability" only in high spin configurations in (π^+, K^+), because its large linear momentum can only be matched to a correspondingly high angular momentum. The same is true for a Ξ^- produced in a (K^-, K^+) reaction. We return later to a discussion of these reactions.

The (K^-, π^-) experiments (10-12) which have been done to date with nuclear targets have had energy resolution in the range $\Delta E \simeq$ 2.5 -6 MeV. With such coarse resolution experiments, one sees only the energy-averaged gross structure of hypernuclei; that is, Λ particle-neutron hole (Λn^{-1}) states. As an example, consider the hypernucleus ${}^{16}_{\Lambda}O$. The ground state doublet of ${}^{16}_{\Lambda}O$ arises by replacing the neutron in the last valence orbit ($nP_{1/2}$) by a Λ in the lowest lying orbit ($\Lambda S_{1/2}$). We thus obtain

$$(\Lambda S_{1/2} \otimes n\, P_{1/2}^{-1})\ 0^-,\ 1^- \tag{5}$$

In hypernuclei, the natural parity member of the ground state doublet ($\pi = (-)^J$) lies lowest in energy. The very small splitting of the doublet cannot be resolved experimentally. Note that the cross section to the 0^- state is essentially zero in the (K^-, π^-) reaction, except for parity violating effects. In general, (K, π) cross sections to other unnatural parity states (1^+, 2^-, 3^+, etc.), which rely on spin-flip amplitudes for their production, are also calculated (13) to be small, even for $\theta_L \neq 0°$. We neglect these in the following. The low-lying excited natural parity states in ${}^{16}_{\Lambda}O$ are

$$\begin{aligned} &(\Lambda S_{1/2} \otimes n\, P_{3/2}^{-1})\, 1^- \\ &(\Lambda P_{1/2,\ 3/2} \otimes n\, P_{1/2}^{-1})\ 0^+,\ 2^+ \\ &(\Lambda P_{1/2,\ 3/2} \otimes n\, P_{3/2}^{-1})\ 0^+,\ 2^+,\ 2^+ \end{aligned} \tag{6}$$

The experimental spectrum for the (K^-,π^-) reaction on ^{16}O and ^{40}Ca targets (10, 11) is shown in figure 5.

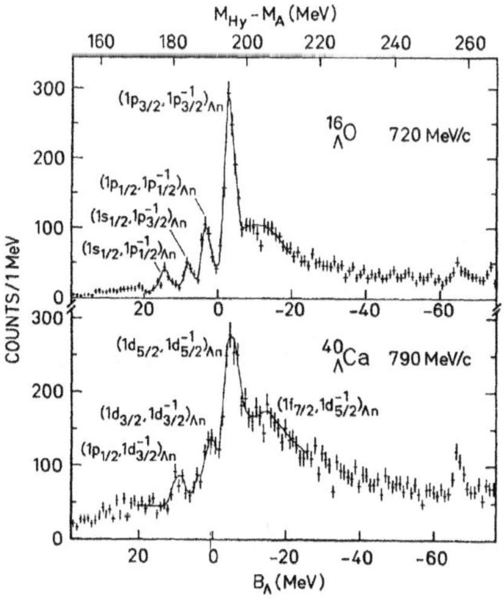

Figure 5

The CERN 0° data for ^{12}C, ^{32}S and ^{209}Bi are given in figure 6. In very light hypernuclei, such as $^{12}_{\Lambda}C$ and $^{16}_{\Lambda}O$, the narrow peaks due to particular Λ particle-neutron hole states (Λn^{-1}) dominate the spectrum.

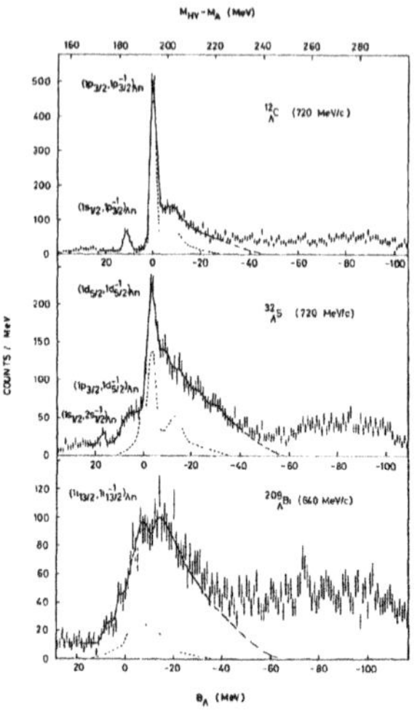

Figure 6

For heavier systems, the quasielastic part of the spectrum becomes relatively more important. This process involves the emission of a Λ into the continuum. This leads to a sizable "background" cross section, for instance in $^{209}_{\Lambda}Bi$, which makes it difficult to resolve individual hypernuclear states.

In the distorted wave impulse approximation (DWIA), the (K^-, π^-) cross section at $\theta_L = 0°$ on a spin zero target takes the form

$$\frac{d\sigma}{d\Omega}(J^\pi) = \left(\frac{d\sigma}{d\Omega}\right)^{0°}_{K^- n \to \pi^- \Lambda} N^J_{eff} \tag{7}$$

where the final Λn^{-1} state has quantum numbers $|J^\pi) = (\Lambda \ell_p j_p \otimes n(\ell_h j_h)^{-1})_{J^\pi}$. In eq. (7),

$$\left(\frac{d\sigma}{d\Omega}\right)^{0°}_{K^- n \to \pi^- \Lambda}$$

is the 0° lab cross section for the elementary $K^- n \to \pi^- \Lambda$ process, and N^J_{eff} is the effective number of neutrons, given by (14)

$$N^J_{eff} = (2J+1)(2j_p+1)(2j_h+1)\begin{pmatrix} j_p & j_p & J \\ \frac{1}{2} & -\frac{1}{2} & 0 \end{pmatrix}^2 F_J(q) \tag{8}$$

Note that at 0°, only natural parity states are excited, so that $\ell_p + \ell_h + J$ must be even. The form factor $F_J(q)$ involves the radial overlap of the Λ and n^{-1} wave functions with kaon and pion distorted waves :

$$F_J(q) \;=\; \left(\int_0^\infty r^2 dr\, R_p(r)\, R_h(r)\, \hat{j}_J\,(r)\right)^2 \qquad (9)$$

In plane wave approximation (PWA), $\hat{j}_J$ reduces to the usual spherical Bessel function $j_J(qr)$. The formulae (7)-(9) can easily be generalized (5) to $\theta_L \neq 0°$.

Near the "magic momentum" for the (K^-, π^-) reaction, $\theta_L = 0°$ corresponds to $q \simeq 0$. In PWA, we see immediately from eq. (9) that only $F_0(q)$ differs from zero for $q \simeq 0$; thus 0^+ states will dominate for $\theta_L = 0°$. As we increase θ_L, q also becomes larger and $F_0(q)$ decreases; other form factors $F_J(q)$ for $J \neq 0$ come into play. The higher spin states not seen at 0° can be revealed by their characteristic angular distributions.

The goal of the theory is to predict the values of N^J_{eff} for various Λn^{-1} configurations, i.e. the (K^-, π^-) "strength function". Some typical results due to Bouyssy (15) are shown in figure 7. The contributions of individual Λn^{-1} states are indicated by dashed lines; they have been spread out in energy to simulate the results of a coarse resolution experiment. The envelope corresponding to the sum of all Λn^{-1} contributions is shown as a solid line, with the corresponding CERN data displayed in the inset.

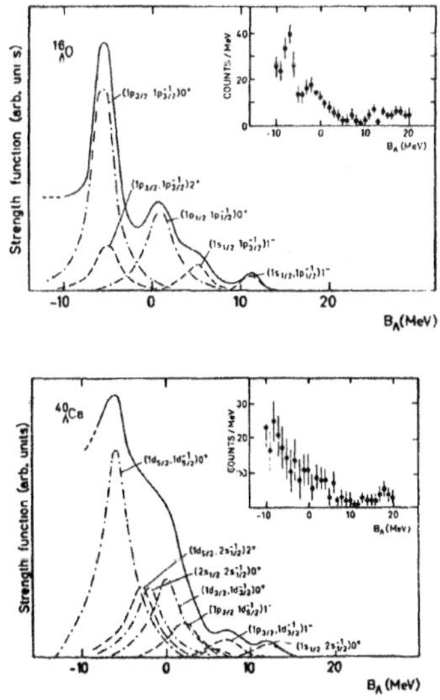

Figure 7

The low spin states are seen to dominate, and the comparison with data enables one to assign experimental peaks to particular Λn^{-1} states, at least in light hypernuclei, where the Λn^{-1} configurations of a given spin J are well spaced and only weakly admixed (14, 15) by the residual ΛN interaction. The positions of the Λn^{-1} peaks are close to the unperturbed particle-hole energies. The n^{-1} single particle energies are taken from observed separation energies, and information on the Λ binding in the $S_{1/2}$ state is available from emulsion data on hypernuclear ground states. To predict the Λ binding energies in excited states, we need a model for the single particle potential $V_\Lambda(r)$ in a hypernucleus. We take a conventional Woods-Saxon form

$$V_\Lambda(r) = -V_o^\Lambda f(r) + V_{LS}^\Lambda \, \underset{\sim}{\ell} \cdot \underset{\sim}{\sigma}_\Lambda \left(\frac{\hbar}{m_\pi c}\right)^2 \frac{1}{r}\frac{df(r)}{dr} \tag{10}$$

with $f(r) = (1 + \exp(r-R)/a)^{-1}$. The geometrical parameters $R = r_o A^{1/3}$, $r_o = 1.1$fm, $a = 0.6$fm are assumed to be similar to those for the nucleon potential. The depth parameters V_o^Λ and V_{LS}^Λ for the central and spin-orbit potentials have been determined (17) by a fit to the observed peak positions in the (K^-, π^-) reactions. The results are (17)

$$\begin{aligned} V_o^\Lambda &\simeq 32 \pm 2 \text{ MeV} \\ V_{LS}^\Lambda &\simeq 4 \pm 2 \text{ MeV} \end{aligned} \tag{11}$$

The depth V_o^Λ is consistent with the emulsion results (16). The new information gained from the CERN (K, π) data is the value of V_{LS}^Λ, which is much smaller than the value $V_{LS}^N \simeq 20$ MeV for nucleons. The fact that the Λ spin-orbit potential is small can be seen, for instance, from the $^{16}_\Lambda O$ spectrum of figure 5. The observed splitting between the $({}_\Lambda P_{3/2} \otimes n\, P_{3/2}^{-1})\, 0+$ and $({}_\Lambda P_{1/2} \otimes n\, P_{1/2}^{-1})\, 0+$ configurations, about 6MeV, can be attributed to the nucleon spin-orbit splitting, with no need for a large value of V_{LS}^Λ. Similar conclusions follow from an inspection of the $^{40}_\Lambda Ca$ spectrum in figure 5.

The Λ spin-orbit potential has been the object of some recent theoretical work (18, 19, 20) based on meson exchange models or the quark-gluon picture. The general conclusion of these authors in that V_{LS}^Λ should be small. Note, however, that the older analysis of p-shell hypernuclei due to Dalitz, Gal and Soper (21) yielded a large spin-orbit potential for the Λ, of opposite sign to that for a nucleon.

The CERN data at 0° are useful for determining the single particle potential of a Λ in a nucleus. These properties can already be determined from the low-spin part of the hypernuclear spectrum. The presence of higher spin states ($J \geq 2$) can only be established by measuring the (K^-,π^-) angular distributions.

In a recent experiment (12) at the Brookhaven AGS, the angular distributions in figure 8 were obtained for the reaction $^{12}C(K^-, \pi^-)^{12}_{\Lambda}C^*$ at an incident momentum of 800 MeV/c.The data have been interpreted in Ref. (13) in DWIA.

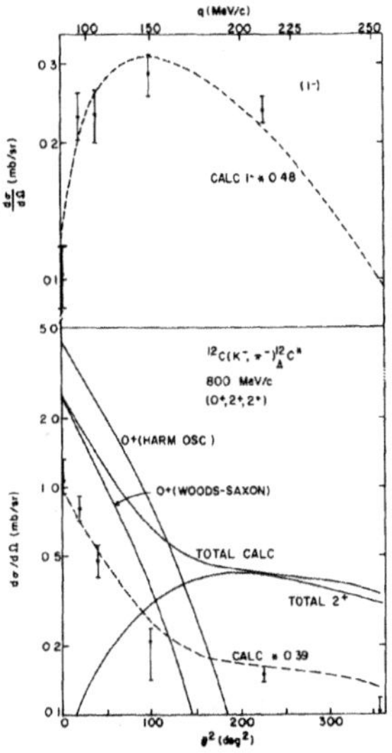

Figure 8

The angular shape for the ^{12}C ground state, with a peak at about 10°, is characteristic of a 1^- state. This is what one anticipates from the $({}_{\Lambda}S_{1/2} \otimes n\, P_{3/2}^{-1})_{1^-}$ configuration. The angular distribution for the group of states at 11 MeV excitation requires more interpretation. One expects the

$$({}_{\Lambda}P_{1/2} \otimes n\, P_{3/2}^{-1})_{2^+} \text{ and } ({}_{\Lambda}P_{3/2} \otimes n\, P_{3/2}^{-1})_{0^+,\, 2^+}$$

states to lie in this region of excitation. A coarse resolution experiment cannot separate these. If the Λ spin-orbit potential is indeed small, the energy splittings of the 0^+, 2^+, 2^+ triplet come only from the residual ΛN interaction, which typically produces splittings of 1 MeV or less. Thus the data shown in the lower half of figure 8 reflect the summed strength for the 0^+, 2^+, 2^+ states. The presence of the 2^+ excitations is revealed as a shoulder in the angular distribution in the region from 10° to 20°. The calculated contribution from the 0^+ state alone is several orders of magnitude below the data in this region. The DWIA theory (13) gives a good account of the shape of the angular distribution. This is sensitive to the relative size of the 0^+ and 2^+ contributions,which the theory gives correctly. Agreement in absolute size of the (K,π) cross sections requires multiplication of the theoretical values by an overall normalization factor of about 1/2. The origins of this discrepancy are discussed in ref. (13). In any case, the data in figure 8 provide clear evidence for the presence of the expected 2^+ strength, not seen in the C^c data.

The CERN group has also investigated the $(K^-, \pi^\pm)$ reactions in the Σ region (22). Some of their results are shown in figure 9 for the $^9Be(K^-, \pi^-)^9_\Lambda Be$ reactions at 720 MeV/c. The incident momentum is chosen to correspond to a peak in the elementary $K^- n \to \pi^- \Sigma^o$ process. The presence of (unbound) Σ states is suggested by the structures located within 20 MeV of the threshold corresponding to zero Σ binding ($B_\Sigma = 0$). Since the momentum transfer q is still reasonably small ($\simeq$ 130 MeV/c), we expect that these states would correspond to 0^+ or 1^- excitations.

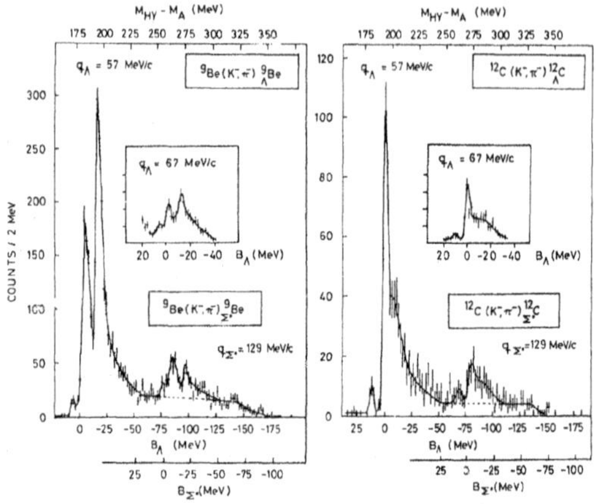

Figure 9

The surprising feature is the narrow width observed for these states ($\Gamma_\Sigma \lesssim 10$ MeV). It had been assumed that the strong conversion mechanism $\Sigma N \to \Lambda N$ would lead to very broad Σ states ($\Gamma \geq 30$ MeV). A rough estimate of the Σ width in nuclei can be obtained from the perturbation theory result

$$\Gamma_\Sigma \simeq (v\sigma)^{AV}_{\Sigma^- p \to \Lambda n} \int_0^\infty \rho_p(r)\, u^2_{n\ell}(r)\,dr \qquad (12)$$

where $u_{n\ell}(r)$ is the Σ radial wave function, $\rho_p(r)$ is the proton density and $(v\sigma)^{AV}$ is the Fermi-averaged $\Sigma^- p \to \Lambda n$ cross section, v being the relative velocity. The experimental data (23) on the $\Sigma^- p \to \Lambda n$ process may be parametrized in the form

$$(v\sigma)_{\Sigma^- p \to \Lambda n} \simeq (v\sigma)_o/(1 + \alpha v) \qquad (13)$$

where $(v\sigma)_o \simeq 65$ mb, $\alpha \simeq 20$. For a Σ^- at rest in the nucleus, the Fermi averaging over the nucleon motion yields

$$(v\sigma)^{AV}_{\Sigma^- p \to \Lambda n} \simeq 14 \text{ mb}.$$

Note that $(v\sigma)^{AV}$ is only about 1/5 of the threshold value $(v\sigma)_o$, which was used in early estimates of the Σ width. Using the smaller value $(v\sigma)^{AV}$, we get typical widths (24)

$$\Gamma_{\Sigma}^{1S} \approx 23\ \text{MeV}, \quad \Gamma_{\Sigma}^{1p} \simeq 13\ \text{MeV} \qquad (14)$$

for $^{12}_{\Sigma}C$. For the 1P Σ-state, the value of Γ_{Σ} in eq. (14) is already tantalizingly close to the width of the structures seen in figure 9. Additional reduction (25) of the Σ width can arise in light nuclei because the low-energy $\Sigma N \to \Lambda N$ conversion mechanism is strongly spin and isospin dependent (the $^3S_1$, I = 1/2 channel dominates). In systems like $^{12}_{\Sigma}C$ and $^{7}_{\Sigma}Li$, the "core" seen by the Σ is not spin-isospin saturated. Thus the matrix element of the $\Sigma N \to \Lambda N$ amplitude between Σ and Λ hypernuclear wave functions will differ from that obtained by neglecting the spin-isospin dependence (as done in eq. (12)). One can thus obtain either larger or smaller widths than from eq. (12). Detailed estimates of Σ widths have been carried out in ref. (25). It was found that 0^+ states with I = 3/2 enjoy an appreciable suppression of the Σ width relative to the limit where spin-isospin effects are ignored. The width suppression effect is limited to only a few states : 1^- and 0^+, I = 1/2 states in $^{12}_{\Sigma}C$ and $^{16}_{\Sigma}O$ are broadened rather than narrowed. Definite predictions for the quantum numbers of narrow Σ states in light nuclei have been made (25). These cannot be firmly established on the basis of the 0° data alone; angular distributions are required.

A particularly interesting case is that of $^{16}_{\Sigma}O$. The two states which are candidates for narrow structure (25) are the

$$I = 3/2,\ ({}_{\Sigma}P_{3/2} \otimes nP_{3/2}^{-1})_{0^+} \text{ and } ({}_{\Sigma}P_{1/2} \otimes n\,P_{1/2}^{-1})_{0^+}$$

configurations. If the spin-orbit potential for the Σ is weak, these two states are predicted to be fairly broad and separated by about 6 MeV. On the other hand, if the Σ spin-orbit potential is comparable to that of the nucleon, the two 0^+ states will strongly mix and <u>one</u> of the diagonalized configurations carries almost all of the (K, π) strength. We then expect <u>one</u> quite narrow 0^+, I = 3/2 state in $^{16}_{\Sigma}O$. No data on $^{16}_{\Sigma}O$ is yet available to test this prediction. For Σ states, it is important to measure both the (K^-, π^-) and the (K^-, π^+) reactions; the latter process filters out the I = 1/2 states.

There have also been some recent measurements of hypernuclear γ rays, in particular for $^{4}_{\Lambda}H$, $^{4}_{\Lambda}He$ (26) and for $^{8}_{\Lambda}Li$ (27). The importance of γ ray measurements has been emphasized by Dalitz and Gal (28). The case of $^{4}_{\Lambda}H$, $^{4}_{\Lambda}He$ is interesting, since it bears on the question of charge symmetry breaking. Earlier experiments were interpreted in terms of a sizable Σ^o - Λ mixing, but the more recent results (26) require much less violation of charge symmetry. The results for $^{8}_{\Lambda}Li$ can be compared with earlier predictions (28). The theoretical results are sensitive to the spin dependence of the ΛN residual interaction, which we hope to test with such data.

The previous sections have been devoted to a brief review of the experimental data on hypernuclei. One may now ask: what are the future prospects for kaon physics ? Several projects for improved kaon beams are in the planning or development stage. At CERN, a K^- beam at 450 MeV/c is under construction (30). The low momentum will be more favorable for the study of Σ hypernuclei, since one is then closer to the "magic momentum" (see figure 4). A workshop group is developing a proposal for a dedicated kaon beam at the Brookhaven AGS with an intensity of perhaps 50 times current values (2×10^4 kaons/burst). Such a proposal could also include an intense pion beam in the 1 - 1.5 GeV/c range. There are also longer range discussions under way, for instance at LAMPF and TRIUMF, concerning the possibility of a "kaon factory". Some of these considerations were reviewed at the recent Kaon Workshop (31) at Vancouver. In view of this discussion, it is important to provide a physics justification for a new kaon facility. I will mention here only a few of the unique possibilities for kaon physics. A number of other topics have also been explored, for which the reader is referred to the literature (31-34).

In heavy hypernuclei, we expect that collective "strangeness analog resonances" (SAR) will be formed (35). As for isobaric analog states, the SAR involve a coherent superposition of Λn^{-1} states coupled to $J^{\pi}=0^{+}$. In a somewhat idealized model (35), the SAR wave function remains antisymmetric with respect to the interchange of the Λ with any neutron, even though the Λ does not need to satisfy the Pauli principle. Theoretical calculations (36) indicate a tendency to develop such collective states in heavy nuclei. Since the quasiclastic contribution to the (K^-, π^-) reaction is large in heavy nuclei (see ${}^{209}_{\Lambda}Bi$ in figure 6), experiments with better energy resolution will be required to see these narrow collective structures. These experiments will become feasible if new intense kaon beams become a reality.

In addition to the SAR, other "supersymmetric" configurations should exist in hypernuclei (37). These states are predicted to lie well below the analog states (by 10-20 MeV) and correspond to symmetries that cannot be realized in ordinary nuclei because of the Pauli principle. For example, consder the $(1p)^5$ configuration outside the closed (1S) shell. The orbital permutation symmetry for the ground state of ${}^{9}Be$ is dominantly [41], which then also holds for the SAR in ${}^{9}_{\Lambda}Be^{*}$. However, "supersymmetric" states of ${}^{9}_{\Lambda}Be^{*}$ also exist with orbital symmetry [5]; these states permit more relative s-state bonds than [41] states, and hence they lie lower in energy than the SAR. The ${}^{9}_{\Lambda}Be^{*}$ excitations may be grouped (37) into SU(6) multiplets, if one neglects the spin dependence of ΛN and NN forces. The SU(6) symmetry would imply that a number of non-analog states would be roughly degenerate with the SAR, leading to an interesting fine structure arising from the distribution of the strength of the SAR over these states. Data with much better energy resolution are required to see this fine structure. The search for such approximate symmetries in hypernuclear spectra is an exciting prospect for kaon physics.

Previously, we have discussed the simple Λ particle-neutron hole model for hypernuclei. Experiments with poor energy resolution are sensitive only to this gross Λn^{-1} structure. The next generation of experiments should be able do detect "core excited" states in hypernuclei. For example, Dalitz and Gal (28) have estimated the (K, π) cross section to the "core excited" states in ${}^{12}_{\Lambda}C$. In particular, a 1^- state consisting of ${}_{\Lambda}S_{1/2}$ coupled to the low-lying $1/2^-$ state in ${}^{11}C$ was predicted to lie about 3-4 MeV above the ${}^{12}_{\Lambda}C$ ground state, and to be populated with appreciable intensity in the (K^-, π^-) reaction. No evidence for this state was seen in the BNL data (12); possible explanations for its absence are discussed in ref. (13). These "core excited" states may also be viewed as arising from ground state correlations in the target, i.e. ${}^{12}_{\Lambda}C$ is a poor "closed shell", containing appreciable two-particle-two-hole correlations. Particle-hole excitations may also be created by inelastic scattering of the kaon or pion. Such dynamical effects can influence cross section estimates based on pre-existing ground state correlations.

The (K, π) reaction has been exploited to produce low spin hypernuclear states. The cross section (π, K) has recently been discussed (38) as a possibility for producing hypernuclei. Since the momentum transfer for $\pi^+ n \rightarrow K^+\Lambda$ is large, as per figure 4, associated production will preferentially populate high spin states, nicely complementing the (K, π) measurements. An attractive experimental feature of the (π, K) process is the availability of large pion fluxes, typically 10^4 π's per K at the Brookhaven AGS. This permits the measurement of much smaller cross sections (1 μb/st is accessible) than for (K, π) with current beam intensities.

The configurations which are predicted (38) to dominate the (π, K) reaction are the natural parity "stretch states" obtained by coupling Λn^{-1} to spin $J = \ell_p + \ell_h$. If we consider nodeless oscillator wave functions for which $R_{1\ell}(r) \sim (r/b)^{\ell} \exp(-r^2/2b^2)$, the form factor $F_J(q)$ of eq. (9) becomes

$$F_J(q) = \frac{z^J e^{-z}}{[(J+1)!!]^2} \tag{15}$$

where $z = (bq)^2/2$. Note that $F_J(q)$ vanishes at $q = 0$ for $J \neq 0$ and assumes its peak value for

$$J = (bq)^2/2 \tag{16}$$

The condition (16) represents the optimum matching of momentum transfer q and total spin J for the oscillator model. Note that $\theta_L = 0°$ corresponds to a large value $q = q_o$; values of $q < q_o$ are not accessible to experiment. The value q_o usually exceeds the optimum value $(2J)^{1/2}/b$ of eq. (16), so that one is already past the peak of $F_J(q)$ for $\theta_L = 0°$, and the (π^+, K^+) cross section is a decreasing function of θ_L.

The question now arises : What is the highest spin Λn^{-1} state that one can manufacture in a given hypernucleus ? If one assumes the potential $V_\Lambda(r)$ of eq. (10), the spectrum of Λ single particle states can be calculated (38). One can also seek guidance from Hartree-Fock calculations (39) for Λ hypernuclei. For systems with $A > 40$ or so, the Λ orbit corresponding to the last bound nucleon orbit is a single particle resonance in the continuum. For typical systems like ^{48}Ca, ^{90}Zr, ^{138}Ba and ^{208}Pb, which have closed $j = \ell + 1/2$ neutron shells with $\ell = 3, 4, 5$ and 6, respectively, the corresponding Λ orbits lie at roughly 5-6 MeV in the continuum. The elastic width of these Λ states becomes narrower as A increases ($\Gamma_{EL} \approx 2.4$ MeV for the 1f in $^{48}_{\Lambda}Ca$, $\Gamma_{EL} \approx 0.3$ MeV for the 1i in $^{208}_{\Lambda}Pb$) so they should be observable as relatively narrow Λn^{-1} states, even though they are unbound. The highest spin narrow Λn^{-1} states that we anticipate are then obtained by coupling these single particle resonances to n^{-1} states : $^{28}_{\Lambda}Si$, $^{48}_{\Lambda}Ca$, $^{90}_{\Lambda}Zr$, $^{138}_{\Lambda}Ba$, and $^{208}_{\Lambda}Pb$ are good candidates, since n^{-1} also has high spin. We can also couple the Λ in the next lowest orbit to n^{-1}, giving states with one less unit of J. We find

$$
\begin{aligned}
^{28}_{\Lambda}Si: \quad & ({}_{\Lambda}d_{3/2,\,5/2} \otimes {}_{n}d^{-1}_{5/2})_{4^+} \\
& ({}_{\Lambda}p_{1/2,\,3/2} \otimes {}_{n}d^{-1}_{5/2})_{3^-} \\
^{48}_{\Lambda}Ca: \quad & ({}_{\Lambda}f_{5/2,\,7/2} \otimes {}_{n}f^{-1}_{7/2})_{6^+} \\
& ({}_{\Lambda}d_{3/2,\,5/2} \otimes {}_{n}f^{-1}_{7/2})_{5^-}
\end{aligned}
\qquad (17)
$$

and so forth. In figure 10, we show these high spin states on a J vs.

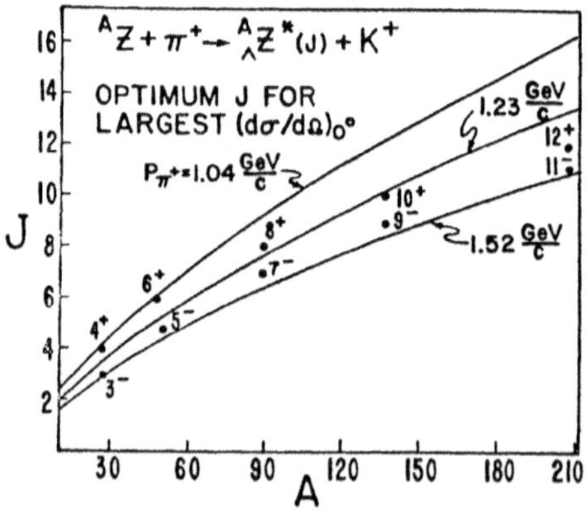

Figure 10

A plot, along with continuous curves corresponding to the optimum matching condition (16). The A dependence of the optimum J arises from the oscillator radius parameter b, which we have taken as proportional to $A^{1/3}$. Note that q decreases with increasing p_π (see figure 4), so that , by varying p_π, we sweep out a band of optimum J values. The elementary $\pi^+ n \to K^+\Lambda$ cross section (40) peaks near p_π = 1.04 GeV/c, but remains sizable up to about 1.5 GeV/c. From figure 10, we see that for light systems such as $^{28}_{\Lambda}$Si or $^{48}_{\Lambda}$Ca, the highest spin state is very well matched at 1.04 GeV/c, so the largest (π^+, K^+) cross sections are to be expected for these states. For heavier hypernuclei, we are no longer well matched at 1.04 GeV/c, so smaller cross sections will result. In ref. (38), estimates of (π^+, K^+) cross sections are given for a range of targets and incident momenta, using PWA, DWIA and eikonal approximation. A typical DWIA theoretical prediction (38) is shown in figure 11 for ^{40}Ca (π^+, K^+) $^{40}_{\Lambda}$Ca* at 1.04 GeV/c.The cross sections for stretch states decrease when J is decreased, but by less than a factor of two per unit of J. Hence several states should have measurable cross sections. The results shown in figure 11 are reduced by only about a factor of 5-10 with respect to PWA estimates, since the K^+ is weakly absorbed in the exit channel.

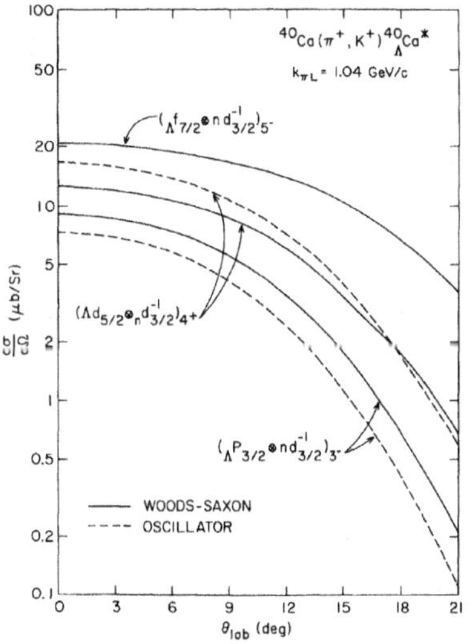

Figure 11

The (π^+, K^+) reaction could open up a new domain of hypernuclear structure physics. The high spin "stretch" states discussed here are of a particularly simple structure, since only one Λn^{-1} configuration of the maximum J exists. In heavy systems, the "spreading width" of very high spin states will be less than that for lower spins, since there is a lower density of compound states (2p 2h, etc.) of the same spin.Thus narrow excitations may exist, even quite high in the continuum. There is as yet no experimental data on the (π^+, K^+) reactions, but an experiment is being planned (41) for the Brookhaven AGS.

As a final topic of the "futuristic" type,we consider the possibility of producing Ξ^- or $\Lambda\Lambda$ hypernuclei via the double strangeness exchange (K^-, K^+) reaction (42). Such studies would be of fundamental interest from several points of view : a) explore a new type of hypernuclear spectroscopy ; b) shed some light on the nature of ΞN and $\Lambda\Lambda$ interaction, thereby extending our knowledge of the SU(3) structure of baryon-baryon forces; c) possibility of narrow Ξ states in nuclei; and d) provide a possible test for Bag model predictions (43) of a strangeness -2 dibaryon.

Two events corresponding to double hypernucleus production are known from emulsion work (44) (${}^{6}_{\Lambda\Lambda}He$ and ${}^{10}_{\Lambda\Lambda}Be$). These are identified by tracks corresponding to two successive weak decays $\Lambda \to p\pi^-$. In the case of ${}^{6}_{\Lambda\Lambda}He$, the first step is a $K^- + p \to K^+ + \Xi^-$ reaction, followed by the slowing down of the Ξ^- in the emulsion, and finally the capture process $\Xi^- + {}^{12}C \to {}^{7}Li + {}^{6}_{\Lambda\Lambda}He$.

Second order processes for the formation of $\Lambda\Lambda$ hypernuclei have been evaluated in ref. (42). These correspond to the graphs of figure 12. The 0° cross section for the $K^-p \to \pi^o\Lambda$ followed by $\pi^o p \to K^+\Lambda$ reaction, summed over all final $\Lambda\Lambda$ hypernuclear states, is given (42) in

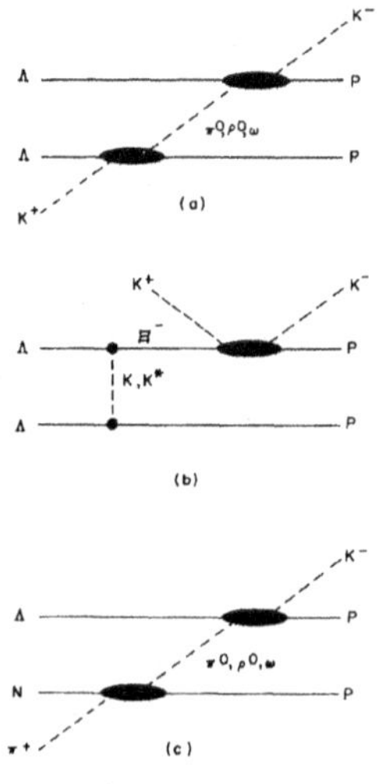

Figure 12

rough approximation by

$$\sum_f \left(\frac{d\sigma}{d\Omega}\right)_f^{0^\circ} = \frac{8\pi^2\alpha_1\alpha_2\xi}{k_\pi^{\;2}} < \frac{1}{r^2} > \left(\frac{d\sigma}{d\Omega_L}\right)_{0^\circ}^{K^-p\to\pi^o\Lambda} \left(\frac{d\sigma}{d\Omega_L}\right)_{0^\circ}^{\pi^o p\to K^+\Lambda} \qquad (18)$$

where

$$\alpha_1 = (1 - q^{(1)}/v_\pi E_\Lambda^{(1)})$$

$$\alpha_2 = (1 - q^{(2)}/v_{K^+} E_\Lambda^{(2)}) \tag{19}$$

$$\xi = \int \frac{d^2\vec{Q}_1}{(2\pi)^2} \left| \frac{t_1(\vec{Q}_1)}{t_1(o)} \right|^2 \left| \frac{t_2(-\vec{Q}_1)}{t_2(o)} \right|^2$$

Here $t_{1,2}$ are t-matrices for processes 1 and 2 and ξ takes account of their angular dependence. A Glauber approximation has been used to reduce the $d^3\vec{Q}$ integral to a two dimensional integral over $\vec{Q}$ (x, y components). The existing experimental data (44) on $\bar{K}N \to \pi\Lambda$ and $\pi N \to K\Lambda$ can be used to obtain

$$\xi \simeq 0.02 \text{ mb}^{-1}$$

$$\left(\frac{d\sigma}{d\Omega_L}\right)^{K^-p \to \pi^o\Lambda}_{0^o} \simeq 0.76 \text{ mb/st} \tag{20}$$

$$\left(\frac{d\sigma}{d\Omega}\right)^{\pi^o p \to K^+\Lambda}_{0^o} \simeq 0.65 \text{ mb/st}$$

for $p_{K^-} = 1.1$ GeV/c, $p_{\pi^o} = 1.02$ GeV/c. This choice of momentum corresponds to the peak in the product of $d\sigma/d\Omega_L$ factors. Using the values in (20), and taking $\langle 1/r^2\rangle \simeq 3p_F^2/20 \simeq 0.028$ mb^{-1} from the Fermi gas model, we obtain

$$\sum_f \left(\frac{d\sigma}{d\Omega_L}\right)^{0^o}_f \simeq 4\mu\text{b/st} \tag{21}$$

for the summed (K^-, K^+) cross section.

Note that the sum over states includes the quasielastic contribution, (one or both Λ's in the continuum) which is a sizable part of the total, since q is large. The contribution of discrete $\Lambda\Lambda p^{-1}p^{-1}$ states to (21) will arise mainly from high spin states. The first Λ, produced in the low q process $K^-p \to \pi^o\Lambda$, would have maximum "sticking probability" in a low spin state, but the second Λ, produced in the high q process $\pi^o p \to K^+\Lambda$, can stick only in a high spin state. Thus the (K^-, K^+) cross sections to discrete states may be too small to measure with current K-beams, although this reaction remains very promising for future facilities.

Rather than considering $\Lambda\Lambda$ hypernucleus formation, we may look at the one step process $K^-p \to K^+\Xi$. The reaction $\Xi^-p \to \Lambda\Lambda$ then serves to provide the Ξ^- state with a conversion width, analogous to the situation for $\Sigma^- p \to \Lambda n$. The two-body threshold for the Ξ^- production is at $p_{K^-} \simeq \simeq 1.05$ GeV/c, so the formation of Ξ hypernuclei is kinematically separated from that of $\Lambda\Lambda$ hypernuclei (the Q value for $\Xi^- p \to \Lambda\Lambda$ is 30 MeV). The summed cross section at 0° for a Ξ hypernucleus in a (K^-, K^+) reaction is (42)

$$\sum_f \left(\frac{d\sigma}{d\Omega_L}\right)^f_{0^\circ} = Z_{eff}\ \alpha\ (d\sigma/d\Omega_L)^{K^- p \to K^+ \Xi^-}_{0^\circ} \qquad (22)$$

where Z_{eff} is the effective proton number

$$Z_{eff} = \int \rho_p(r)\ |\chi^{(-)}_{K^-}(r)|^2\ \ |\chi^{(+)}_{K^+}(r)|^2\, dr \qquad (23)$$

and $\alpha = (1 - q/v_{K^+} E_\Xi)$.

In eq. (23), $\chi^{(\pm)}$ are distorted waves obtained from the optical model. We estimate (42) Z_{eff} to be of the order of 1 and take

$$\left(\frac{d\sigma}{d\Omega_L}\right)^{K^- p \to K^+ \Xi^-}_{0^\circ} \simeq 30\ \ \mu b/st \qquad (24)$$

at $p_{K^-} \gtrsim 1.26$ GeV/c, from experiment (45). This gives

$$\left(\frac{d\sigma}{d\Omega_L}\right)^f_{0^\circ} \simeq \ \ \mu b/st \qquad (25)$$

for Ξ hypernucleus formation. Since $q \simeq 400$ MeV/c, only discrete Ξ states of high spin will be populated in the (K^-, K^+) reaction. The quasielastic process represents much of the sum of eq. (25).

The Hartree potential for a Ξ in the nucleus and the width Γ_Ξ arising from $\Xi^- p \to \Lambda\Lambda$ have been estimated in ref. (42). Because of the smaller phase space for the conversion process, Ξ widths are intrinsically smaller than Σ widths. The quenching mechanism discussed earlier for Σ widths (25) also operates for the Ξ, since at low energies, the process $\Xi^- p \to \Lambda\Lambda$ takes place only from the I = 0, $^1S_0$ channel. Very narrow Ξ states may thus be anticipated (42). Intense kaon beams are required to find them !

There are also a number of unique possibilities for the K^+ as a probe of nuclear structure. The K^+ does not transfer strangeness to the nucleus, but its weak absorption makes it distinctive among hadrons. Its features have been discussed in some detail in the literature (5 ,46-48) so we only give a skeleton outline here. The K^+ should in principle be useful for: a) extraction of information on neutron densities (5, 47) from elastic scattering, as well as spin flip from factors (47). This requires a good knowledge of the elementary K^+N amplitudes; b) K^+ inelastic scattering studies of giant isoscalar 0^+ and 2^+ resonances and high spin unnatural parity states. It would be interesting to compare these results to π, p and e^- inelastic scattering to the same class of states; c) charge exchange (K^+, K^o) reactions (5) to look at the isospin structure of giant resonances; d) investigation of nucleon quasiparticle properties via the (K^+, K^+p) knockout reaction (49), and comparison with (e, e'p) and (p, 2p) results. There is some recent data (50) on $K^\pm$ elastic scattering on ^{12}C and ^{40}Ca at 800 MeV/c. These data have been discussed in the context of the optical model (48). A first order potential (linear in the nuclear density $\rho(r)$) with no free parameters gave excellent agreement (48) with the K^+ elastic scattering data.

I have tried to indicate some of the interesting questions which arise in kaon-nucleus physics, but the list has been far from complete. There are many open questions and opportunities for advances in our understanding of how strange mesons and baryons interact with nuclei.

References and notes

*. Supported by the U.S. Department of Energy under Contract No. DE-AC02-76CH00016.

1. Gal, A.: 1975, in 'Advances in Nuclear Physics' 8, Eds. M. Baranger and E. Vogt, Plenum Press, New York, 1-120.

2. Dover, C.B.: 1979, Proceedings of the Kaon Factory Workshop, Vancouver, B.C., Canada, August. Ed. M.K. Craddock, TRIUMF Report TRI-79-1, 4-11.

3. Povh, B.: 1976, Reports on Progress in Physics 39, 824; 1978, Ann. Rev. Nucl. Part. Sci. 28, 1.

4. Povh, B.: 1979, "Nuclear Physics with Hyperons", Max Planck Institute Preprint MPI H-1979-V25.

5. Dover, C.B., and Moffa, P.J. : 1977, Phys. Rev. C16, 1087.

   Dover, C.B., and Walker, G.E.: 1979, Phys. Rev. C19, 1393.

6. Li, K.K.: 1976, Proceedings of the Topical Conference on Meson-Nuclear Physics, Carnegie Mellon University, May.

   Kycia, T., and Li, K.K., Private communication.

7. Li, K.K., Private communication; the data points are from a Brookhaven experiment; the solid lines are based on the analysis of B.R. Martin: 1975, Nucl. Phys. B94, 413.

8. Dover, C.B.: 1979, Proceedings of the International Topical Conference on Meson-Nuclear Physics, Houston, March; AIP Conf. Proc. No. 54, Ed. E.V. Hungerford III, 634-657.

9. Feshbach, H., and Kerman, A.K.: 1965, Preludes in Theoretical Physics (Amsterdam, North Holland), 260.

10. Bonazzola, G.C., et al.: 1975, Phys. Rev. Lett. 34, 683.

Brückner, W., et al.: 1976, Phys. Lett. 62B, 481.

11. Brückner, W., et al.: 1978, Phys. Lett. 79B, 157.

12. Chrien, R., et al. : 1979, Phys. Lett. 89B, 31.

13. Dover, C.B., Gal, A., Walker, G.E., and Dalitz, R.H. : 1979, Phys. Lett. 89B, 26.

14. Hüfner, J., Lee, S.Y., and Weidenmüller, H.A.: 1974, Nucl. Phys. A234, 429.

15. Bouyssy, A.: 1977, Nucl. Phys. A290, 324.

16. Pniewski, J., and Zieminska, D.: 1977, Proceedings of the Seminar on Kaon-Nuclear Interaction and Hypernuclei, Zvenigorod, Russia, September, 33-50.

17. Bouyssy, A.: 1979, Phys. Lett. 84B, 41.

18. Brockmann, R., and Weise, W.: 1977, Phys. Lett. 69B, 167.

19. Pirner, H.J.: 1979, Phys. Lett. 85B, 190.

20. Noble, J.V.: 1980, Phys. Lett. 89B, 327.

21. Gal, A., Soper, J.M., and Dalitz, R.H. : 1971, Ann. Phys. 66, 63; ibid : 1972, 72, 445.

22. Bertini, R.: 1979, Proceedings of the Second International Conference on Meson-Nuclear Physics, Houston, March; AIP Conf. Proc. No. 54, Ed. E.V. Hungerford III, 703-713.

23. Nagels, M.: 1975, Nijmegan Thesis.

Nagels, M., Rijken, T.A., and deSwart, J.J.: 1977, Phys. Rev. D15, 2547.

24. These Σ widths were actually obtained by solving for the complex eigenvalues of the Schrödinger equation, using a first order ($t\rho$) approximation to the Σ optical potential.

25. Gal, A., and Dover, C.B.: 1980, Phys. Rev. Lett. 44, 379.

26. Bedjidian, M., et al.: 1979, Phys. Lett. 83B, 252.

27. Bedjidian, M., et al., Phys. Lett., to be published.

28. Dalitz, R.H., and Gal, A.: 1978, Ann. Phys. 116, 167.

29. Bamberger, A., et al.: 1973, Nucl. Phys. B60, 1.

30. Birien, P.: 1979, in Proceedings of the International Conference on Hypernuclear and Low Energy Kaon Physics, Jablonna, Poland, September; to appear in Nukleonika (1980).

31. Proceedings of the Kaon Factory Workshop, Vancouver, B.C., Canada, August 1979, Ed. M.K. Craddock, TRIUMF Report TRI-79-1.

32. Proceedings of the Summer Study Meeting on Nuclear and Hypernuclear Physics with Kaon Beams, Brookhaven National Laboratory, July 1973; Ed. H. Palevsky, BNL 18335.

33. Proceedings of the Summer Study Meeting on Kaon Physics and Facilities, Brookhaven National Laboratory, June 1976; Ed. H. Palevsky, BNL 50579.

34. See ref. 2.

35. Kerman, A.K., and Lipkin, H.J.: 1971, Ann.Phys. 66, 738.

36. Kisslinger, L.S., and Giai, N.V.: 1977, Phys. Lett. 72B, 19.

37. Dalitz, R.H., and Gal, A.: 1976, Phys. Rev. Lett. 36, 362.

38. Dover, C.B., Ludeking, L., and Walker, G.E.: 1980, Preprint.

39. Rayet, M.: 1976, Ann. Phys. 102, 226.

40. Baker, R.D., et al. : 1977, Nucl. Phys. B126, 365; 1978, Nucl.Phys. B141, 29.

Saxon, D.H., et al.: 1979, Rutherford Lab Preprint, July.

41. Thiessen, A., Private communication.

42. Dover, C.B.: 1979, Proceedings of the International Conference on Hypernuclear and Low Enregy Kaon Physics, Jablonna, Poland, September; Dover, C.B., and Gal, A.: in preparation.

43. Jaffe, R.: 1977, Phys. Rev. Lett. 38, 195.

44. Van Horn, A.J., et al.: 1972, Phys. Rev. D6, 1275.

Keren, J.: 1964, Phys. Rev. 133, B457.

45. Burgun, G., et al.: 1968, Nucl. Phys. B8, 447.

46. Dover, C.B., and Walker, G.E., Phys. Reports C, in preparation.

47. Cotanch, S.R.: 1978, Nucl. Phys. A308, 253; 1978, Phys. Rev. C18, 1941.

48. Rosenthal, A.S., and Tabakin, F., Phys. Rev. C, to be published.

49. Koshel, R.D., Moffa, P.J., and Redish, E.F.: 1977, Phys. Rev.Lett. 39, 1319.

50. Eisenstein, R., ref. 2, 75-81.

DISCUSSION

- J. Bowcock :

The two particular nuclei ^{12}C and ^{16}O are believed to have a considerable amount of α-particle clustering in their ground-state wave functions. If the Λ substitutes into one of these clusters would not this upset the interpretation of the experimental results in terms of single particle states and the single particle potential?

- C. B. Dover :

Apparently not. Antisymmetrized cluster models and the single particle shell model (in the SU(3) basis) are equivalent representations of ^{12}C and ^{16}O. The data on the (K,Π) reaction with these targets have been successfully (quantitatively) interpeted in terms of weakly mixed Λ particle-neutron hole configurations. States involving "core excitations" are only weakly excited with closed-shell targets.

KAON-NUCLEUS SCATTERING, REGENERATION AND ANALYTICITY*

O. Dumbrajs
Institute of Physics - Aarhus University

1. Introduction

The use of the analytic properties of scattering amplitudes is of increasing importance in nuclear physics. By applying analyticity arguments to the scattering amplitude one can correlate existing experimental data, predict the real parts of the amplitude and determine coupling constants. Much work has already been done along these lines in pion-nucleus, nucleon-nucleus and nucleus-nucleus scattering (see e. g. reviews 1-3).

In my talk I will discuss possibilities of applying the concept of analyticity to the kaon-nuclear scattering amplitudes.

2. Forward dispersion relations

The most interesting dispersion relation is, perhaps, the dispersion relation for the antisymmetric amplitude $f^-(\omega)$

$$\mathrm{Re}f^-(\omega) = \frac{2\omega r}{\omega^2-\omega_p^2} + \frac{2\omega}{\pi} P\int_{\omega_o}^{\infty} \frac{\mathrm{Im}f^-(\omega')}{\omega'^2-\omega^2} d\omega' \tag{1}$$

Here ω is the kaon laboratory energy, r is the residue of the effective pole, ω_p is the position of the effective pole and ω_o is the beginning of the unphysical cut. This dispersion relation is convenient for two reasons: i) the pole term contribution is not suppressed, ii) both the real and the imaginary parts of the amplitude are directly measured in regeneration experiments :

$$\begin{aligned} \mathrm{Re}f^- &= \left|\frac{f-\bar{f}}{K}\right| \; p\cos\phi, \\ \mathrm{Im}f^- &= \left|\frac{f-\bar{f}}{K}\right| \; p\sin\phi. \end{aligned} \tag{2}$$

E. Ferrari and G. Violini (eds.), Low and Intermediate Energy Kaon-Nucleon Physics, 187–190.

Here $f(f^-)$ is the $K^o(\bar{K}^o)$ scattering amplitude in the c.m. system, K is the c.m. momentum, p is the laboratory momentum and ϕ is the phase of the antisymmetric amplitude directly related to the phase of the regeneration amplitude

$$\phi = \pi + \arg(f-f^-). \tag{3}$$

What experimental data are available for analysis ? Regeneration data on several nuclei, e.g. on d (4) and on ^{12}C (5) are at our disposal. In the near future the coherent regeneration of K^o_S from K^o_L on several other nuclei will be measured (6). Total cross sections are well known only for $K^\pm d$ scattering (7) and, in a limited energy region, also $K^\pm$ ^{12}C scattering (8, 9). The real parts have been measured only for the $K^\pm$ ^{12}C forward scattering amplitude (10). However, it has been reported recently (11) that a systematic study of the energy dependence of the ratio, $\alpha = \mathrm{Re}f(0)/\mathrm{Im}f(0)$, determined from Coulomb interference, and of the total cross sections below 1 GeV/c on several nuclei is in progress at the 12 GeV proton synchrotron of KEK in Japan. At threshold, the K^--nucleus scattering amplitude (the scattering length) has been extracted from all available K^--mesonic atom data (12).

In the case of the pion-nucleus scattering the unphysical cut is due to pion absorption and has little influence (1, 13). In the kaon-nucleus dispersion relation calculations one can expect more difficulties in handling the unphysical region. In fact, already in kaon-nucleon dispersion theory (14) the presence of the unphysical region is the main source of uncertainties in spite of the existence of some well developed models, like the K-matrix parametrization (15). Moreover, it is known (16) that the K^--nucleon and K^--nucleus interactions near the threshold are dominated by different physical mechanisms, in particular, the role of the $\Lambda(1405)$-resonance could be quite different in the two cases. This would make simple adoption of the existing unphysical region parametrizations impossible.

Two attempts (17, 18) have been made so far to parametrize the unphysical region in the kaon-nucleus dispersion relation calculations.

In (17) the contributions of the lowest intermediate states ΛN and ΣN to the scattering amplitude for elastic K^-d forward scattering were investigated. The analytic continuation of these contributions with respect to the kaon mass was performed by making use of the impulse approximation for the K^-d adsorptive process $K^- + d \rightarrow Y + N$. It was discovered that in the unphysical region in addition to anomalous logarithmic branch cuts, anomalous poles play an important role. Using the information on the deuteron wave function the following sum-rule has been deduced :

$$\mathrm{Re}f^-(\omega) = \Pi_{\Lambda N}(\omega) + \Pi_{\Sigma N}(\omega) + C_{\Lambda N}(\omega) + C_{\Sigma N}(\omega)$$
$$+ C^{(3,\,4)}(\omega) + \frac{2\omega}{\pi} P \int_{m_K}^{\infty} \frac{\mathrm{Im}f^-(\omega')}{\omega'^2 - \omega^2}\, d\omega' , \tag{4}$$

where $\Pi_{\Lambda N}$ and $\Pi_{\Sigma N}$ are the anomalous pole contributions, $C_{\Lambda N}$ and $C_{\Sigma N}$ are the continuum contributions composed of the contributions of the anomalous and the normal unphysical continuum and $C^{(3,\ 4)}$ represents the integral contribution of the three or four particle intermediate states $\Lambda N\pi$, $\Sigma N\pi$ and $\Lambda N2\pi$ in the unphysical region. In the first approximation it can be shown that $\Pi_{\Lambda N}$, $\Pi_{\Sigma N}$, $C_{\Lambda N}$ and $C_{\Sigma N}$ depend only on the elementary coupling constants g^2_{Λ} and g^2_{Σ} and that the contribution of the three and four particle intermediate states is expected to be small. I would like to note that the relation (4) has not been tested numerically.

A much simpler approach was adopted in (18) where the elastic $K^{\pm\,12}C$ scattering has been considered. An effective pole was introduced to represent the pole due to the ground and the excited states of the $^{12}B_{\Lambda}$-hypernucleus and the unphysical region due to the $^{11}B_{\Lambda}$-cut and due to the anomalous pole and the anomalous cut. The discrepancy function

$$\Delta(\omega) = \mathrm{Re}f^{-}(\omega) - \frac{2\omega}{\pi} P \int_{m_K}^{\infty} \frac{\mathrm{Im}f^{-}(\omega')}{\omega'^2 - \omega^2} d\omega' \qquad (5)$$

has been evaluated at those values of ω where experimental data on $\mathrm{Re}f^{-}$ exist and a fit with a pole term

$$\Delta(\omega) = \frac{2\omega r}{\omega^2 - \omega_p^2} \qquad (6)$$

has been made. The residue of the effective pole r was found to be stable with respect to the change of the pole position ω_p. The numerical value turned out to be larger than the coherent sum of the elementary $KN\Lambda$ and $KN\Sigma$ residues. It is interesting to note that in the case of pion-nucleus scattering the residues of the effective poles as a rule are always smaller than the elementary πNN residue (3, 19).

3. Analyticity in the cosθ-plane

The simplest way of using analyticity in the cosθ-plane would be to extrapolate a differential cross-section to a pole in the cosθ-plane at fixed energy. By doing this one could determine a coupling constant. However, it was shown (18) that one can hardly expect to benefit from this method in the case of the kaon-nucleus scattering. Due to the very unfavourable mass ratios, the corresponding poles in the cosθ variable are very far from the physical region !

4. Conclusions

I think that already now, with a limited amount of experimental data, analyticity in the form of forward dispersion relations can be used not only for correlating the existing pieces of experimental information on kaon-nucleus scattering and regeneration but also for getting an insight into the unphysical region. For the latter purpose one could use the simplest possibility : to represent all the unknown contributions by one effective pole and to calculate its residue. This has been done in the case of the $K^{\pm 12}C$ forward scattering, it can be done already now for the $K^{\pm}d$ forward scattering and it will be possible for other nuclei in the nearest future. Even before developing some "microscopic" models for the unphysical region one could hope to learn more about the kaon interaction with nuclei by comparing different effective residues.

References and note

*. Work supported by a grant from the Danish Natural Science Research Council.

1. Ericson, T.E.O., and Locher, M.P.; 1970, Nucl. Phys. A148, 1.
2. Dubnicka, S., and Dumbrajs, O: 1975, Phys. Reports 19, 142.
3. Locher, M.P., and Mizutani, T.: 1978, Phys. Reports 46, 43.
4. Albrecht, K.F. et al.: 1979, Nucl. Phys. B158, 29.
5. Roehrig, J., et al.: 1977, Phys. Rev. Lett. 38, 1116.
6. Diu, B., and Ferraz de Camargo filho, A.: 1978, Nuovo Cimento 47A, 495.
7. Caroll, A.S., et al.: 1976, Phys. Lett. 61B, 303.
8. Bugg, D.V., et al.: 1968, Phys. Rev. 168, 1466.
9. Abrams, R.J., et al.: 1971, Phys. Rev. D4, 3235.
10. Gobbi, B., et al.: 1972, Phys. Rev. Lett. 29, 1278.
11. Endo, I., and Sumi, Y.: 1979, Proceedings of the TRIUMF Kaon Factory Workshop, Vancouver, Ed. Craddock, 101.
12. Seki, R.: 1972, Phys. Rev. Lett 29, 240
13. Pilkuhn, H., Zovko, N., and Schlaile, H.G.: 1976, Z. Phys. A279, 283.
14. Queen, N.M. : Contribution to this Workshop.
15. Martin, A. : Contribution to this Workshop.
16. Ericson, T.E.O. : 1971, Proceedings of the IV International Conference on High-Energy Physics and Nuclear Structure, Dubna, September 7-11.
17. Bebel, D.: 1965, Nucl. Phys. 66, 577.
18. Dumbrajs, O.: 1979, Nuovo Cimento 54A, 155.
19. Dumbrajs, O.: 1980, J. Phys. A6, L9.

MEASUREMENT OF THE POLARIZATION PARAMETERS FOR $K^+n \rightarrow K^+n$ AND K^op AT THE MOMENTUM OF 1.06,1.28, 1.39 AND 1.49 GeV/c*

F. Takasaki
National Laboratory for High Energy Physics
Oho-machi, Tsukuba-gun, Ibaraki-ken 305, Japan

According to the standard quark model, mesons are made of a quark and an antiquark pair, while baryons consist of three quarks.

Since K^+ meson is supposed to be made of an Up-quark (U) and a Strange-antiquark ($\bar{S}$), K^+ can make a bound state with a baryon only with configurations of at least five quarks, namely, four quarks and one strange antiquark. If such a bound state is proved to exist, it may add quite a new aspect to the fundamental interaction of quarks.

Search for such exotic bound states has been made many times in the last decade (1). Although no conclusive evidence has been found yet, there are some analyses which claim the possible existence of Z^* states. For example, BGRT group made a phase shift analysis of the strangeness +1 Kaon Nucleon system and found a solution which has a resonant like behaviour in the Argand Plot (2), together with non resonant like solutions. It has been pointed out that the discrimination of their solutions could be best done by the measurement of the P-parameters both for the K^+n elastic and charge exchange reactions at the momentum region around 1 GeV/c.

The measurement of the K^+n elastic and the charge exchange reactions is also interesting from the view point of studying the reaction dynamics of hadrons. Regge theory is considered to work even at low energy region down to around 1 GeV/c for the exotic channels. It makes some predictions as to the values of P-parameters for the K^+ nucleon reactions (3). For example, the sign of the P-parameters for the K^+n elastic scattering process is opposite to those for the K^+p elastic scattering process and their magnitude for the K^+n elastic is smaller than those for the K^+p elastic process. It also predicts that the magnitude of the P-parameters for the charge exchange process is small due to the exchange degeneracy.

There has been no experimental attempt (4) to measure the P-parameters for these reactions around 1 GeV/c region where the possible existence of the Z^* resonances is claimed to exist.

E. Ferrari and G. Violini (eds.), Low and Intermediate Energy Kaon-Nucleon Physics, 191–206.

We made measurements of these parameters at the momenta 1.06, 1.28, 1.39 and 1.49 GeV/c. The experiment was done at the K2 beam line (5) of KEK, Japan.

The K^+'s produced off the production target placed in the extracted 12 GeV proton beam from the KEK proton synchrotron were guided by a series of magnets, a DC static separator and a mass slit to the experimental target. The distance from the production target to the experimental target was about 29 m. The beam size was defined to be about 2 ^{cm} x 1.5^{cm} at the target.

The target we used was made of the deuterated propanediol $C_3D_6(OD)_2$ contained in a teflon capsule of the dimension of 6.5 ^{cm} long, 2^{cm} wide and 1.5^{cm} high. This target was cooled down to about 100 m°K by the H_e^3-H_e^4 dilution method (6). Typical magnitude of the target polarization of the neutron was ranged from 30 to 40%. The correction due to the D-wave state component of the deuteron was applied for the calculation of the final reaction asymmetry.

To pick up the K^+'s in the beam, we used a DISC type liquid Cherenkov counter (7) placed about 70 cm upstream the target. The contamination of the other particles is less than one per cent at all the measured momenta.

The scattered charged particles, namely, K^+'s, protons and pions, were detected by the magnetic spectrometer TELAS (8). We show in figure 1 the top view of the experimental setup.

The signature of the event was formed by the coincidence of a series of signals from the scintillation counters and the Cherenkov counter (S3, S3', BD, C, AL, BUL and TL) and the anti-coincidence of other scintillation counter signals (AR, BUR, TR and TV). Symbols S3, S3' and BD mean signals of the scintillation counters defining the beam, while C stands for that of the Cherenkov counter. AL and BUL (AR and BUR) denote the scintillation counters placed downstream the target covering the left-(right-)hand hemisphere (about 90° from the incident beam). TL (TR) means a signal formed when any one of the T-counters (Scintillation counters surrounding the magnet) placed on the left (right) hand side from the beam orbit produce signals. TV stands for a signal made from the three T-counters in the beam.

The information of the neutron counters was not required for forming the trigger signature at momenta 1.06 and 1.39 GeV/c. However, if any of the neutron counters produce signals in coincidence with the trigger signal, they were taken into our data through the CAMAC coincidence register modules.

To improve the separation of scattered K^+ from proton, we installed 15 liquid Cherenkov counters at the extreme forward angles.

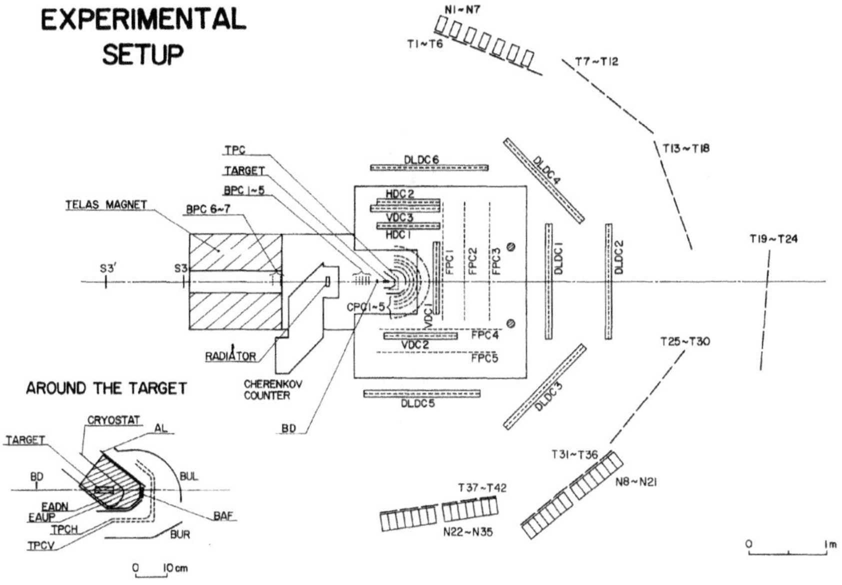

Figure 1 - Experimental setup

Typical beam intensity of K^+ was ranged from 4 K to 15 K K^+'s per burst depending upon the beam momentum. The momentum acceptance was set to be ± 3%.

The event rate was about 1 event per 1 K K^+'s.

The momentum of the scattered particles was measured by a series of proportional wire chambers and drift wire chambers placed in the field volume of TELAS.

Since the field distribution of the spectrometer shows a rapid change in the field volume of the magnet, from 25 KGauss to about 1 KGauss in about 1.5 m, careful field mapping and careful momentum calibration were required. (On the field mapping, see reference 9).

To know the momentum resolution of our spectrometer, we have measured several different beam momenta with both signs of charges of the beam particle. Beam momentum were also measured independently by the time of flight technique. In figure 2, we show a distribution of the beam momentum measured by our system at the beam momentum of 1.28 GeV/c. It tells that the momentum resolution of our system is about ± 3%.

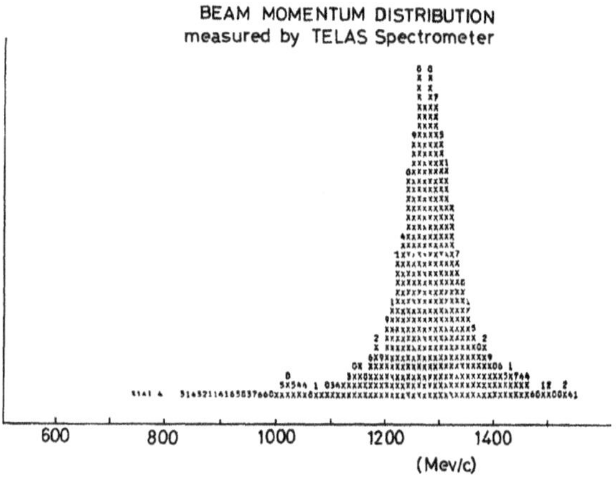

Figure 2 - Beam momentum distribution measured by TELAS spectrometer

In figure 3, we made a comparison of the momenta measured by TOF technique and by this spectrometer. They show a good agreement over the momentum range.

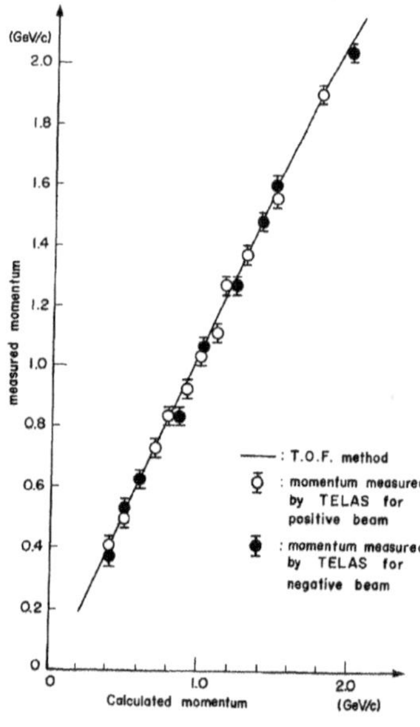

Figure 3

To see where the scattering occurs, we make a plot of the crossing point of the incident beam trajectory and the scattered particle trajectory, which were traced by the Spline-fit method (10). In figures 4-a and 4-b, we show the distribution of the scattering points of the beam particles projected to the axis perpendicular to the incident beam direction for the full and the empty target. If we compare the two figures, we can see a clear peak corresponding to the target.

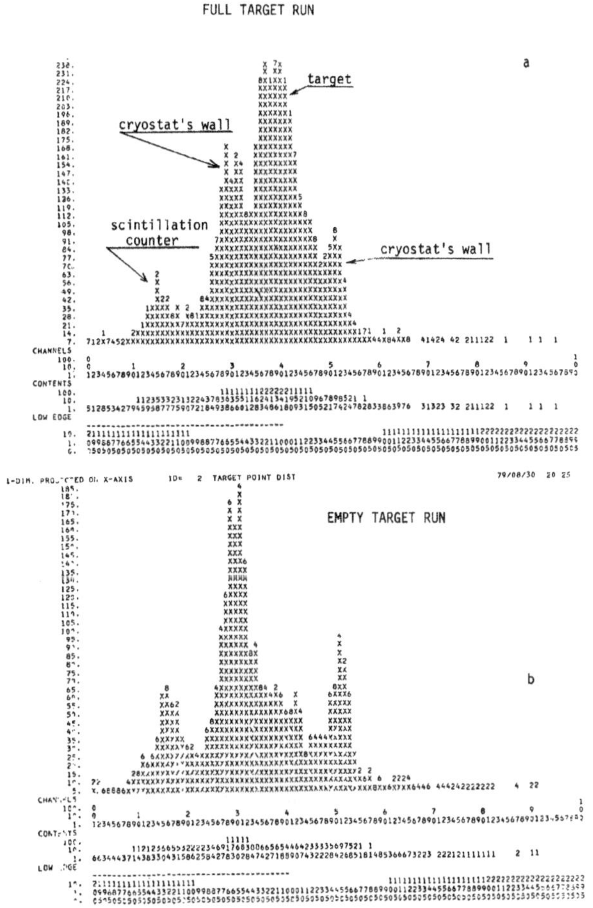

Figure 4 a), b)

The identification of the scattered particles was made in the off-line analysis by observing the correlation between the time of flight and the momentum of the scattered particle. In figure 5, we show a typical example of the TOF/momentum distribution. It shows a clear separation of pions, kaons and protons.

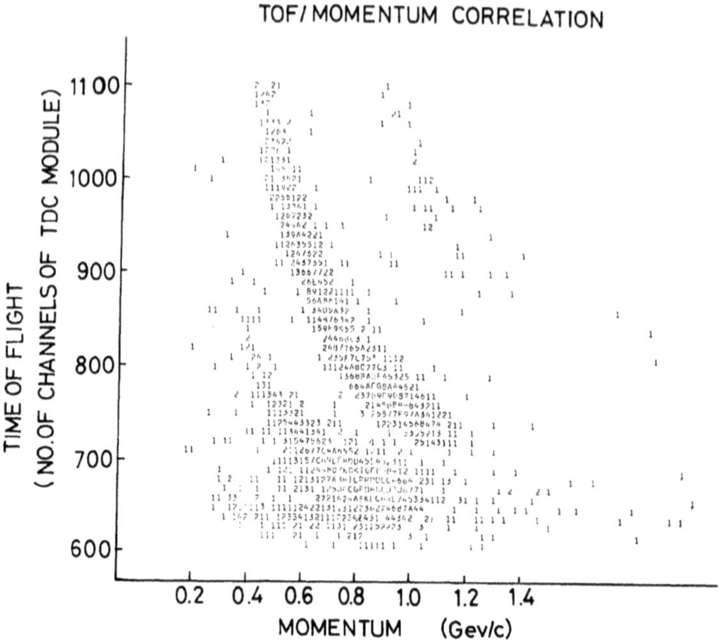

Figure 5 - TOF/Momentum correlation

We give the mass plot calculated from the velocity and the momentum of the scattered particle in figure 6.

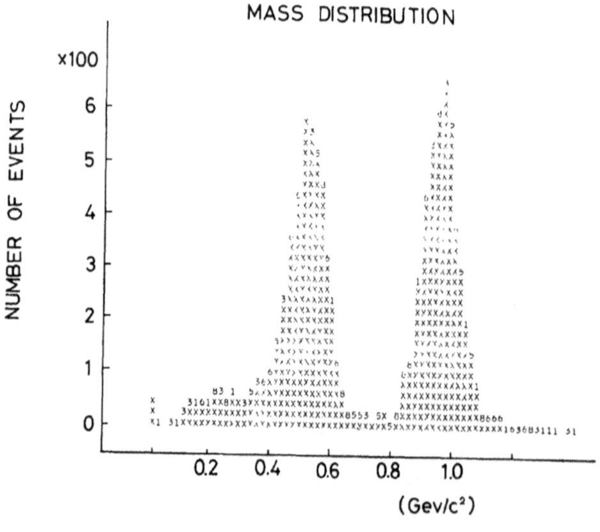

Figure 6

To ensure that we measure the elastic kaon nucleon scattering, we calculate the missing mass using the three momenta and the mass of the scattered particles obtained in the above procedure. In figure 7, we show the missing mass spectrum at the beam momentum of 1.28 GeV/c in the case when the detected particles are kaons. We see a peak at the nucleon mass and a bump at the higher mass region. The former peak can be attributed to the elastic K^+n scattering and the latter bump to the inelastic processes.

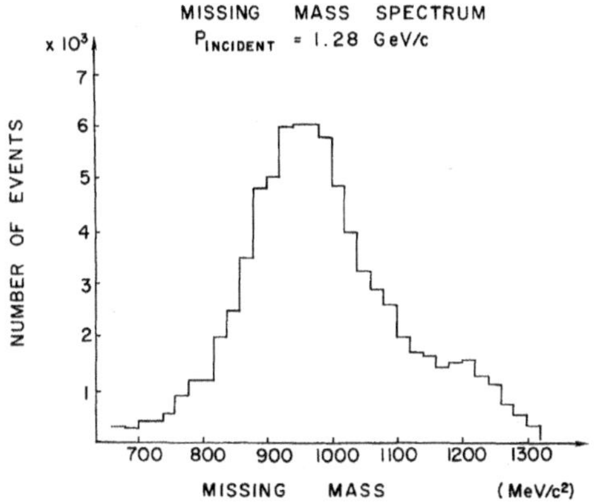

Figure 7

To select the events in the elastic peak region, we applied a cut in the TOF/momentum plot. How to make a cut in the plot was studied by a Monte Carlo method which will be discussed later.

Due to the Fermi motion of the nucleon in the deuteron, the missing mass peak becomes broad and we can not separate the events originating from the deuteron from those coming from the other complex nuclei, like carbon and oxygen contained in the propanediol target. Therefore, to extract the events produced from the deuteron, we have to make an extra measurement using the carbon target, in which the same amount of the complex nuclei is contained as in the propanediol target, to subtract the amount of the events off the complex nuclei.

Similar analysis was made for the charge exchange reaction, where protons were detected in the final state. In this case, relative amount of the background mainly coming from the one pion production process is larger than the case of the elastic scattering.

To see the amount of the contamination from the other processes, we made a Monte Carlo simulation both for the K^+n elastic and charge exchange reactions.

For the elastic scattering, the following processes which may contaminate should be examined: 1) K^+p and K^+D elastic scattering in which rec protons and deuterons stop in the target and, therefore, escape from being vetoed. 2) Single π^0 production processes. As far as the former case is concerned, the Monte Carlo analysis tells that their contributions are negligible at the scattering angle of K^+ larger than 30° in the center of the mass system above the beam momentum of 1 GeV/c. On the single π^0 production processes, we considered the process $K^+n \rightarrow K^+n\,\pi^0$. In the Monte Carlo simulation, we assumed that pions were produced through the decay of excited nucleons and kaons, namely, Δ^0 and K^{*0}. The differential cross sections for these processes were taken from the compilation of Giacomelli et al. (11). In the simulation, we assumed the Fermi motion of the nucleon in the deuteron described by the Hulthén wave function (12). For the Fermi motion in the complex nuclei, we took the momentum distribution given by the formula $1/(1+e^{(p/p_0)^2})$, in which we used the cutoff momentum of 200 MeV/c.

In figure 8, we show an example of a missing mass distribution simulated by the Monte Carlo method when scattered K^+'s are detected in the final state. In this figure, a similar spectrum, obtained by applying a momentum cut for the present experimental data, is presented, in which the unhatched (hatched) area corresponds to the elastic (inelastic) events. The two spectra agree well with each other. The contamination of the inelastic events is considered to be small.

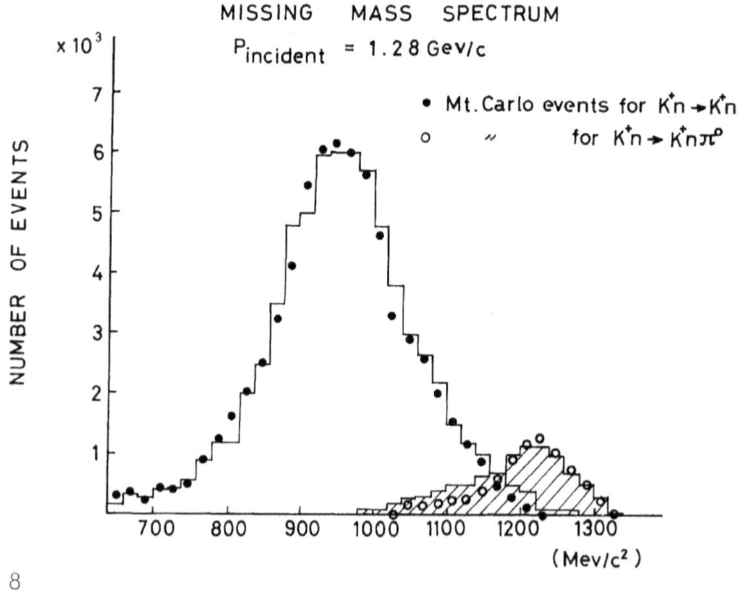

Figure 8

For the charge exchange process, the main background is $K^+n \rightarrow K^0p\pi^0$. We made a similar Monte Carlo simulation for this process to know the contamination of the inelastic processes.

In figure 9, we give an example of a momentum spectrum for recoil protons at the beam momentum of 1.49 GeV/c. In this figure, we plot also points obtained by the Monte Carlo method. A good agreement between two values assures that the contamination of the inelastic processes can be made negligible by applying an appropriate cut in the momentum spectrum without seriously losing charge exchange events.

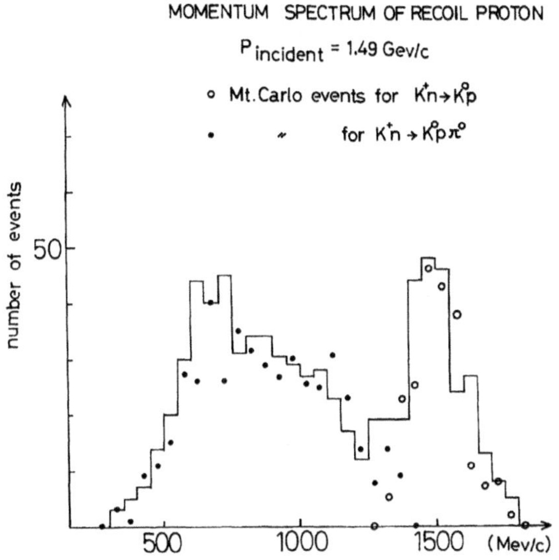

Figure 9

In figure 10, we show a missing mass spectrum for the case when recoil protons are detected in the final states. We see a peak at the mass of kaon, together with the inelastic backgrounds.

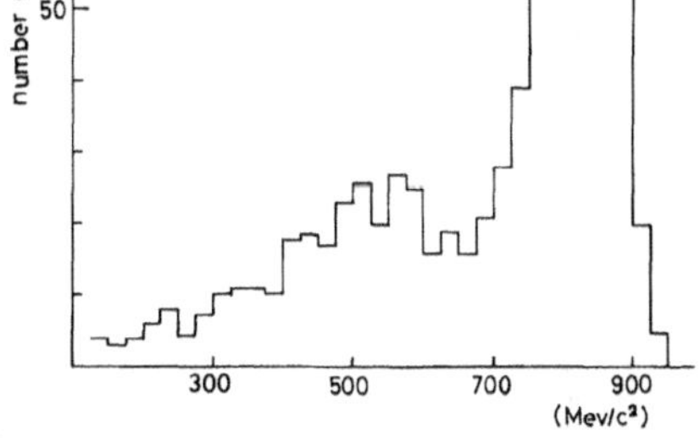

Figure 10 .

We present the mass spectrum after applying the momentum cut in figure 11. Unhatched area corresponds to the charge exchange process, while the hatched area is interpreted to be the inelastic events.

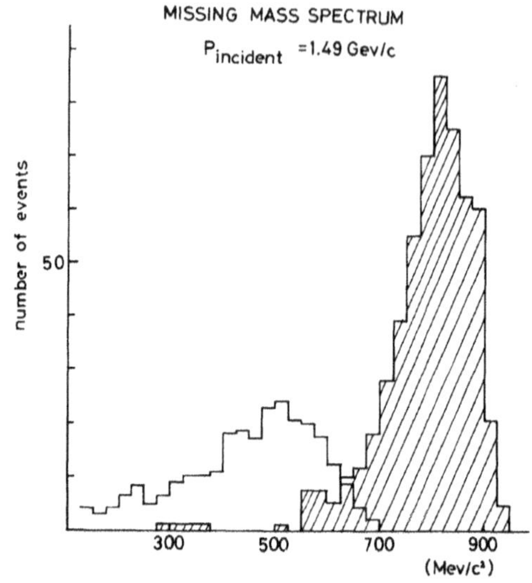

Figure 11

We deduced the differential cross sections for the K^+n elastic process at the beam momentum of 1.49 GeV/c. The result is presented in figure 12 together with the experimental data taken from the data compilation (11). Results of this experiment show good agreement with the previous data.

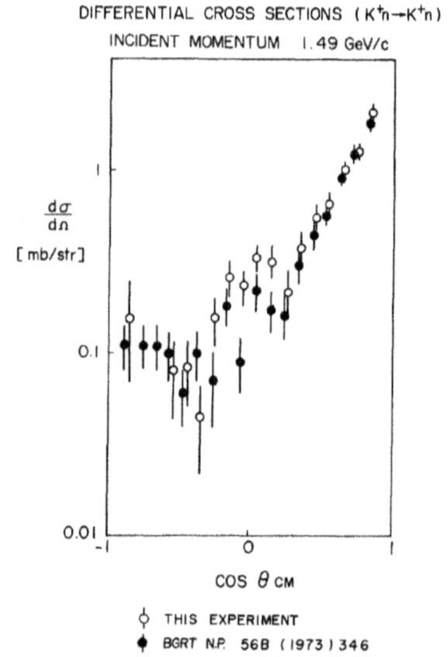

Figure 12 - Differential cross sections for $K^+n \rightarrow K^+n$ at 1.49 GeV/c.

We took about 300 K events for the spin up and down configuration using the deuterated propanediol target and also for the carbon target at all four momenta.

To calculate the P-parameter, we define the following quantities.

$N_{\uparrow(\downarrow)}$: the normalized yield for the spin up (down) target

N_c : the same quantities for the carbon target

$P_{\uparrow(\downarrow)}$: the magnitude of the target polarization with the spin directing up-(down-)ward

$\bar{N}_{\uparrow(\downarrow)} = N_{\uparrow(\downarrow)} - N_c$

Then, the P-parameter is given by the following equation,

$$P = (\bar{N}_\uparrow - \bar{N}_\downarrow)/(P_\downarrow \ \bar{N}_\uparrow + P_\uparrow \ \bar{N}_\downarrow)$$

Since no experimental data is available for the K^+n elastic and the charge exchange reactions to compare results of this experiment, we measured the pn elastic scattering with the same method to check our way of experimentation.

In figure 13, we show our results at the beam momenta of 1.06 and 1.49 GeV/c together with the experimental data of Cheng et al. (13). Reasonable agreement between the two sets of data supports our way of experimentation.

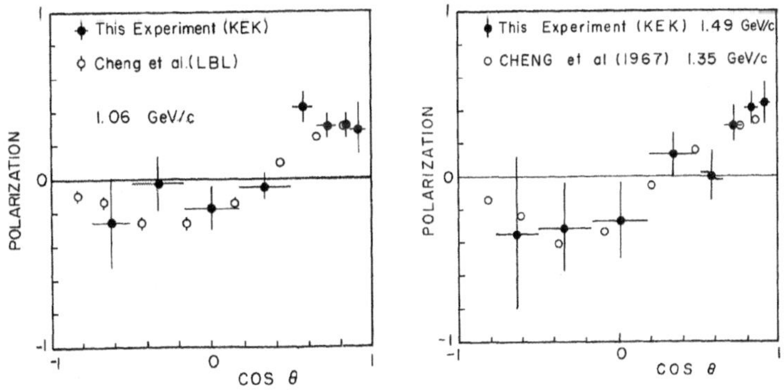

Figure 13

To check the intrinsic asymmetry of our detection system, we tried to calculate the asymmetry using the events produced off the walls of the cryostat.

In figure 14, obtained P-parameters are presented. They are consistent with zero at all momenta and angles.

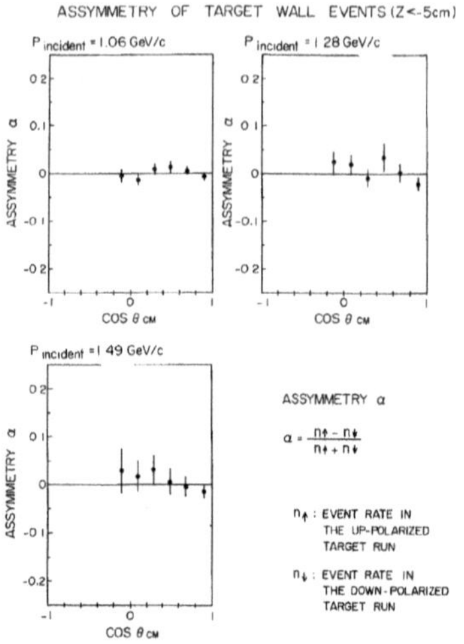

Figure 14 - Asymmetry of target wall events (Z < -5 cm)

In figures 15 and 16, we present preliminary results of the P-parameters for the K^+n elastic and charge exchange processes, respectively. The error bars quoted in the figures are statistical error only.

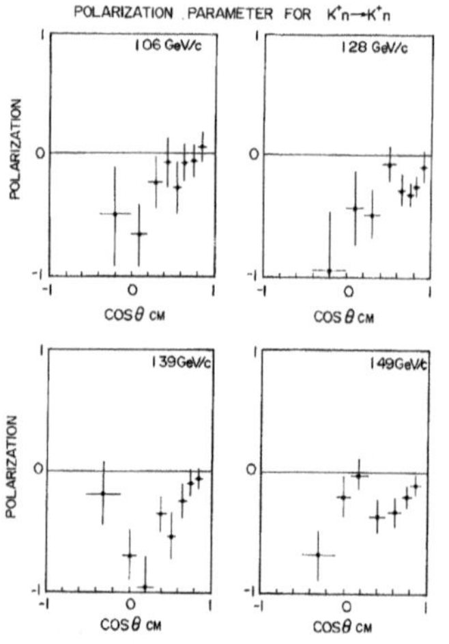

Figure 15 - Polarization parameter for $K^+n \to K^+n$

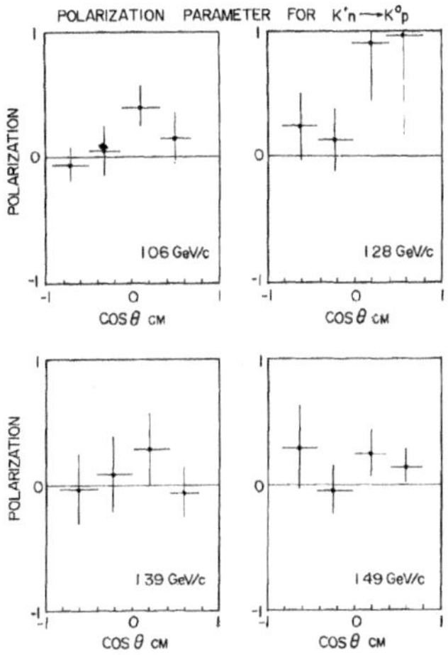

Figure 16 - Polarization parameter for $K^+n \rightarrow K^0p$

Characteristic features of the experimental results are summarized. 1) The sign of P-parameters for the K^+n elastic scattering is mostly negative at all four momenta. It makes sharp contrast with that for the K^+p elastic scattering, which is positive at this momentum region. This feature is quite consistent with the prediction of the Regge pole theory. 2) Although the error bars are large, P-parameters for the charge exchange process are mostly positive and their magnitude is not so large, which is again consistent with the Regge pole theory.

We compared present results with the prediction of the phase shift analyses of BGRT group (2) and B. Martin (14). Their predicted values are displayed in figures 17 and 18 for the elastic and charge exchange scattering, respectively, together with present results. None of their solutions reproduce present results well all through the momentum and angular range. Among these solutions, B. Martin's solution predicts correctly the sign of P-parameters for both processes.

In conclusion, it is noted that we could not find the positive evidence of the existence of Z_0^* states from the present data by comparing it with the previous analyses. Since at the time of the previous analysis, no experimental data on the P-parameters for K^+n reactions was available, it is quite interesting to make similar phase shift analyses including present data. We have started this kind of attempts.

We also try to improve the statistics by a factor of two in the coming experimental machine time. It is also stated that the predictions of the Regge pole theory are quite consistent with the present experimental results.

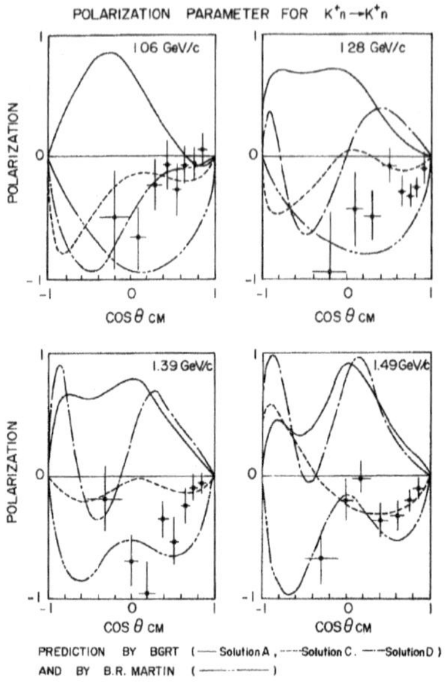

Figure 17 - Polarization parameter for $K^+n \to K^+n$.

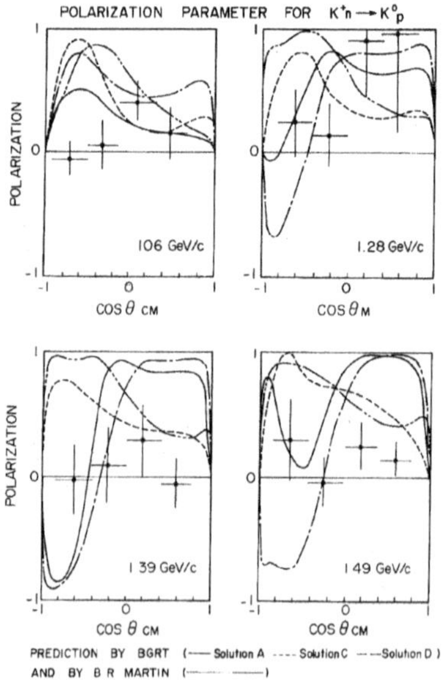

Figure 18 - Polarization parameter for $K^+n \to K^op$.

References and notes

*. This experiment was made by the following people; S. Isagawa, S. Ishimoto, A. de Lesquen, M. Morimoto, A. Masaike, K. Ogawa, S. Kabe, M. Suetake, F. Takasaki, Y. Watase (KEK), N. Kim, K. Nakajima (University of Tokyo), M. Sakuda (University of Kyoto), S. Kobayashi, A. Murakami (Saga University).

1. See the compilation of the Particle Data Group.
2. Giacomelli, G., et al.: 1974, Nucl. Phys. 1371, 138.
3. Fukugita, M.: 1974, KEK Lecture note, unpublished.
4. There is only one experiment of the P-parameter measurement for $K^+n \rightarrow K^0p$ at 600 MeV/c by Ray et al.: 1969, Phys. Rev. 183, 1183.
5. Hirabayashi, H. et al.: 1979, IEEE Transactions on Nucl. Science, NS-26, n° 3, June.
6. Ishimoto, S., et al.: 1979, KEK Preprint-79-9/1979 to be published in Nucl. Instr. Methods.
7. Takasaki, F.: 1979, KEK-EXP FACILITIES-79-3, Sept.
8. Kim, N., et al., to be published in KEK Report.
9. Nakajima, K., et al., to be published in KEK Report.
10. Wind, H.: 1974, Nucl. Instr. Methods 115, 431.
11. Giacomelli, G., et al.: 1976, CERN-HERA 76-01/1976.
    Casadei, U., et al.: 1975, CERN-HERA 75-1/1975.
    Bracci, E., et al.: 1972, CERN-HERA 72-2/1972.
12. Hulthén, L., and Sugawara, M.: 1957, Handbuch der Physik 39, 1.
13. Cheng, D., et al.: 1967, Phys. Rev. 163, 1470.
14. Martin, B.: 1975, Nucl. Phys. B94, 413.

DISCUSSION

- D. Miller :

Many of the ambiguities between solutions could be resolved if you could get closer to $\cos\theta = -1$. What is the problem? Would the slow K^+ not get out of your target?

- F. Takasaki :

Our detector covers the centre of mass angle of about 130 degrees for the reaction $K^+n \rightarrow K^+n$. The limitation of the acceptance is mainly due to the limited angular coverage of the inner cylindrical propanediol chamber.

- S. Watts :

Have you a distribution of reconstructed Fermi momentum?

- F. Takasaki :

We detect only one charged particle in the final state.(We detect neutrons and K_L^o's but we do not require their signals for trigger.) So we cannot get the Fermi motion of the nucleon for most events.

- S. Watts :

What typical polarizations for the target did you achieve?

- F. Takasaki :

About 40%.

POLARIZATION IN K^-p ELASTIC SCATTERING BETWEEN 0.955 GeV/c and 1.272 GeV/c

H.C. Bryant[*], A.A. Carter, M. Coupland[**], E. Eisenhandler,
W.R. Gibson, P.I.P. Kalmus, H. Sandhu and S.J. Watts
Queen Mary College - London

A. Astbury, J.K. Davies[***], G.H. Grayer, A.W. Robertson[+],
T.P. Shah and C. Sutton[++]
Rutherford Laboratory - Chilton, Didcot

Presented by S.J. Watts

An experiment on the above interaction was carried out at the Nimrod proton synchroton as part of a programme of kaon-nucleon polarization. The experimental layout is shown in figure 1.

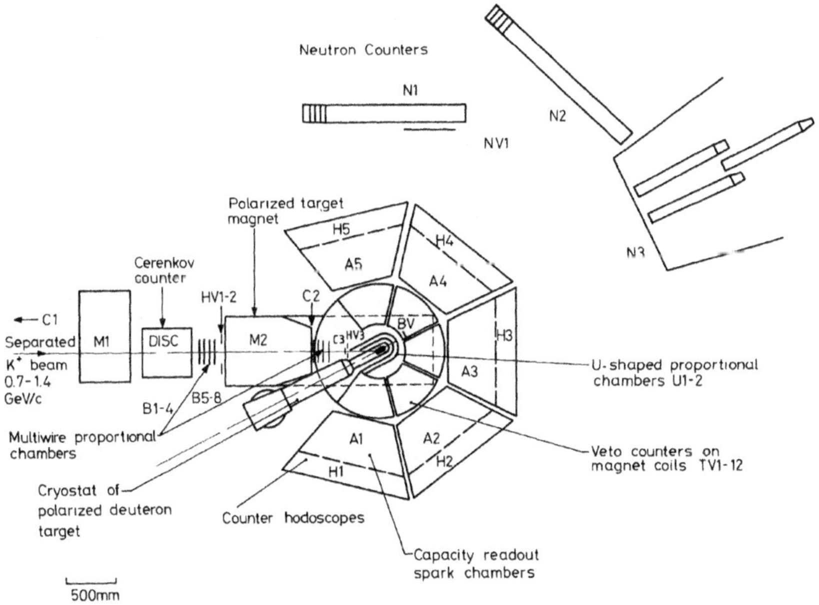

Figure 1

E. Ferrari and G. Violini (eds.), Low and Intermediate Energy Kaon-Nucleon Physics, 207–213.

For the K^-p experiment the neutron counters were not used. An electrostatically separated K^- beam was scattered from a polarized proton target. The incident kaon and both scattered particles were detected by an arrangement of multiwire proportional chambers, wire spark chambers and associated scintillation counters.

The kaon beam was produced at zero degrees from an external 4 x 4 x 100 mm^3 Cu target using a 7 GeV proton beam. The 24 metre long beam line had a useful momentum range of 0.7 GeV/c to 1.4 GeV/c. The beam momentum was calibrated to an accuracy of ± 0.5% using time-of-flight. The momentum acceptance was 2% FWHM which implied a centre of mass energy resolution of ± 10 MeV at 1 GeV/c. A differential Cerenkov counter, DISC, was used to identify kaons. The contamination of the kaon beam by other particles was less than 0.5%. The beam intensity varied from 6×10^3 K^-/burst at 0.955 GeV/c to 1.6×10^4 K^-/burst at 1.272 GeV/c for 3.4×10^{12} extracted protons.

The beam entered the magnet of the polarized target, a 2.5 Tesla C-type dipole magnet M2, through a hole in the return yoke. This magnet supplied the polarizing field, and was also used to determine the momentum of the outgoing particles. The cylindrically symmetric magnetic field was uniform to ± 0.5 mT over a radius of 90 mm about the centre of the pole faces, which were 102 mm apart. The magnet M1 was used to steer the beam on to the centre of the target. The beam dimensions at the centre of the target were typically 26 mm horizontally by 32 mm vertically and the divergence in both planes was ± 15 mrad. Since the incident kaon trajectory varied with beam momentum, the position of the beam defining counter C3 had to be correspondingly adjusted in the horizontal plane.

The trajectory of the incident beam particle was measured by two sets of multiwire proportional chambers, B(1-4) positioned just before the magnet and B(5-8) close to the target inside the magnetic field region. Each set gave two horizontal and two vertical coordinates.

The kaon beam was scattered from a polarized target consisting of beads of propanediol doped with chromium V complexes. The target was 25 mm in diameter and 100 mm long. It was inclined in the horizontal plane at an angle of 11° to the axis of M2 in order to match the beam trajectories. The target polarization was monitored by an arrangement of four small NMR coils positioned in the volume of the target. A proton polarization of typically 90% was obtained using a He^3 evaporation cryostat which produced a temperature of 0.4°K.

The counters HV(1-3) were used to reject particles outside the main beam. Unscattered beam particles were vetoed using BV, whose position was adjusted for each incident momentum. The pole faces and coils of M2 were lined with scintillation counters TV(1-12) which were used to reject unwanted scatters.

The scattered particles were detected in five arrays of wire spark chambers A(1 - 5), with capacity read-out, which were placed around the magnet to give a large acceptance. These chambers were fabricated from wires supported on a plastic frame to give low mass and minimum loss of acceptance due to edge supports. Each array consisted of 4 vertical, 4 horizontal, and 2 diagonal sets of wire planes. The distance between the wires in each plane was 1.5 mm. The arrays were placed outside the stray field of the magnet and tracks in these gave the direction of the scattered particle with an angular resolution of $\pm$ 3 mrad. Sets of scintillation counters H(1-5), each containing 5 vertical elements, were placed at the back of the arrays. These were used in forming the event trigger.

Two U-shaped multiwire proportional chambers (U1, U2) with 2 mm wire spacing surrounded the target and supplied position coordinates in the horizontal plane in the central magnetic field region.

The trigger for an elastic event formed by a coincidence between the beam, which was defined as C1. C2. C3. $\overline{\text{HV}}$ (1-3). DISC, and a matrix coincidence of hodoscope counters chosen to impose loose K^-p elastic kinematic constraints at each momentum. Only events in which two hodoscope counters fired were accepted. Background was reduced by putting the signals from the TV(1-12) and BV counters in veto. A PDP9/L computer monitored the performance of the apparatus and wrote the event formation on to the magnetic tape.

In order to minimize the effects of any drifts in the efficiency of the equipment, the sign of the polarization of the target was changed approximately every eight hours. Triggers from 2 x 10^6 incident kaons were recorded at each momentum. To enable background corrections to be made for events produced by interaction from bound protons, data were recorded from a carbon target whose mass was equivalent to the non-hydrogenous content of the propanediol.

The analysis was performed on the IBM 360/195 at the Rutherford Laboratory in the following stages.

a) The raw data tapes were copied and decoded on to high density tapes in a more convenient form for analysis.

b) The main analysis program reconstructed the events and wrote a data summary tape of acceptable K^-p candidates.

c) The polarization parameter was calculated using the events of the data summary tape, after applying suitable cuts and background subtraction.

Details of the data analysis can be found in ref. 1, which also shows how the polarization parameter was calculated. An estimation of the background and the evaluation of target polarization are also given. These two effects are largely responsible for systematic errors in the data.

The uncertainty in the target calibration is estimated to be ± 3%. Because of the very good signal to background ratio, the final results on the polarization parameter are relatively insensitive to small errors in the estimate of the background. An error of ± 5% in the level of the carbon background, would imply a $\Delta P/P$ of ± 0.25%. A fuller account of this work is given in ref. 1.

The results are shown in figure 2 where the continuous curve is the prediction of the partial wave analysis of Alston-Garnjost et al. (2). This analysis includes the recent polarization data of Ehrlich et al. (3) and also the latest charge exchange data but does not include the polarization data given in this paper. The agreement is good.

Our polarization data, and the differential cross sections taken from Conforto et al. (4) and interpolated to our momenta, were parametrized as

$$P(\cos\theta^*)\ \frac{d\sigma}{d\Omega}(\cos\theta^*) = \lambda\!\!\!{}^{-2} \sum_{\ell=1}^{\ell_{max}} B_\ell\ P_\ell^1(\cos\theta^*)$$

$$\frac{d\sigma}{d\Omega}(\cos\theta^*) = \lambda\!\!\!{}^{-2} \sum_{\ell=0}^{\ell_{max}} A_\ell\ P_\ell(\cos\theta^*)$$

where P_ℓ and P_ℓ^1 are the normal and associated Legendre polynomials. The values of A_ℓ and B_ℓ were determined simultaneously by an optimization of the χ^2 for the overall data set, with P obtained by

$$P(\cos\theta^*) = \frac{\sum B_\ell P_\ell^1(\cos\theta^*)}{\sum A_\ell P_\ell(\cos\theta^*)}$$

The order of the fit was increased until the χ^2 per degree of freedom became stable. This was the case for $\ell_{max} = 5$ below 1.1 GeV/c and $\ell_{max} = 6$ at higher momenta. The value of A_ℓ were consistent with the results of ref. (4).

Figure 3 shows the values of the B coefficients as a function of the momentum for this and 3 previous experiments (3, 7, 8). The curves are the predictions of the phase shift analysis of Alston-Garnjost et al. (2). This analysis predicts the following baryon states in our energy region-S01(1725), S11(1770), P01(1703), P03(1908), D05(1825), D15(1777), F05(1819), F15(1937).

Some of the resonance parameters coming from this analysis differ significantly from previous ones (5, 6).

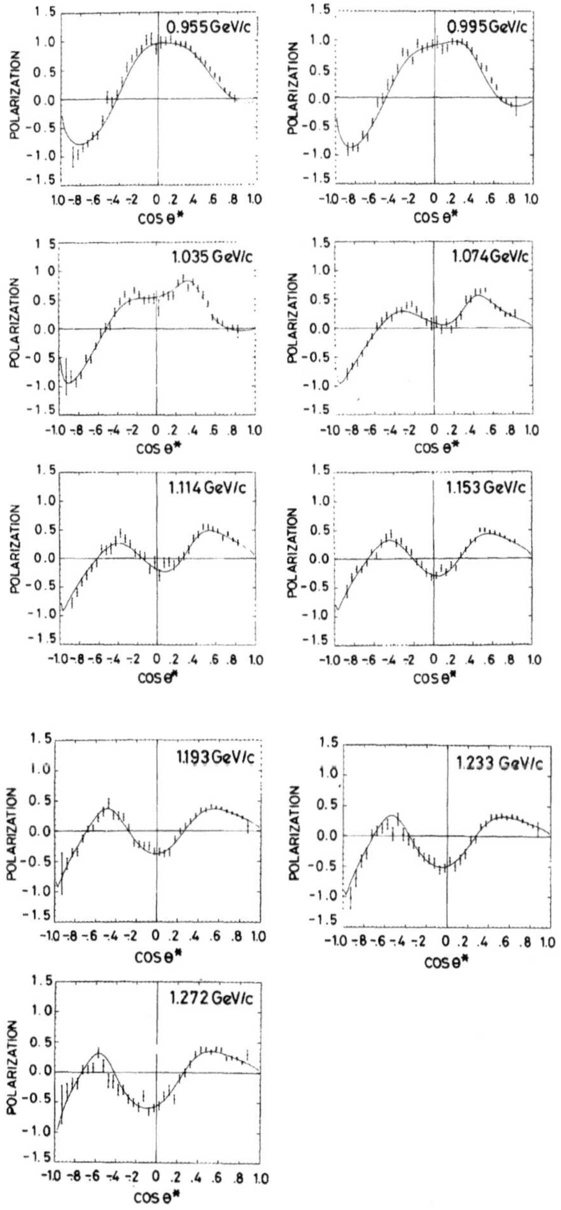
0.955 GeV/c
0.995 GeV/c
1.035 GeV/c
1.074 GeV/c
1.114 GeV/c
1.153 GeV/c
1.193 GeV/c
1.233 GeV/c
1.272 GeV/c
POLARIZATION
cos θ*

Figure 2

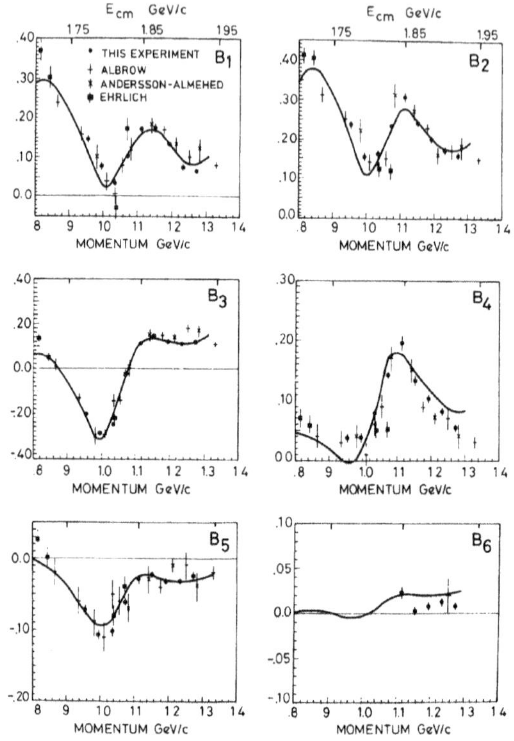

Figure 3

Since our results appear to support this recent analysis, inclusion of our data in a future phase shift analysis should help to clarify the resonance structure in the $\bar{K}N$ system.

* Visitor from University of New Mexico, Albuquerque, USA.
** Now at Birkbeck College, London, England.
*** Now at Department of Nuclear Physics, Oxford University, England.
+ Now with Marconi Ltd., Camberley, England.
++ Now with New Scientist, London, England.

References

1. Bryant, H.C., et al., Rutherford Lab. Report RL 79-078, to be published in Nucl. Phys. B.

2. Alston-Garnjost, A., et al.: 1978, Phys.Rev. 18D, 182.

3. Ehrlich, R.D., et al.: 1977, Phys. Letters 71B, 455.
4. Conforto, B., et al.: 1976, Nucl. Phys. B105, 189.
5. Gopal, G.P., et al.: 1977, Nucl. Phys., B119, 362.
6. Martin, B.R., and Pidcock, M.K.: 1977, Nucl. Phys. B126, 285.
7. Albrow, M.G., et al.: 1971, Nucl. Phys. B29, 413.
8. Andersson-Almehed, S., et al.: 1970, Nucl. Phys. B21, 515.

POLARIZATION IN K^+n ELASTIC AND CHARGE EXCHANGE SCATTERING BETWEEN 0.851 GeV/c AND 1.351 GeV/c

S.J. Watts, D.V. Bugg, A.A. Carter, M. Coupland[+],
E. Eisenhandler, W.R. Gibson, P.I.P. Kalmus and H. Sandhu
Queen Mary College - London

A. Astbury, G.H. Grayer, A.W. Robertson[++], T.P. Shah and
C. Sutton[+++]
Rutherford Laboratory - Chilton, Didcot

Presented by S.J. Watts

A resonance formed in the K^+nucleon system cannot be constructed out of 3 quarks. Structures found previously in K^+p and K^+d total cross sections (1) can be interpreted as resonances in both the I = 1 and I = 0 isospin states. Information on the I = 1 part of the KN system can be obtained directly from K^+p scattering, but the I = 0 part requires data involving neutral kaons or neutrons. In this experiment information on two different combinations of the isospin states was obtained by measuring simultaneously the elastic and charge exchange scattering of K^+ mesons from neutrons in a polarized deuteron target.

The apparatus was almost identical with that used for the previous study of polarization in K^-p elastic scattering, and is shown in figure 1 of the accompanying paper (2). Kaons in the beam were identified with a DISC differential Cerenkov counter except at the lowest momentum where time of flight was used.

The target material consisted of beads of deuterated propanediol doped with chromium V complexes (3). To obtain high polarization and neutron density a mixture of $C_3H_2D_6O_2$ and $C_3D_8O_2$ in the ratio 1:2 was used. The deuteron polarization was monitored with an NMR coil. A dynamic deuteron polarization of typically 32% was obtained.

For the elastic channel the neutron was detected using counters N1-3. The N1-2 counters (4) each consisted of 30 elements of plastic scintillator, 1.5 m high, 50 mm wide and 150 mm in depth, viewed by photomultipliers at the top and bottom. Since neutrons are detected by rescattering they may be observed in more than one counter. A maximum of 5 adjacent elements was allowed to determine the position of the neutron. The time-of-flight relative to the beam was measured for each photomultiplier. The average of these gave the time-of-flight of the

E. Ferrari and G. Violini (eds.), Low and Intermediate Energy Kaon-Nucleon Physics, 215–222.

neutron to an accuracy of $\pm$ 1 ns and their difference gave the vertical coordinate with a spatial resolution of $\pm$ 50 mm. The vertical position was calibrated by extrapolating charged particles from the spark chambers and the absolute time-of-flight to the counters was calibrated using X-rays from the reaction $\pi^- p \rightarrow \pi^o n$.

The N3 counter (5) consisted of a set of 1 m long perspex cells containing liquid scintillator. The effective cell face was approximately 10 x 10 cm^2 at 3 metres and up to 3 by 3 elements were combined to allow for cross-talk. Time-of-flight was not measured for the N3 counters since it would have had inadequate resolution even to differentiate neutrons from γ-rays. The complete apparatus was surrounded by concrete blocks to reduce the background of low energy neutrons.

The elastic trigger was formed by putting the beam, given by $C(1-3).\overline{HV}(1-3).DISC$, in coincidence with the neutron counters and the counter hodoscope arrays H(1-3), excluding the beam region. Only events in which one hodoscope counter fired were accepted. Background was reduced by putting the signals from the counters H(4,5), NV1, TV(1-12) and BV in anticoincidence.

The charge exchange trigger was a coincidence between an incident K^+ and three outgoing charged particles. A multiplicity requirement was demanded from hodoscope counters and the proportional chamber surrounding the target. This looked for events in which the K^o decayed into two charged pions.

Triggers from 6.10^8 incident kaons were recorded at each momentum for each sign of target polarization and from a carbon target whose effective neutron density was equivalent to that of the non-deuterated content of the propanediol.

The incident kaon track was determined using the proportional chambers B(1-8). The scattered kaon vector outside the magnetic field was obtained from the spark chamber arrays, and this was then tracked backwards into the magnetic fiels by an iterative fitting routine using coordinates from the U1 and U2 chambers. The direction of the neutron was inferred from its measured position in the neutron array, and in some cases time-of-flight gave a determination of neutron velocity. Details of the analysis chain for the elastic channel are given in (6).

For the charge exchange channel three charged particle vectors were obtained outside the field. Although the particles were not identified, their charge could be determined from the curvature in the field of the polarizing magnet. The two positive tracks were tried as both a π^+ and a proton in the subsequent analysis. The K^o decay vertex was reconstructed first. The interaction vertex was then reconstructed using the incident beam, the outgoing proton and the K^o direction. Further details can be found in (7).

The data have been corrected for the D-state of the deuteron (8) (9 $\pm$ 2%), and the mean value of cos ϕ (2%), where ϕ is the azimuthal angle. No correction has been applied for neutron binding in the deuteron; such a correction would be greatest at forward cos θ^*. There are systematic uncertainties of ± 5% in the target polarization, and ± 2% from the beam normalization.

Results for the elastic channel are shown in figure 1. The Basel convention has been used to determine the sign of the polarization. The curves are the predictions of the phase shift analyses of BGRT (9) labelled A, C, D and of Martin (10) labelled M. Results for the charge exchange channel are shown in figure 2.

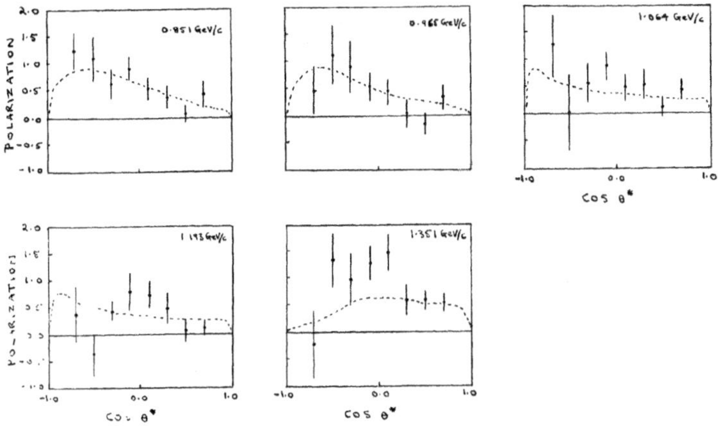

Figure 1

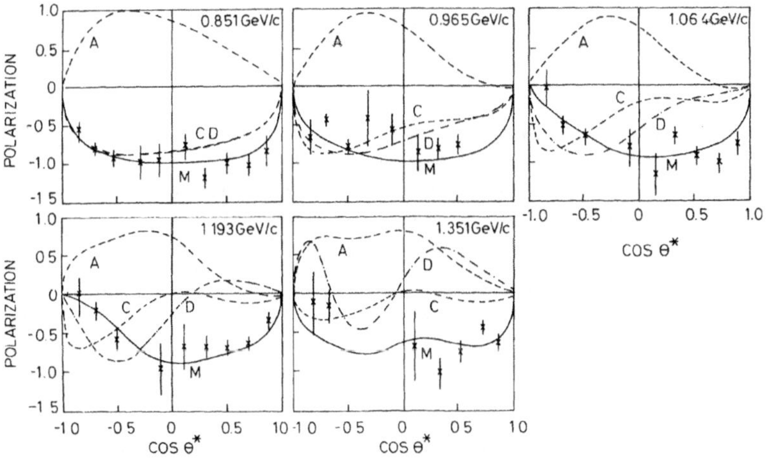

Figure 2 - Preliminary results for the polarization in K^+n charge exchange scattering. There is no cos ϕ correction. This would scale numbers by 1.02

The error bars represent statistics. An estimate of the uncertainty in subtracting background results in an extra Δp of about ± 0.03 at the top two momenta.

A phase shift analysis incorporating this and other data has been performed. This analysis used both K^+p and K^+n data. The data set is that of Martin (10), with the addition of data listed in table 1. Data of Armitage et al. require a renormalization varying strongly with momentum; they have been given a floating normalization. In order to analyse the data of Corden et al. (12), $\bar{K}N$ phase shifts of Gopal et al. were used. There were no significant changes in KN phase shifts when the $\bar{K}N$ phase shifts of Alston-Garnjost et al. (13) were used instead. The phase shift analysis requires a renormalization of the data of Corden et al. by a factor 1.62 ± 0.05.

Table 1 - Data used in the phase shift analysis additional to those by Martin (10)

Data type		Authors	Ref.
$K^+p \to K^+p$	$d\sigma/d\Omega$	Albrow et al.	18
	$d\sigma/d\Omega$	Giacomelli et al.	19
	$d\sigma/d\Omega$	Charles et al.	20
	P	Zeller et al.	17
	Ref(0)	Martin	21
$K^+n \to K^+n$	P	Robertson et al.	6
	Ref(0)	Martin	21
$K^+n \to K^op$	$d\sigma/d\Omega$	Sakitt et al.	22
	P	This paper	
$K^op \to K^+n$	$d\sigma/d\Omega$	Armitage et al.	23
$K^o_Lp \to K^o_Sp$	$d\sigma/d\Omega$	Corden et al.	11

Earlier phase shift analyses (10, 14) of K^+p data have found a poorly determined solution above the inelastic threshold. The problem is that there is one direction in parameter space in which the minimum is very shallow. This uncertainty has been eliminated in the present analysis using a dispersion relation. The phase shift showing the largest uncertainty is $\delta(P_{11})$ and several other phases are strongly correlated with it, notably $\delta(P_{13})$. These two move in opposite directions; $\delta(P_{11}) + 2\delta(P_{13})$ is well determined, but $\delta(P_{11}) - \delta(P_{13})$ is not. This uncertainty can be resolved by using the fixed-t dispersion relation for the invariant amplitude B. The logic of this is that B is dominated by $g/\sin\theta$, and the latter is dominated by P_{11}-P_{13}. The value of this latter quantity at 851 MeV/c is used as a subtraction constant, and P_{11}-P_{13} at higher energies is calculated from the dispersion relation. All phase shifts are now well determined and satisfy partial wave dispersion relations closely; in particular the relation for S_{11} helps strongly in determining $\eta(S_{11})$.

The I = 0 phase shifts are now unique from data alone up to 1205 MeV/c. A second solution appears at 1355 MeV/c with $\delta(P_{01}) = 90^o$.However this solution can be ruled out because it does not continue smoothly through data around 1200 MeV/c to lower momenta. It also fails dramatically to satisfy the backward dispersion relation for the B amplitude.

Table 2 shows the phase shift solution from 800 to 1400 MeV/c.

Table 2 - The phase shift solution at 100 MeV/c steps of lab momentum. Units are degrees. G waves are taken from Alcock and Cottingham (21). Those η parameters are not tabulated are 1. Errors are very approximate

Lab Momentum (MeV/c)	800	900	1000	1100	1200	1300	1400	Typical error
δ S_{11}	-39.9	-41.0	-41.0	-42.4	-44.9	-47.5	-52.5	±1.0
P_{11}	- 9.8	-11.0	-14.0	-17.5	-22.6	-27.0	-30.0	±1.0
P_{13}	8.2	10.6	12.1	12.6	10.5	4.2	- 3.0	±0.5
D_{13}	- 3.0	- 4.1	- 5.4	- 6.5	- 7.5	- 8.6	- 9.6	±0.4
D_{15}	0.20	0.41	0.65	0.75	0.64	0.58	0.56	±0.2
F_{15}	- 0.40	- 0.48	- 0.64	- 0.80	- 0.93	- 1.09	- 1.26	±0.2
F_{17}	0.19	0.28	0.39	0.60	0.85	1.02	1.03	±0.2
η S_{11}	1	0.984	0.892	0.755	0.645	0.575	0.520	±0.03
P_{11}	1	0.950	0.872	0.820	0.785	0.775	0.780	±0.03
P_{13}	0.989	0.930	0.845	0.735	0.605	0.525	0.520	±0.04
D_{13}	0.990	0.967	0.949	0.940	0.940	0.940	0.940	±0.02
D_{15}	1	0.999	0.992	0.980	0.973	0.972	0.970	±0.02
F_{15}	1	1	1	1	1	0.994	0.980	±0.02
F_{17}	1	1	1	1	0.993	0.975	0.958	±0.02
δ S_{01}	-11.2	-13.8	-16.5	-19.2	-22.0	-24.3	-26.8	±2.0
P_{01}	50.9	55.9	59.2	59.1	58.2	57.0	55.0	±2.0
P_{03}	- 7.1	- 8.1	- 9.2	-10.4	-12.0	-12.9	-13.7	±1.5
D_{03}	8.8	10.9	13.0	13.9	13.6	12.5	10.2	±1.0
D_{05}	- 0.60	- 0.80	- 0.98	- 1.22	- 1.55	- 1.90	- 2.25	±0.3
F_{05}	1.80	2.20	3.80	4.90	6.05	7.25	8.40	±1.0
F_{07}	- 0.21	- 0.40	- 0.58	- 0.86	- 1.14	- 1.48	- 1.83	±0.4
η D_{05}	0.997	0.945	0.885	0.790	0.715	0.669	0.653	±0.03
F_{05}	1	1	0.995	0.969	0.928	0.884	0.835	±0.03
G_{07}	1	1	1	1	0.996	0.986	0.967	±0.02

A rough estimate of the errors is shown in the final column, but these can only be prorpagated approximately through the dispersion relations. The curves in figure 2 show the polarization given by this solution.

Fitted elasticities agree quite closely with solution B of BGRT (15) for $K^+p = K\ \Delta$. A conspicuous feature of P_{13} is a distinct loop in the Argand diagram. However, it is almost certainly not a resonance. If it were, it would be highly inelastic, and strongly coupled to $K\Delta$. But solution B of BGRT shows no large phase variation for P_{13}. We interpret the behaviour of P_{13} as a classic 'woolly cusp' (16) due to the opening of the strong inelastic channel. There have been suggestions (17) that there may exist a P_{01} resonance. Our solution shows that P_{01} reaches a broad maximum at 60°, and does not resonate. Hence this analysis provides evidence against the existence of a Z_o^* resonance.

\+ Now at Birkbeck College, London.
++ Now with Marconi Ltd., Camberley.
+++ Now with New Scientist, London.

References

1. Carroll, A.S., et al.: 1973, Phys. Lett. 45B, 531.
   Bowen, T., et al.: 1973, Phys. Rev. D7, 22.
2. Bryant, H.C., et al., accompanying paper
3. Cash, A.R., et al.: 1979, Journal of Phys. E (Scientific Inst.), and: 1978, RL-78-110.
4. Strong, J.A., et al.: 1978, Nucl. Inst. and Meth. 156, 411.
5. Brown, R.M., et al.: 1976, Nucl. Inst. and Meth. 136, 307.
6. Robertson, A.W., et al.: 1980, Rutherford Laboratory RL-80-01.
7. Watts, S.J., et al.: 1980, Rutherford Laboratory, RL-80-
8. Daum, C.: 1973, Rutherford Laboratory RL-73-088, 172.
9. Giacomelli, G., et al.: 1974, Nucl. Phys. B71, 138.
10. Martin, B.R.: 1975, Nucl. Phys. B94, 413.
11. Corden, M., et al.: 1979, Nucl. Phys. B155, 13.
12. Gopal, G.P., et al.: 1977, Nucl. Phys. B119, 362.
13. Alston-Garnjost, M., et al.: 1978, Phys. Rev. D18, 182.
14. Kelly, R.L.: 1979, AIP Conf. Proc. 17, 501 (High Energy Physics with Polarised Beams and Polarised Targets, Argonne 1978), ed. G.H.Thomas, American Institute of Physics, New York.
    Arndt, R.A., Roper, L.D., and Steinberg, P.H.: 1978, Phys. Rev. D18, 3278.

Cutkosky, R.E., et al.: 1976, Nucl. Phys. B102, 139.

15. Giacomelli, G.: 1976, Nucl. Phys. B110, 67-80.
16. Nauenberg, M., and Pais, A.: 1962, Phys. Rev. 126, 360.
17. Zeller, M.E., et al.: 1978, quoted by R.L. Kelly at the Meeting on Exotic Resonances, Hiroshima, Sept.; and preprint LBL-7976, 1978.
18. Albrow, M.G., et al.: 1971, Nucl. Phys. B30, 273.
19. Giacomelli, G., et al.: 1970, Nucl. Phys. B20, 301.
20. Charles, B.J., et al.: 1977 , Nucl. Phys. B131, 7.
21. Martin, A.D.: 1976, Phys. Lett. 65B, 346; and private communication.
22. Sakitt, M., Shelly, J., and Thompson, J.: 1977, Phys. Rev. D15, 1846.
23. Armitage, J.C.M., et al.: 1977, Nucl. Phys. B123, 11.

DISCUSSION

- G.C. Oades :

Does your phase shift analysis make significant renormalizations of the $K_L^o p \rightarrow K^+n$ data of Armitage et al. and are you worried by the large renormalization of the regeneration data of Corden et al. ?

- D.V. Bugg* :

Two data sets required significant renormalization. Scaling factors found in single energy phase shift solutions are shown in table 1.

Table 1 - Scaling factors required for the data of (a) Armitage et al., ref. (1), and (b) Corden et al., ref. (3).

momentum (MeV/c)	(a)	(b)
860	0.652	1.642
970	0.690	1.585
1075	0.735	1.542
1205	1.067	1.779
1365	1.393	- -

The renormalization of the $d\sigma/d\Omega$ ($K_L^o p \rightarrow K^+n$) data of Armitage et al. (1) is not surprising. One of the difficulties of such an experiment is establishing the spectrum of the K_L^o beam. Figure 24 of the original publication shows that integrated cross-sections differ from those of Giacomelli et al. (2) by factors agreeing closely with table 1.

The second large renormalization is of data of Corden et al. (3) on $d\sigma/d\Omega$ ($K^o_Lp \rightarrow K^o_Sp$). The experimental normalisation was made indirectly by measuring $K^op \rightarrow \Lambda^oK^+$ and relying on charge independence to normalize against $K^-p \rightarrow \Lambda^o\pi^o$, which has been measured well. One would expect this procedure to be reliable, and Corden et al. quote errors of typically $\pm$ 9%. The scaling factors shown in table 2 are surprisingly large, but are reasonably consistent in the four momentum ranges where the data are available. Without them, the fits are terrible. They change little according to the choise of $\bar{K}N$ phase shifts. Note that the momentum range covers several $\bar{K}N$ resonances, so the stability of the scaling factor in table 1 is really quite surprising.

At the meeting it was suggested that such a renormalization would create problems with forward dispersion relations. In defence, may I remark that my phase shift analysis fits accurately total cross-sections and real parts of forward K^+p and K^+n amplitudes supplied by A.D. Martin in January 1980. So if there is a problem with the dispersion relations, it must be in the $\bar{K}N$ sector.

One should also be aware that $d\sigma/d\Omega$ ($K^o_Lp \rightarrow K^o_Sp$) is strongly peaked close to $\cos\theta = \pm 1$. This could lead to experimental problems with normalization, although the authors of the data express surprise at the magnitude of the renormalization I require. Alternatively, the normalization may be sensitive to assumptions about high partial waves; this requires further study.

1. Armitage, J.C.M., et al. : 1977, Nucl. Phys. B123, 11.

2. Giacomelli, G., et al. : 1972, Nucl. Phys. B42, 437.

3. Corden, M., et al. : 1979, Nucl. Phys. B155, 13.

* This comment has been sent by dr. Bugg to the editors and it expands the original answer by dr. Watts.

EXOTIC ATOMS AND THE KAON-NUCLEON INTERACTION

C.J. Batty
Rutherford Laboratory - Chilton, Didcot

1. Introduction

An exotic atom is formed when an electron in an "ordinary" atom is replaced by a heavier particle such as a muon, π or K-meson, antiproton or a Σ-hyperon. Studies of these exotic atoms cover a whole range of disciplines from atomic and solid-state physics to nuclear and elementary particle physics. From the nuclear point of view, the strongly interacting pions, kaons, antiprotons and sigma-hyperons give information about the very low energy hadron-nucleus interaction and can in principle give further information about the nucleus such as the matter distribution. Information about the properties of the hadron, such as its mass or magnetic moment, and about the hadron-nucleon interaction can also be obtained.

Much of the early interest in kaonic atoms had as its objective the study of the matter distribution in the periphery of the nucleus. These studies assumed some simple form for the kaon-nucleon interaction, which is required in the energy region below threshold since the nucleons are bound in the nucleus. Unfortunately, for the kaon-proton system this is the energy region dominated by the presence of the Λ(1405) resonance close to the K^- -p threshold. As a result realistic models must include the dynamics of the Λ(1405) inside the nucleus and hence become rather complicated. However as we shall see, analyses using a simple model with a phenomenological <u>effective</u> kaon-nucleon interaction have had considerable success in fitting the experimental data. Unfortunately it is still a considerable task to relate such effective parameters to those of the free kaon-nucleon interaction.

More recently the interest has turned to the study of kaonic atoms, particularly kaonic-hydrogen, to obtain information about the low energy kaon-nucleon interaction. As discussed later a measurement of strong interaction effects in kaonic-hydrogen can give values for the (complex) kaon-proton scattering length. The experiments are technically very difficult but some progress has been made over the last 2 or 3 years and interesting results have been obtained. The extension of these studies

E. Ferrari and G. Violini (eds.), Low and Intermediate Energy Kaon-Nucleon Physics, 223–242.

to K^--d atoms has not so far been attempted but is possible at least in principle.

After the kaonic atom has been formed, the kaon cascades down through the sequence of atomic levels until finally nuclear absorption takes place. A reasonable fraction of these reactions go through the $\Sigma^{\pm}\pi^{\mp}$ channel (∿ 45% in the case of kaonic-hydrogen) and can result in the formation of Σ^- atoms in the same target. These Σ^- atoms have been observed in nuclear targets but not so far in hydrogen. However, if observed these Σ^--p atoms could give extremely useful information about the low energy Σ^- - nucleon interaction.

The above topics will be discussed in more detail in the later sections of this paper. However first the features of the kaon-nucleon interaction of particular relevance to exotic atoms are briefly reviewed and the relevant characteristics of exotic atoms discussed.

2. The kaon-nucleon interaction

In kaonic atom studies we are principally interested in the characteristics of the $\bar{K}$-N interaction at threshold and in the energy region below threshold. Analyses of the low energy $\bar{K}$-N interaction use data available from experiments on the interactions of K^- and K^o_2 with protons at momenta less than ∿ 300 MeV/c. However in general the data is rather limited because of the experimental difficulties associated with the use of very low momenta kaons and mostly dates from before 1970. In addition the analysis is complicated by a number of other factors.

1. Even at threshold (1432 MeV total energy in the centre of mass) the channels $K^-p \rightarrow K^-p$, $\Lambda\pi^o$, $\Sigma^o\pi^o$, $\Sigma^{\pm}\pi^{\mp}$, $\Lambda\pi\pi$ are open although for momenta below ∿ 200 MeV/c there are no significant contributions from three-body final states.

2. The charge exchange threshold occurs at low momenta (∿ 90 MeV/c) for the channel $K^-p \rightarrow \bar{K}^on$.

3. The s-wave resonance Λ(1405) occurs just below threshold.

4. Resonances occur in other partial waves above (Λ(1520) in d-wave) and below (Σ(1385) in p-wave) threshold.

Rewiews of the present status of analyses of the low-energy $\bar{K}$-N interaction have been given recently (1) and will be discussed elsewhere (2) in this Kaon Workshop, so we will only briefly comment on those aspects which are of direct relevance to kaonic-atom studies.

Analyses have usually been made using a K-matrix formulation to describe the multichannel problem. A constant scattering length (CSL) approach was used (3, 4) in early analyses of K^-p data, equivalent to neglecting the variation in the final state momentum. Energy dependences were then allowed in subsequent zero-range (4, 5) (ZR) and finite range (6) (FR) K-matrix analyses and in some cases (4, 6) partial waves higher than s-waves were included so as to exploit higher energy data. Chao et al. (7) pointed out that these various analyses gave rather differing values for the parameters of the $\Lambda(1405)$ resonance just below threshold and added to the K-matrix analysis the constraint that the $\pi\Sigma$ mass spectrum obtained from the $\pi^- p \rightarrow \pi\Sigma K$ reaction should be fitted simultaneously with the data above threshold. Their analysis showed that without the constraint provided by the $\pi\Sigma$ mass spectrum in the energy range between the $\pi\Sigma$ and $\bar{K}N$ thresholds little information about the parameters of the $\Lambda(1405)$ could be determined. For example for a solution where the $\pi\Sigma$ mass distribution was not included they obtained M = 1420 MeV and Γ = 20 MeV for the mass and width of the $\Lambda(1405)$ whilst with the extra data they obtained M = 1406 MeV and Γ = 55 MeV. (The latest values given in the particle data compilation are M = 1402 ± 4 MeV and Γ = 38 ± 4 MeV from production experiments). This is a result of particular interest for kaonic-atom studies in view of the dominant effect of this resonance in the region just below threshold.

More recently Martin (8) has shown how measurements of the real parts of kaon-nucleon amplitudes can be used with dispersion relations to provide powerful constraints on the analysis of low energy K^-p data. Indeed dispersion relations had already shown significant discrepancies between the predicted scattering at low energies and that measured in $\bar{K}N$ experiments.

Values for the I = 0 and 1, $\bar{K}$-N scattering amplitudes a_0, a_1 from a variety of analyses are given in table 1 with references to the method of analysis in table 2.

Table 1 - Values of $\bar{K}$-N scattering lengths

Fit	a_0	a_1	a_s
1	-1.67 + i 0.71	-0.07 + i 0.68	-0.87 + i 0.70
2	-1.69 + i 0.72	-0.04 + i 0.70	-0.87 + i 0.71
3	-1.74 + i 0.70	-0.05 + i 0.63	-0.90 + i 0.67
4	-1.66 + i 0.69	-0.09 + i 0.54	-0.88 + i 0.62
5	-1.65 + i 0.73	-0.13 + i 0.51	-0.89 + i 0.62
6	-1.70 + i 0.71	0.00 + i 0.61	-0.85 + i 0.66
7	-1.60 + i 0.76	0.10 + i 0.69	-0.75 + i 0.73
8	-1.66 + i 0.75	0.35 + i 0.66	-0.66 + i 0.71

Table 2 - Fits to $\bar{K}$-N data

Fit	Ref	Method
1	(3) Kim 65	Constant scattering length
2	(4) A.Martin 70	Constant scattering length 0-280 MeV/c
3	(4) A.Martin 70	Zero Range 0-280 MeV/c
4	(5) B.Martin 69	Zero Range 0-280 MeV/c
5	(6) Kim 67	Finite Range 0-550 MeV/c
6	(7) Chao 73	Zero Range 0-280 MeV/c
7	(7) Chao 73	As 6 with fit to $\pi\Sigma$ mass spectrum
8	(8) A.Martin 76	Includes dispersion relations

Also given is $a_s = (a_0 + a_1)/2$ which is the amplitude relevant to K^-p interactions. It is seen that there are significant differences between the results of the various analyses and in particular the real part of a_1 is seen to change quite dramatically when the dispersion relation constraints are included (8) in the analysis. On the other hand the two analyses by Chao et al. (7) show rather smaller changes than might have been expected at first sight from the very different predictions for the mass and width of the $\Lambda(1405)$ which have been discussed earlier. This point is emphasised by the dispersion relation analysis of Martin (8) who shows that the $\pi\Sigma$ mass spectrum is very sensitive to the energy dependence of the scattering matrix elements. Indeed he had previously suggested (9) that it may be possible to obtain additional information about the energy dependence from the study of data from K^- absorption in nuclei. This is interesting since a recent experiment on kaonic-He atom X-rays, to be discussed later, where the nucleons in the nucleus are very tightly bound, has apparently shown much larger stronger interaction effects than expected.

Finally we should comment on the nature of the $\Lambda(1405)$ resonance. Continuation of the low energy analyses below K^--p threshold indicates a virtual $\bar{K}N$ bound state resonance with parameters similar to those of the $\Lambda(1405)$. However Cline et al. (10) using K^- scattering off virtual nucleons in the deuteron have suggested that there is only weak coupling of $\bar{K}N \rightarrow \Lambda(1405) \rightarrow \Sigma\pi$ which could imply that the resonance was not a local phenomena caused by forces in the $\bar{K}N$ system but a more global effect, such as produced say by a three quark system in a unitary singlet configuration. In view of its importance in the sub-threshold region it seems rather important to properly understand the nature of the $\Lambda(1405)$.

3. Characteristics of Exotic Atoms

When a negatively charged hadron, such as a kaon, is slowed down in a target to sufficiently low energy, it can be captured by the Coulomb field of the nucleus to form a highly excited hadronic atom. This "exotic atom" will then de-excite, with the hadron cascading between the various atomic levels, first by Auger transitions with the ejection of atomic electrons and later by radiative transitions with the emission of X-rays. As the hadron becomes more tightly bound, reaching states of relatively low principal quantum number n, the overlap between the hadron and nuclear wave function increases and absorption of the hadron by the nucleus may take place. At some final n-value, nuclear absorption dominates, all hadrons reaching the level are captured and the X-ray series terminates.

The strong interaction between the hadron and the nucleus causes a shifting of the energy of this last X-ray transition from the purely electromagnetic value whilst the nuclear absorption reduces the lifetime of the final atomic state and so the X-ray line is broadened. Precise measurements of the X-ray energy spectrum then allow the energy shift (ε) and width (Γ) associated with this final level to be determined. As the atomic number and size of the nucleus increase the absorption occurs from higher n-values. For example in kaonic-hydrogen transitions to n = 1 states have been observed; in light nuclei such as He, Li, Be etc. the transitions terminate at n = 2 whilst in heavy nuclei such as Pb or U the X-ray series finish at n = 7.

Although many of the hadrons reach this final level, some capture occurs from the next higher level but the effects on the level energy and width are generally too small to be measured directly. However the width (Γ_u) of this upper level can frequently be deduced indirectly by measuring the decrease in intensity of the final X-ray transition. The relative yield Y of this transition ($n + 1, \ell + 1 \rightarrow n, \ell$) is obtained by comparing the intensity of the transition with the sum of the intensities feeding the upper ($n + 1, \ell + 1$) level.

$$Y = \frac{I_X\,(n + 1,\ \ell + 1 \rightarrow n,\ \ell)}{\sum_{i=n+2}^{\infty} \left[I_{XA}\,(i,\ \ell + 2 \rightarrow n + 1,\ \ell + 1) + I_{XA}\,(i,\ \ell \rightarrow n + 1,\ \ell + 1) \right]}$$

where I is the intensity, the index XA refers to the sum of X-ray and Auger transitions and X refers to the X-ray intensity alone.

For K^- and Σ^- the interaction is sufficiently strong that levels of angular momentum ℓ are fully absorbed and the second term in the denominator can be neglected. Contributions from Auger transitions and from transitions with large Δn are not observed experimentally and corrections for these must be obtained from a calculation of the cascade process (11). Parallel transitions such as ($i, \ell + 2 \rightarrow n, \ell + 1$) and ($i, \ell + 1 \rightarrow n, \ell$) have very similar energies and are difficult to separate. Again a correction for these latter transitions must be obtained from a cascade calculation.

The yield Y is related to Γ_u, the strong interaction width of the upper state by $Y = \Gamma_x/(\Gamma_u + \Gamma_x)$ where Γ_x is the width for the X-ray transitions which can be calculated (12). Measurement of the X-ray spectra thus allows, in favourable cases, three quantities, ε, Γ and Y or Γ_u, which are relevant to the strong interaction to be determined.

In order to determine the energy shift ε, the electromagnetic energy of the unshifted line must first be calculated. This is usually obtained by solving the Klein-Gordon wave equation using the Coulomb potential obtained from realistic nuclear charge distribution. Other corrections can be calculated using second-order perturbation theory and include first and higher-order vacuum polarization, electron screening, relativistic corrections to the reduced mass approximation, nuclear polarization and the Lamb shift. Depending on the energy of the level to be calculated and on the magnitude of the various corrections (see for example table 1 in ref. 13) the electromagnetic energy can usually be calculated to an accuracy of a few electron volts.

4. Strong interaction effects for Z > 1

There are now quite a large number of measurements of strong interaction effects in kaonic atoms for a range of nuclei over the periodic table. A compilation of these various values is given in ref. 14 and the results for the shifts, widths and yields are displayed in figures 1, 2 and 3.

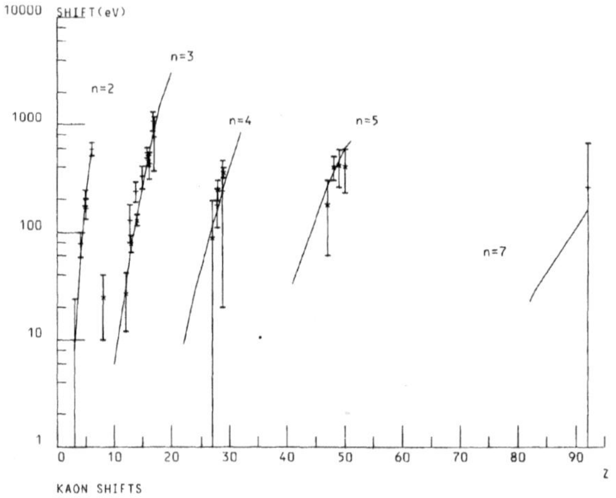

Figure 1 - Measured shifts in kaonic atoms. The continuous lines are theoretical fits to the data (see text)

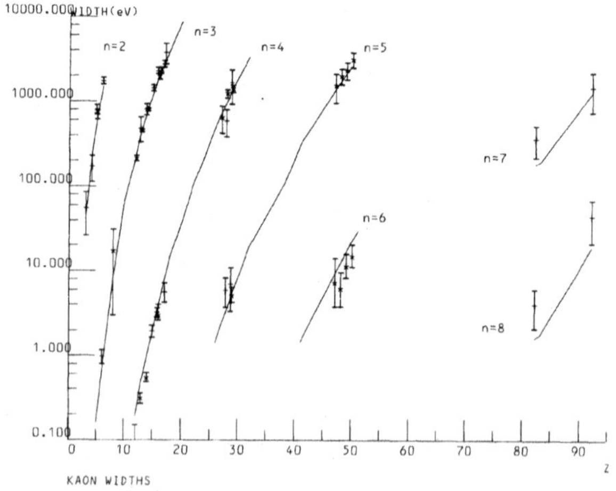

Figure 2 - Measured widths in kaonic atoms. The continuous lines are theoretical fits to the data (see text)

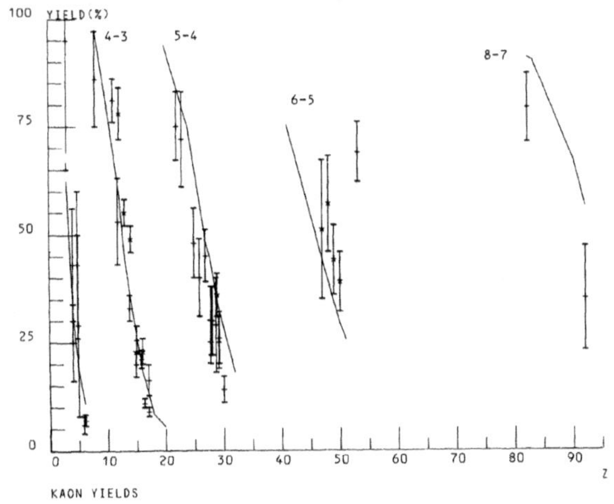

Figure 3 - Measured yields in kaonic atoms. The continuous lines are theoretical fits to the data (see text)

The region of elements up to Z = 50 is seen to be reasonably well covered but there are few measurements available for n = 7 levels in heavy nuclei or for n = 6 levels in the Z $\approx$ 70 region where the nuclei are deformed.

Since kaons are predominantly absorbed on single nucleons it seems reasonable to describe the kaon-nucleus interaction in terms of an optical potential with both the real and imaginary parts proportional to the nuclear density and this was the form used in very early analyses. Using more recent notation this gives a potential of the form

$$V = -\frac{2\pi}{\mu}\left(1 + \frac{m_K}{m_N}\right)\bar{a}\ \rho(r)$$

where μ is the reduced kaon-nucleus mass, m_K the mass of the kaon, m_N - the mass of the nucleon, $\rho(r)$ is the nucleon density distribution and $\bar{a}$ is the average kaon-nucleon scattering length. This potential is then used, together with the Coulomb potential, in the Klein-Gordon equation to calculate the (complex) energy of the bound atomic states and hence the strong interaction level shifts and widths. The formula can easily be generalised to the case where the neutron and proton density distributions are taken to be different but with the present precision of the data this has not so far been necessary.

A detailed analysis of kaon data using this form of the potential has been made recently (14). There are problems in fitting the shift, width and yield measurements simultaneously which may be due to the particular form used for the nucleon density distribution. However fitting all the available shift and width measurements gives $\bar{a}$ = (0.34 ± .03) + i(0.84 ± .03) fm with χ^2 = 35.6 and a χ^2 per degree of freedom of 0.98 which is very satisfactory. These fits to the data are also shown in figures 1, 2 and 3.

It was assumed in the early work that $\bar{a}$ would be the average of the kaon-proton and kaon-neutron free scattering lengths. However this was soon found to be invalid. For example the parameters obtained by Martin and Sakitt (5) (fit 4 in table 1 and 2) would give $\bar{a}_{free}$ = -0.49 + i 0.58 whilst the more recent parameters of Martin (8) (fit 8 in tables 1 and 2) give $\bar{a}_{free}$ = -0.16 + i 0.69 fm.

The difference in sign between the real part of the free scattering length ($\bar{a}_{free}$) and the real part of the effective scattering length ($\bar{a}$) has given rise to much confusion. The effective $\bar{a}$ has a positive real part which corresponds to an attractive K^--nucleus interaction in the sign convention used here. The free $\bar{a}$ has a negative real part but the free K^--nucleon potential is attractive and strongly absorptive (15). It is the strong absorption which causes the real part of the scattering length to be negative, a fact which was neglected in the original derivations of the above equation where it was assumed that the real parts of the potential and its scattering length would have the same sign.

Another point to be considered in the derivation of the optical potential is the binding of the nucleons in the nucleus which changes the effective energy of the kaon-nucleon interaction to the region below threshold, which, as we have already discussed, is dominated in the K^-p channel by the Λ(1405) resonance.

The effect of this resonance was included in an analysis of kaon data by Bethe and Siemens (16) who replaced the threshold value of a_o, the I = 0 scattering length by Breit-Wigner form with parameters corresponding to the Λ(1405). The scattering amplitudes for neutrons and protons were then calculated at the appropriate effective energies after taking into account the nucleon binding energy. However, in common with a number of early papers, Bethe and Siemens used perturbation theory to calculate the kaonic atom level shifts and widths. This is now known to be invalid (17) so that care must be used in interpreting results of analyses in the literature, particularly those prior to 1972.

More recently (18, 19) calculations have been carried out, again invoking the role of the Λ(1405) resonance inside the nucleus and also including the different binding of the various nuclear shells. Using amplitudes from the latest analysis due to Martin (8) they find that a fit to the shift and width values for kaonic - C and S systematically suggests a repulsive shift and increased width for the Λ(1405) inside the nucleus.

An alternative approach which has been considered (15, 20) is to construct a kaon-nucleus potential by folding a finite-range complex kaon-nucleon potential into the nuclear density distribution

$$V(r) = \int \nu(\underline{x}) \ \rho \ (\underline{r} - \underline{x}) \ d\underline{x}$$

In the work of Deloff and Law (20) a Gaussian form was used for the (complex) kaon-proton and kaon-neutron potentials and the strengths adjusted to fit the corresponding scattering length values. The range was obtained by fitting low energy K-He scattering data. The resulting K-N potential was then used, via the folding procedure to calculate the kaon-nucleus potential and hence the kaonic atom level shifts and widths. Good agreement was obtained with the data then available. Atarashi et al. (21) in a similar method use for the real part a Yukawa type potential with a range corresponding to heavy meson exchange and get improved agreement with experimental shift and width measurements.

This brief discussion of the analysis of kaonic atom data in terms of realistic models has illustrated the complexity of the problem. The kaon-nucleon interaction at threshold has many open channels whilst the scattering amplitudes vary rapidly in the energy region below threshold. The presence of the $\Lambda(1405)$ resonance close to the K^--p threshold implies that realistic models must include non-locality effects, nucleon-correlations and the dynamics of the $\Lambda(1405)$ inside the nucleus. Whilst some progress has been made in the analysis of data using a variety of models, the results have generally only been applied to a few nuclei. The very wide range of nuclei now available presents a very serious challenge to such models. Nevertheless it is amusing to note that a very simple potential model for the kaon-nucleus interaction using only two adjustable parameters (the real and imaginary parts of $\bar{a}$) seems to be capable of giving an excellent fit to the data.

Finally we should comment that recent data (22) for kaonic-He do not fit in with this simple phenomenological picture. Values obtained for the energy shift and width of the 2p level are $\varepsilon = -35 \pm 12$eV and $\Gamma = 30 \pm 30$eV. These values are much larger than those predicted by the simple model which gives $\varepsilon \sim 0.1$eV and $\Gamma \sim 2$eV for He but which gives values for Li and Be in good agreement with measurements (23). Assuming that the experimental measurements are not in error then it is interesting to speculate as to possible reasons for the discrepancy. Because the nucleons in ^{4}He are tightly bound by $\sim$ 20MeV the effective energy ($\sim$1411 MeV) for kaon-nucleon interactions in ^{4}He is very close to the energy of the $\Lambda(1405)$ resonance. As a result it may not be too surprising that parameters which fit data for heavier nuclei where the interaction energy is further away from the resonance do not fit the Helium values. A rather wilder speculation (14) is that the results could be an indication of a kaon-nucleus bound state as discussed by Seki (17) and others. A macroscopic description for a nucleus containing only four nucleons is in any case very questionable and the explanation, if one is needed, may perhaps be found in some form of microscopic model.

5. Kaonic-Hydrogen X-rays

5.1 Experimental

Studies of kaonic-hydrogen atoms are of particular interest since the shift and width of the 1s level can be related in a fairly model independent way to the real and imaginary parts of the s-wave, K^--p scattering length a_s. As discussed already, the free particle scattering lengths have so far been obtained from low energy kaon-nucleon scattering data. A measurement of kaonic-hydrogen X-rays to the 1s state should determine the scattering length directly at threshold. However there are two particular problems which make the detection ot these X-rays difficult.

Firstly the Stark effect causes strong absorption of the kaon from high-n atomic s-states during collisions between the electrically neutral K^--p atoms and other protons in the target. This effect was first calculated by Leon and Bethe (24) for liquid hydrogen who obtained a 2p-1s X-ray yield of about 0.1%/stopped kaon in liquid hydrogen. More recently the calculations have been extended by Borie and Leon (25) who have also considered hydrogen gas at various pressures. A second problem arises due to the very low energy of the X-rays to be detected. The calculated energy of the 2p-1s X-ray is 6.482 keV. As a result several attempts to detect kaonic hydrogen X-rays have been unsuccessful.

The experiments have used either gas or liquid hydrogen targets. Gas targets have the advantage that the yield of X-rays per stopped kaon is relatively high. However, the target must be relatively long ($\sim$ 1m) if a reasonable number of kaons are to be stopped and this then implies that large area proportional X-ray counters must be used. As a result background problems become more severe and the energy resolution of the detectors ($\sim$ 1.5 keV fwhm at 6 keV) is relatively poor. The long target also means that a significant number of the kaons decay in the target region so that the usual anti-coincidence counter at the rear of the target cannot be used to define a stopped kaon trigger. A target system of this type has been used (26) to detect $\bar{p}$-p X-rays but similar experiments with kaons have failed completely.

At first sight experiments with a liquid target look less attractive due to the low X-ray yield per stopped kaon. However much larger numbers of kaons can be stopped in a reasonable target volume so that the total X-ray yield can be comparable in the two cases. This is illustrated in table 3 where we show the yield of 2p-1s K^--p X-rays for a 1m long gas

Table 3 - K^--p X-ray yields for different targets. The 2p-1s X-ray yields are taken from ref. 25

Target	2p-1s X-ray yield (%)	No. of stopping kaons	Total X-ray yield
Gas (1 atm)	10	38	4
Gas (10 atm)	5	860	43
Liquid	0.1	10,000	10

target at either 1 or 10 atmospheres pressure and for comparison the yield for a 10 cm long liquid H_2 target. The results are normalised to 10,000 kaons stopping in the liquid target, whilst the yield values have been taken from ref. 25. The smaller volume of the liquid target also allows Si(Li) X-ray detectors to be used, which, whilst they have a much smaller area than proportional counters, do have considerably better energy resolution ($\sim$ 250eV at 6 keV). This is particularly important if shift and width values are to be measured and should improve the signal to noise ratio for detecting the K^--p X-ray.

An experiment of this latter type, using liquid hydrogen has been carried out at the Rutherford Laboratory (27) and is now being repeated and extended at CERN. The experiment uses a specially constructed liquid-hydrogen target fitted with an integral, high resolution Si(Li) detector. The 12 litres of liquid hydrogen are contained in a pure Al vessel (figure 4) which is designed so that the low energy X-rays from kaonic hydrogen can be detected against a minimum of discrete and continuous backgrounds.

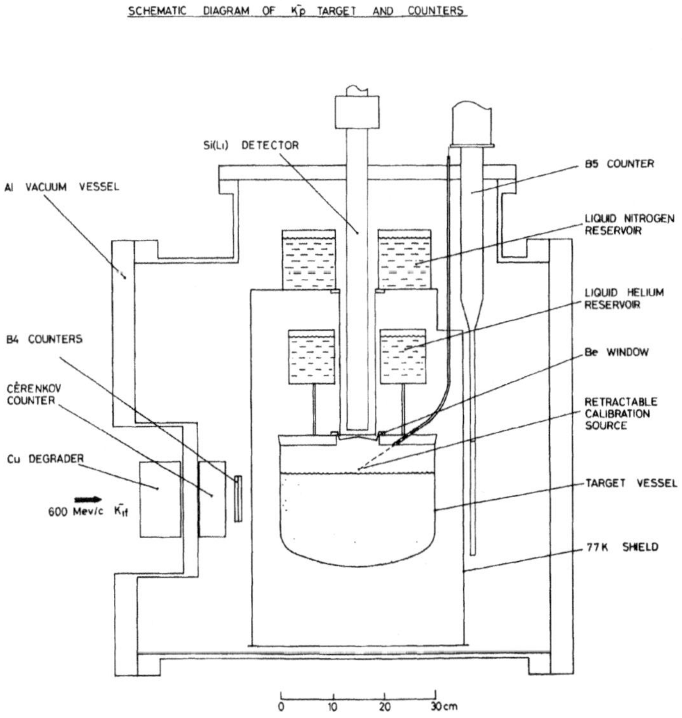

Figure 4 - Layout of special hydrogen target for kaonic X-ray experiment

The X-rays are detected using a 300 mm^2 x 5 mm thick Si(Li) detector, mounted immediately above the target, which views the liquid hydrogen in the target vessel through a 0.5 mm thick Be window. The hydrogen is condensed directly in the target vessel using liquid He for cooling. The detector is calibrated, and the gain stability monitored, using a ^{57}Co source which can be lowered remotely into the target vessel.

In the Nimrod experiment (27) carried out at the Rutherford Laboratory the detector had a resolution of 260eV at 6.50 keV and the X-ray spectrum in the detector was recorded as a function of the arrival time of the detector signal relative to the stopped kaon trigger. Spectra obtained for a total of 1.13 x 10^8 kaon stop triggers and for incident pions with 2.8 x 10^8 pion stop triggers are shown in figure 5.

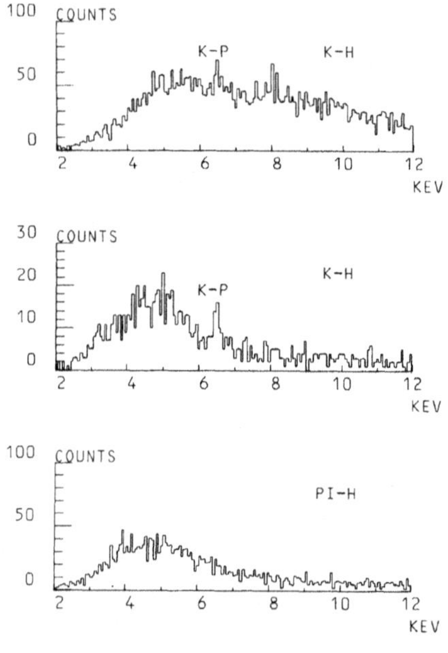

Figure 5 - X-ray spectra measured (27) with kaons and pions stopping in liquid hydrogen. The upper spectrum is for a 120 ns wide timing gate whereas for the other two spectra this was reduced to the last 60 ns period

The upper spectrum for kaons was obtained with a wide timing gate and shows weak peaks at 6.5 and 8.0 keV on a large background mainly originating from high-energy photons Compton scattering in the bulk of the detector. However these Compton electrons produce events with the same timing characteristics as high-energy photons and the middle and lower spectra in the figure show K-H and π-H data where the time gate has been selected to optimize the signal-to-background ratio in the energy region around 6.5 keV. The peak at 6.5 keV is now seen more clearly in the kaon data and is not present in the pion results. The 8 keV peak is attributed to π-Be X-rays from the window materials and possible Cu K_α fluorescence in the Si(Li) crystal mounting. No other peaks appear in the low energy pion or kaon spectra.

Detailed analysis of the data shows that the energy of the X-ray line is 6.52 ± .06 keV giving an energy shift ε = 40 ± 60eV. The peak shape is consistent within rather large errors with the detector resolution and although fits to the data give $\Gamma = 0.0^{+0.23}_{-0.0}$ keV the quoted errors ignore severe correlations with the functional form assumed for the background. The area of the peak is 45 ± 23 counts corresponding to an X-ray yield per stopped kaon of 0.11 ± .06% in good agreement with the theoretical calculations (24, 25) discussed earlier.

Running of the experiment at the Rutherford Laboratory was abruptly terminated by the closure of Nimrod and the experiment is now being repeated at CERN, with some improvements to the apparatus, in an attempt to confirm and improve the earlier results. In the new experiment an improved on-line data recording and display system is being used. As well as recording a two-dimensional spectrum of the energy pulse from the Si(Li) X-ray detector as a function of its time delay relative to the incident kaon, the system also separately measures the time-of-flight of the incident particle, the pulse heights in two beam counters immediately before the hydrogen and also in a further (veto) counter after the target. Each event is also individually recorded on magnetic tape which can be replayed with selection criteria applied to the various signals so as to better define the accepted events. In this way it is hoped to improve the signal to noise ratio of the wanted events and to improve the shift and width measurement.

5.2 Interpretation

The energy shift (ε) and width (Γ) of the 2p-1s X-ray are related (28, 29) to the complex s-wave K^--p scattering length a_c. To first order in ($\mu\alpha a_c$)

$$\varepsilon + i\ \Gamma/2 = 2\mu^2\alpha^3 a_c = 412.11\ a_c\ \text{eV fm}^{-1}$$

where μ is the reduced mass of the kaon, α the fine structure constant and a_c the scattering length in fm. Using the values of ε and Γ given above, the value

$$a_c = (0.10 \pm .15) + i(0.0^{+\ .28}_{-\ .0})\ \text{fm}$$

is obtained in strong disagreement with values for the s-wave isospin averaged scattering length a_s given in table 1.

It is however important to note that a_c and a_s are not the same quantity. The value a_c is the scattering length for the K^--p system and includes Coulomb effects whereas a_s refers to the (neutral) $\bar{K}$-N system. This point has recently been discussed in detail by Deloff and Law (30) following the early work of Dalitz and Tuan (31).

There are two principal corrections. The first takes into account the threshold difference between the $\bar{K}$ p and $\overline{K^o}$n channels and gives

$$a_t = \frac{a_s + |k_o|\, a_o a_1}{1 + |k_o|\, a_s}$$

where a_o, a_1 are the isospin 0 and 1 s-wave scattering lengths, $a_s = (a_o + a_1)/2$ and $|k_o| = 0.2970\ fm^{-1}$ is the $\overline{K^o}$ n centre of mass momentum evaluated at the K^-p threshold. Deloff and Law show that the Coulomb correction, again following the work of Dalitz and Tuan, gives

$$a_{DT} = \frac{a_t}{1 + \dfrac{2(2\gamma + \ell n(2R/B))a_t}{B}}$$

where B is the Bohr radius (83.6 fm) for kaonic hydrogen, $\gamma = 0.57722$ is Euler's constant and R is a cutoff radius which is usually taken to be 0.4 fm.

However Deloff and Law argue that the treatment of the Coulomb correction given by the above formula is incomplete and following the work of Sauer (32) obtain a considerably more complicated expression including an additional quantity Δ which depends on the behaviour of the strong interaction force at small distances. The quantity Δ then has to be calculated using some particular form for the strong interaction and so is model dependent. If Δ is very small then the above formula for the Coulomb correction holds. On the other hand if Δ is large then Deloff and Law show that the Coulomb corrected scattering length and hence the X-ray shift and width could be very small in agreement with the present experimental values. However they are only able to obtain this result by taking a rather extreme value for Δ.

An alternative treatment of the Coulomb correction has been given by Stepien-Rudzka and Wycech (33). Using a separable form for the strong interaction and with Yamaguchi form-factors $v \sim e^{-\chi r}/r$ they obtain (usign our sign convention for the scattering length)

$$\frac{1}{a_w} = \frac{1}{a_t} + \frac{2}{B}\left(-\frac{1}{a_t\chi} + \ell n\left(\frac{4}{\chi B}\right) + 0.577\right)$$

In table 4 we illustrate the size of these various corrections using as an example the scattering lengths found in the most recent analysis by Martin (8). (This is fit 8 in tables 1 and 2). We also show in table 4 the corresponding shifts (ε) and widths (Γ) calculated using the first order formula (28, 29) given above.

Table 4 - Corrections to scattering lengths. Values of ε and Γ given in parentheses were calculated using the formulae correct to second order in the scattering length. For the other values the simple first order formula was used.

	Scattering Length (fm)	ε (eV)	Γ (eV)
Isospin average $a_s = (a_0 + a_1)/2$	-0.655 + i 0.705	-270 (-268	581 611)
Including threshold effects a_t	-0.995 + i 0.827	-410 (-414	682 736)
Coulomb Correction (Dalitz) a_{DT}	-0.963 + i 0.702	-397 (-404	579 620)
Coulomb Correction (Wycech) a_W	-0.961 + i 0.683	-396 (-402	562 606)

The inclusion of threshold effects is seen to give a marked change in the value of the scattering length, the real part increasing by about 50% and the imaginary part by about 20%. Inclusion of Coulomb corrections to the threshold corrected value using either the Dalitz or Wycech formulae given above gives very similar results. The real part decreases by only a small amount (∿ 3%) whilst the imaginary part is reduced by 17%. However the Coulomb correction could be much larger if the suggestion of Deloff and Law (30) is correct.

An alternative suggestion has been made by Kumar and Nogami (34) to explain the apparent difference between the scattering lengths obtained by an extrapolation of low-energy scattering data and the value obtained from the X-ray experiment. They postulate that the I = 0 part of the K^-p scattering amplitude varies rapidly with energy in the threshold region such that it is very small at threshold and increases rapidly as the energy increases. They use a specific model with the Λ(1405) considered as a three quark system in a unitary singlet configuration and are able to obtain agreement with the X-ray results. It remains to be seen how successful such a model would be in reproducing the details of the $\bar{K}N$ scattering data and the πΣ spectrum below threshold.

Finally a comment on the relationship between the shift, width and K^-p scattering length a_c given previously. The formula

$$\varepsilon + i\,\Gamma/2 = 2\mu^2\alpha^3 a_c$$

is correct to first order in $(\mu\alpha a_c)$. Higher order terms have been considered by Trueman (28) and by Stepien-Rudzka and Wycech (33). To second order they obtain

$$\varepsilon = 2\mu^2\alpha^3 \left(a_{cR} - 3.077 \frac{(a_{cR}^2 - a_{cI}^2)}{B} + \ldots \right)$$

$$\frac{\Gamma}{2} = 2\mu^2\alpha^3 a_{cI} \left(1 - \frac{2 a_{cR}}{B} \times 3.077 + \ldots \right)$$

where a_{cR} and a_{cI} are the real and imaginary parts of a_c. Values of the shift and width calculated using these formulae are given in parentheses in table 4. The effect on the calculated shift ε is seen to be small but the values of the width Γ are typically increased by about 10%.

6. Other experiments

When a kaon stops in hydrogen, the following reactions may occur

$$\begin{aligned} K^- p &\rightarrow \Sigma^-\pi^+ && 46\% \\ &\rightarrow \Sigma^+\pi^- && 20\% \\ &\rightarrow \Sigma^0\pi^0 && 27\% \\ &\rightarrow \Lambda^0\pi^0 && 7\% \end{aligned}$$

The Σ^- formed in the first reaction have an energy of $\sim$10 MeV and a range of about 6 mm in liquid hydrogen. Only about 50% of the Σ hyperons will decay in this distance so that $\sim$ 25% of kaons stopped in liquid hydrogen produce Σ^- which can in principle form Σ^- - atoms.

The formation of Σ-atoms by a similar mechanism in nuclear targets has been observed and shifts and widths have been measured (35). In the case of Σ^--p atoms the 2p-1s transition energy is predicted to be 10.8 keV. Again Stark mixing effects will be important and whilst no detailed calculations appear to have been made, they are likely to be as severe as in the kaon case.

As the Σ hyperon has spin $\frac{1}{2}$, the Σ-p system may exist in a singlet or triplet state. Predictions of expected shifts and widths have been made by Wycech (36) who obtains $\varepsilon = -0.59$ keV, $\Gamma = 6.6$ keV for the triplet state and $\varepsilon = 0.86$ KeV, $\Gamma = 0.5$ keV for the singlet state. The very large width will make transitions to the triplet state difficult to observe but a search for transitions to the singlet state would be interesting and could yield valuable information about the Σp interaction.

Another interesting experiment would be to look for X-rays from kaons stopped in deuterium. Such a study would complement the data from hydrogen and could permit deduction of the scattering lengths below the $\bar{K}N$ threshold and in combination with good K^-p data might allow the K^-n scattering length to be determined. The 2p-1s X- ray energy is predicted to be 7.78 keV and again Stark mixing effects are expected to be large although no detailed calculations are available. The energy shift and width for the 1s state predicted by Wycech (36) are $\varepsilon = -1.0$ keV and $\Gamma = 0.62$ keV.

7. Conclusions

In this review we have discussed the relationship between kaonic atom studies and the kaon-nucleon interaction and found that it is a two-way dialogue. Information about the kaon-nucleon interaction, particularly in the subthreshold region is required for exotic atom studies involving nuclei. On the other hand experiments involving exotic atoms formed in hydrogen or deuterium can in principle give information about the kaon-nucleon interaction. Finally there may be some exciting prospects for gaining information about the Σ-nucleon interaction at threshold.

Acknowledgments

I wish to thank my colleagues in the Birmingham, Rutherford, Surrey Collaboration for many discussions about exotic atoms and to J. Law and S. Wycech for keeping me informed about their calculations.

References

1. Miller, D.: 1979, Talk given at the Conference on Hyperfragment and K^- Physics, Jablonna, Poland. To be published in Nukleonika.
2. Martin, A.D., Invited talk 'The kaon-nucleon interaction near threshold' at this meeting.
3. Kim, J.K.: 1965, Phys. Rev. Lett. 14, 29.
4. Martin A.D., and Ross, G.C. : 1970, Nucl. Phys. B16, 479.
5. Martin B.R., and Sakitt, M.: 1969, Phys. Rev. 183, 1345.
6. Kim, J.K.: 1967, Phys. Rev. Lett. 19, 1074.
7. Chao, Y.A., et al. : 1973, Nucl. Phys. B56, 46.
8. Martin, A.D.: 1979, Phys. Lett. 65B, 346.
9. Martin A.D.: 1970, Springer Tracts in Modern Physics 55, 141.
10. Cline, D., Laumann, R., and Mapp, J.: 1971, Phys. Rev. Lett. 26, 1194.
11. Hufner, J.: 1966, Zeit. für Physik 195, 365.
12. Bethe, H.A., and Salpeter, E.E. : 1957, Quantum Mechanics of one-and two-electron atoms (Springer-Verlag).
13. Batty, C.J., Biagi, S.F., Blecher, M., Hoath, S.D., Riddle, R.A.J., Roberts, B.L., Davies, J.D., Pyle, G.J., Squier, G.T.A., Asbury, D. M., and Clough, A.S.: 1979, Nucl. Phys. A329, 407.
14. Batty, C.J.: 1979, Talk given at the Conference on Hyperfragment and K^- Physics, Jablonna, Poland. To be published in Nukleonika.
15. Koch, J.H., and Sternheim, M.M.: 1972, Phys. Rev. Lett. 28, 1061.
16. Bethe, H.A., and Siemens, P.J.: 1970, Nucl. Phys. B21, 589.

17. Seki, R.: 1972, Phys. Rev. C5, 1196.

18. Weise, W., and Tauscher, L.: 1976, Phys. Lett. 64B, 424.

19. Brockman, R., Weise, W., and Tauscher, L.: 1978, Nucl. Phys. A308, 365.

20. Deloff, A., and Law, J.: 1974, Phys. Rev. C10, 1688.

21. Atarashi, M., Hira, K., and Namuri, H.: 1978, Prog. Theor. Phys. 60, 209.

22. Batty, C.J., Biagi, S.F., Hoath, S.D., Sharman, P., Davies, J.D., Pyle, G.J., and Squier, G.T.A.: 1979, Nucl. Phys. A326, 455.

23. Batty, C.J., Biagi, S.F., Blecher, M., Riddle, R.A.J., Roberts, B. L., Davies, J.D., Pyle, G.J., Squier, G.T.A., and Asbury, D.M.:1977, Nucl. Phys. A282, 487.

24. Leon, M., and Bethe, H.A.: 1962, Phys. Rev. 127, 636.

25. Borie, E., and Leon, M., Phys. Rev. A , to be published.

26. Auld, E.G., Averdung, H., Bailey, J.M., Beer, G.A., Dreher, B., Drumm, H., Erdman, K., Gastaldi, U., Klempt, E., Merle, K., Neubecker, K., Sabev, C., Schwenk, H., Walther, V.H., Wendling, R.D., White, B. L., and Woodrich, W.R.: 1978,Phys. Lett. 77B, 454.

27. Davies, J.D., Pyle, G.J., Squier, G.T.A., Batty, C.J., Biagi, S.F., Hoath, S.D., Sharman, P., and Clough, A.S.: 1979, Phys. Lett. 83B, 55.

28. Trueman, T.L.: 1961, Nucl. Phys. 26, 57.

29. Deloff, A.: 1976, Phys. Rev. C13, 730.

30. Deloff, A., and Law, J.: 1979, Phys. Rev. C20, 1597.

31. Dalitz, R.H., and Tuan, S.F.: 1960, Ann. Phys. 10, 307.

32. Sauer, P.U.: 1974, Phys. Rev. Lett. 32, 626.

33. Stepien-Rudzka, W., and Wycech, S.: 1977, Nukleonika 22, 929.

34. Kumar, K.S., and Nogami, Y. Preprint.

35. Batty, C.J., Biagi, S.F., Blecher, M., Hoath, S.D., Riddle, R.A.J., Roberts, B.L., Davies, J.D., Pyle, G.J., Squier, G.T.A., and Asbury, D.M.: 1978, Phys. Lett. 74B, 27.

36. Wycech, S. Preprint.

DISCUSSION

- G.C. Oades :

How many channels would the 2p-1s hydrogen peak shift and how much wider would it become if you were to input A. Martin's scattering lengths into the simple formula?

- C. Batty :

The peak would shift by about 3 channels whilst the width of the peak would change from 3 channels to 6 channels. These effects would be easily seen.

- P. Gensini :

I am puzzled by the use of the simple linear formula to extract the K^-p scattering length from the complex energy shift. Since the works by Kroll and Seki (to quote only those I recall best) it is known that strong absorption gives rise both to effective repulsion and highly non-linear effects in the (relativistic) wave equation describing the one-particle problem. Since as the wave equation is concerned this problem for K^-p system is not different from the one for $K^- - {}^{32}S$ or $K^- - {}^{12}C$ systems (once radii are properly scaled down), I cannot see any reason why we should not expect the same to happen here, as for the heavier nuclei. Thus the use of formulae suitable only for small absorption would be completely unwarranted here. I believe somebody ought to make some Klein Gordon equation evaluation, similar to Seki's ones, for this case as well.

- C. Batty :

The work of Kroll and Seki is related to two particular problems which arise for kaonic atoms formed in nuclear targets. Firstly they showed that the use of simple perturbation theory to calculate shifts and widths is invalid and secondly that an effective scattering length and not the free scattering length must be used in the optical model discussed earlier. The simple equation relating the complex K^-p scattering length a_c to the measured shift and width is based on an expansion in powers of $(\mu\alpha\, a_c)$ and, as the work of Trueman has shown, the next term in the expansion is small but not negligible. However, further work on the problem of relating the measured shift and width to the $\bar{K}N$ scattering length would certainly be worthwhile.

- A.D. Martin :

Interpretation of the K^- hydrogen mesic atom results

The observed K^-p branching ratios at rest give constraints which should be respected in the interpretation of the K^- hydrogen mesic atoms results of Batty et al. (1). Two of the three observed at-rest ratios (2)

$$\frac{\Lambda\pi}{\Sigma^0\pi + \Lambda\pi} \sim 0.2 \qquad \frac{\Sigma^+\pi^- + \Sigma^-\pi^+}{\Sigma^0\pi^0 + \Lambda\pi} \sim 1.9 \tag{1}$$

can be solved to give directly

$$\frac{\sigma_0}{\sigma_1} = \frac{b_0}{b_1} \left| \frac{(1 + |k_0|\, A_1)}{(1 + |k_0|\, A_0)} \right|^2 \sim 4 \tag{2}$$

where σ_I are the $\bar{K}N \to \pi Y$ absorption cross-sections and $A_I = a_I + ib_I$ are the scattering lengths for isospin $I = 0, 1$; k_0 is the c.m. momentum in the $\bar{K}^0n$ channel, and so $|k_0| \sim 0.3\ \mathrm{fm}^{-1}$ at the K^-p threshold.

The present indications are that the K^- hydrogen atom experiment (1) requires a complex scattering length $A(K^-p) \sim 0$. If we take $b_0 \sim b_1$, then the most physically reasonable solution of this result together with the result of eq. (2), is

$$a_0 \sim -a_1 \sim -1\ \mathrm{fm}$$

How sensitive is this estimate to the assumption $b_0 \sim b_1$? The preliminary mesic data also imply small values of b_I at threshold, whereas the K^-p scattering data require $b_0 \sim b_1 \sim 0.7$ fm. Due to the $\Lambda(1405)$ it might be conceivable to have a strong energy dependence in the $I = 0$ channel such that b_0 is smaller at threshold. From eq. (2) this would require even larger values of $|a_I|$.

As indicated by this simplified discussion, in practice it is not found possible to simultaneously describe the low energy K^-p scattering data, the at-rest ratios, and the mesic atom result $A(K^-p) \sim 0$.

The above simple discussion also suggests that modifications of the K^-p Coulomb corrections (3) are not going to resolve the uncompatibility, since the corrections should cancel in the at-rest ratios of eq. (1).

1. Batty, C.J., these Proceedings.

2. Miller, D., these Proceedings

3. References given in the contribution of C.J. Batty.

THE Λ^o AND Σ^o PRODUCTION RATIOS FOR $\Sigma^- p$ AND $K^- p$ INTERACTIONS AT REST

M. Goossens* and G. Wilquet
Inter-University Institute for High Energies (ULB-VUB), Brussels

J.L. Armstrong** and J.H. Bartley
Department of Physics and Astronomy - University College London

1. The Experimental Sample

The 1.5 m British Bubble Chamber, equipped with a plexiglass Track Sensitive Target (TST) was exposed to a 600 MeV/c negative kaon beam from the NIMROD accelerator. Directing the kaons through a 30 cm aluminum degrade installed inside the hydrogen filled TST brought the kaon momentum down to approximately 250 MeV/c at the beginning of the fiducial volume, so that most of the kaons came to rest near the middle of the TST. The TST was surrounded by a Ne $-H_2$ mixture to increase the γ conversion probability. About 25,000 pictures were double scanned for V^o configurations, which could be due to one of the following processes (neglecting 3 particle final states)

$$K^- p \rightarrow \Lambda \pi^o \qquad (1)$$

$$K^- p \rightarrow \Sigma^o \pi^o \qquad (2)$$

$$K^- p \rightarrow \bar{K}^o n \qquad (3)$$

$$K^- p \rightarrow \Sigma^- \pi^+$$

$$\quad \vert\!\rightarrow \Sigma^- p \rightarrow \Sigma^o n \qquad (4)$$

$$\quad \vert\!\rightarrow \Sigma^- p \rightarrow \Lambda n \qquad (5)$$

(The presence or otherwise of a converted γ-ray from Σ^o or π^o decay was also noted, though in the present study no direct use is made of this γ-ray information).

Preliminary results for channel (3), which at our energies and in our TST can be clearly separated from the other processes, have already been reported elsewhere and final results have been published on the ratio of Σ^+ to Σ^- production and on the charged to neutral hyperon production ratio for K^- at rest (1). This paper is only concerned with

E. Ferrari and G. Violini (eds.), Low and Intermediate Energy Kaon-Nucleon Physics, 243–249.

processes (1), (2), (4) and (5) for K^- or Σ^- interacting at rest.

Events were only retained for measurement when there was a clear separation between the K^- (Σ^-) production point and the Λ hyperon decay vertex. After measurement events for which the kaon or sigma particles entered a plexiglass wall were eliminated from the sample. All remaining candidates were geometrically and kinematically reconstructed. A fiducial volume guarenteeing a minimal distance of 1 cm to the nearest plexiglass wall was introduced for the primary and decay vertices to reduce the effect of the presence of the walls.

When an event failed in the reconstruction program, its characteristics were scrutinized on a scantable and recorded. It was found that failures were almost exclusively due to short Λ decay paths, short Λ decay proton ranges, large Λ decay pion dips or large Λ decay opening angles. Also events for which the Λ decay proton entered a plexiglass wall were poorly reconstructed, since the energy loss of the proton inside the perspex, and hence the proton momentum, is largely unknown. Furthermore the typical range of the proton being about 10 cm, its momentum cannot reliably be determined from its curvature in the 1.2 Tesla magnetic field.

The influence of all these geometrical effects will be studied below.

2. Analysis

The kinematical ranges for the Λ hyperon kinetic energy T in processes (1), (2), (4) and (5) are given in figure 1.

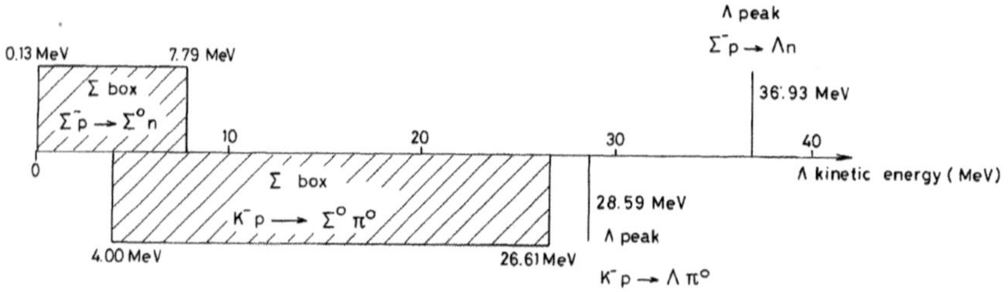

Figure 1 - Definition of the Λ kinematical region for the Λ production processes (1), (2), (4) and (5)

However, due to the finite experimental resolution of the reconstructed particle parameters and to the biases described above, the experimental T_Λ distributions for the processes (1) and (2) are not clearly separated (figure 2). To study the way the experimental uncertainties influence the production ratios S and R_N, a Monte Carlo simulation program was developed, which tracked every particle through the composite bubble chamber (2).

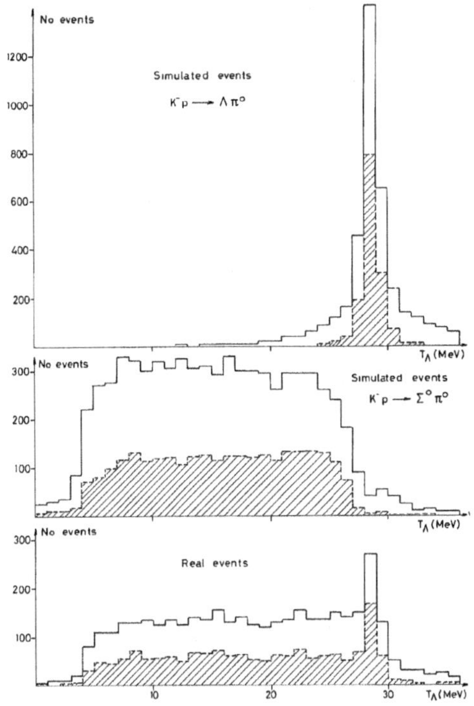

Figure 2 - Λ kinetic energy plots for the R_N ratio sample for simulated and real events, before and after (hatched part) the application of the cuts described in the text

To test the Monte Carlo program the S-ratio sample was used (processes (4) and (5)) since the kinematical separation between the directly and indirectly produced Λ hyperons is, in this case, trivial. A suitable number of Σ^- interaction events of types (4) and (5) was generated and, from the knowledge gained on the scantable looking at problematic events, cuts were introduced on the real and simulated samples. These cuts correspond to values where it was accepted that the geometrical reconstruction program becomes almost 100% efficient, so that the geometrical losses from the experimental sample are negligible.

In table 1 the cuts introduced are defined, while in table 2 and in figure 3 their effect on the sample can be found.

Table 1 - Definition of the cut values

Cuts		Fraction of events eliminated (*)	
		Σ box	Λ peak
Λ opening angle	160°	.36	.07
Λ decay path	2 mm	.40	.13
Λ proton length	2 mm	.26	.0
Λ pion dip	60°	.12	.13
Fiducial volume cut	present	.09	.10
Λ proton in perspex	eliminate	.05	.56
Fraction left after all cuts		0.22	0.29

(*) To give some indication of the influence of the introduction of the cuts on the studied samples, we quote in the last two columns the fraction of events left after each cut for the simulated events for processes (4) and (5), respectively labelled "Σ box" and "Λ peak"

Table 2 - Results for the Σ^- interaction sample

Number of Σ^- interaction events			
	Before cuts	After cuts	Corrected by Monte Carlo simulation
Σ box	462	106	476 ± 37
Λ peak	558	196	667 ± 29

From the figures in table 2 the Σ^- production ratio can be determined to be

$$S = \frac{\sigma(\Sigma^- p \rightarrow \Lambda n)}{\sigma(\Sigma^- p \rightarrow \Lambda n) + \sigma(\Sigma^- p \rightarrow \Sigma^o n)} = 0.58 \pm 0.06$$

This value is stable against variations in the cuts and compares well with two previous determinations, namely S = 0.56 ± 0.03 and S = 0.61 ± 0.03 (3).

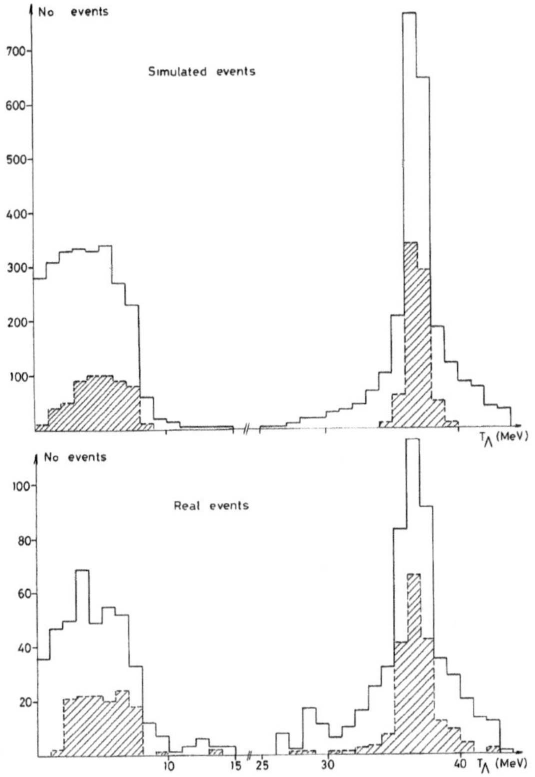

Figure 3 - Λ kinetic energy plots analogous to the ones of figure 2 but for the Σ^- interaction sample

Then the same Monte Carlo program was used to study reactions (1) and (2). Because the experimental T_Λ plot (figure 2, last part) does not show a clear separation between the directly and indirectly produced Λ kinetic energy regions, we developed a maximum likelihood method to extract the R_N ratio. Again a large number of events of types (1) and (2) were generated and the cuts mentioned in table 1 were introduced on the real and simulated samples. This yielded the numbers quoted in table 3 (see also figure 2).

Table 3 - Results for the R_N ratio sample

	Number of events before cuts	T_Λ < 25 MeV	25 < T_Λ < 26	26 < T_Λ < 27	27 < T_Λ < 28	T_Λ > 28 MeV
Real events	3925	1166	60	48	66	248
Simulation						
process (1)	4100	42	17	41	193	1186
process (2)	7140	2469	111	72	16	18

The simulated T_Λ plot was divided into 5 regions and the probabilities for a Λ^o from processes 1 or 2 being found in a particular region i (p_i^1, p_i^2) were calculated. The likelihood of detecting N_i events in kinetic energy region i is therefore

$$L(R_N) = \prod_{i=1}^{5} \left(\frac{p_i}{\sum_{i=1}^{5} p_i} \right)^{N_i}$$

where $p_i = R_N p_i^1 + (1-R_N)\, p_i^2$.

Using the experimental number N_i of events in the ith bin (cf. table 3), and varying the ratio R_N to maximise L, the maximum-likelihood value of R_N is determined.

Checks on the Monte Carlo simulation indicate possible systematic errors of the same order as the statistical error obtained from the likelihood analysis. For this reason the error has been increased by 50% to give a final result of

$$\underline{R_N = 0.206 \pm 0.015}$$

This is to be compared with the two previously published results which gave $R_N = 0.192 \pm 0.015$ (4) and $R_N = 0.189 \pm 0.017$ (5).

* Navorser I.I.K.W. Present address: E.P. Division, CERN, Geneva.

** SRC Research Student.

References

1. Bedford, N.H., Evans, E., Major, J.V., Goossens, M., Wilquet, G., Bartley, J.H., Davis, D.H., Tymieniecka, T., Nowak,R.J., Rondio, E., and Zakrzewski, J.A.: 1979, 'Charged Σ and K^o production in K^-p Collisions below 250 MeV/c'. Data presented to the Conference on Hyperfragments and K^- physics, Jablonna, Warsaw 11-15 September;(to be published in Nukleonika).

   Nowak, R.J., Armstrong, J., Davis, D.H., Miller, D.J., Tovee, D.N., Bertrand, D., Goossens, M., Vanhomwegen, G., Wilquet, G., Abdullah, M., Bedford, N.H., Ciborowski, J., Major, J.W., Zakrzewski, J.A., Kielcewska, D., Tymieniecka, T., and Rondio, E. : 1978, ' The Ratios $(\Sigma^-\pi^+/\Sigma^+\pi^-)$ and (Charged/Neutral) from K^-p Interactions at Rest '. Nucl. Phys. B139, 61.

2. Goossens, M.: 1978, Thesis, Brussels University, unpublished.

3. Dosch, H.G., et al.: 1964, Proceedings of the 12th International Conference on High Energy Physics, Dubna, 679.

4. Kim, J.K. : 1966, Thesis Columbia, Nevis 149.

5. Humphrey, W.E.: 1961, Thesis, University of California, UCRL 9752.
Humphrey, W.E., and Ross, R.R.: 1962, Phys. Rev. 127, 1305.

SLOW AND STOPPING K^-p INTERACTIONS IN A BUBBLE CHAMBER

D.J. Miller
Department of Physics and Astronomy - University College London

R.J. Nowak and T. Tymieniecka
Institute of Physics - University of Warsaw

Presented by D.J. Miller

1. Status of Experiment

The British 1.5 m bubble-chamber, with a hydrogen-filled Track-Sensitive-Target (TST) surrounded by a neon-hydrogen mixture, was exposed in 1973 to a K^- beam of between 250 and 580 MeV/c. 900,000 pictures were taken and the analysis is still continuing in two collaborations; the "TST collaboration" (whose progress is reported here) Brussels, Durham, UCL, Warsaw (with earlier participation from Birmingham) below 450 MeV/c and a Durham, Rutherford collaboration above ∿ 450 MeV/c.

Results have been published so far on the at-rest ratios ($\Sigma^-\pi^+$ to $\Sigma^+\pi^-$) and (charged to neutral) (1, 2). Preliminary results were reported in 1979 (3) on the $\Sigma^+\pi^-$, $\Sigma^-\pi^+$ and $\bar{K}^0n$ production cross-sections up to 250 MeV/c. A further sample of film is now being scanned for $\bar{K}^0n$ below 250 MeV/c, since the preliminary results show a considerably higher cross-section than had been observed in previous experiments (see below for further discussion). New results are also reported at this meeting on the ratio of ($\Lambda^0\pi^0$ to $\Sigma^0\pi^0$) from K^-p and also from Σ^-p interactions at rest (4). The other significant analysis still to be completed in the sample below 450 MeV/c is the determination of the $\Lambda^0\pi^0$ and $\Sigma^0\pi^0$ cross-sections upwards from ∿ 150 MeV/c. Unpublished theses (5) also contain values for charged and neutral hyperon production cross-sections between 200 and 450 MeV/c. The statistics in these theses are smaller than those of the Berkeley group but the cross-sections are in good agreement with those from Berkeley (6).

2. Discussion of Results

Forward Dispersion Relations have been used by A. Martin and many others to correlate parametrisations of the low energy K^-p with KN and

E. Ferrari and G. Violini (eds.), Low and Intermediate Energy Kaon-Nucleon Physics, 251–257.

$\bar{K}N$ elastic scattering data (7). The results suggest that the older low energy K^-p data (8) used in the effective-range or zero-range parametrisations (9) were inadequate and possibly biased (10). That is why the TST collaboration has attempted to repeat or improve the measurements in the hyperon production channels below 250 MeV/c. The chief concern in this analysis has been to understand the systematic errors and biases which arise in even so "ideal" a detector as a bubble chamber.

Table 1 shows the latest values for the at-rest K^-p production ratios. It will be noted that the value of the γ ratio from an earlier experiment disagrees by many standard deviations with the emulsion and the TST collaboration's values. Because of the brevity of the account given in the original paper it has been impossible to understand exactly how this early result was calculated but a Monte Carlo study has shown that biased corrections, after cutting events with short Σ tracks, could give rise to just such a reduction in γ as there appears to have been.

Table 1 - K^-p at-rest branching ratios; comparison of results

Experiment	$R_C = \frac{\Sigma^+\pi^- + \Sigma^-\pi^+}{\Lambda^o\pi^o + \Sigma^o\pi^o + \Sigma^\pm\pi^\mp}$	$R_N = \frac{\Lambda^o\pi^o}{\Lambda^o\pi^o + \Sigma^o\pi^o}$	$\gamma = \frac{\Sigma^-\pi^+}{\Sigma^+\pi^-}$
H_2 Bubble Chamber Humphrey & Ross (8)	0.655(± 0.011)	0.186(± 0.017)	
H_2 Bubble Chamber Kim (8)		0.192(± 0.015)	2.06(± 0.06)
Emulsion Tovee et al. (11)			2.34(± 0.08)
T.S.T. Collaboration (1) Nowak et al. & Marx et al.	0.664(± 0.011)		2.38(± 0.03)
T.S.T. Collaboration Goossens et al. (4)		0.206(± 0.015)	

Similar problems of brevity in the original papers make it impossible to be sure where the disagreement arises between the earlier workers (12) and the TST collaboration (3) on the $K^-p \rightarrow \bar{K}^on$ cross-sections. But the techniques used are very similar so one can expect that many of the biases must be the same. In particular, the TST collaboration has found a clear and significant loss of events due to scanners missing the decay of slow $\bar{K}^o \rightarrow \pi^+\pi^-$ because the pion tracks are almost collinear and look like a single continuous track. No mention is made of such an effect in either of the two earlier accounts, and it was even assumed that any losses could be taken as independent of the K^- beam momentum.

Figure 1 shows a comparison between the preliminary TST cross-sections for $\bar{K}^0n$ production and those from the earlier experiments. The heavy bars represent "floor" values for the TST cross-sections, based only on the observed numbers of events corrected for scanning efficiency and for $\bar{K}^0$ decay but not corrected for the more subtle geometrical biases which have been proved to be present and which must have reduced the total number of events. It is therefore hard to escape the conclusion that the cross-sections must rise, and the Forward Dispersion Relation fits (7) seem to confirm this.

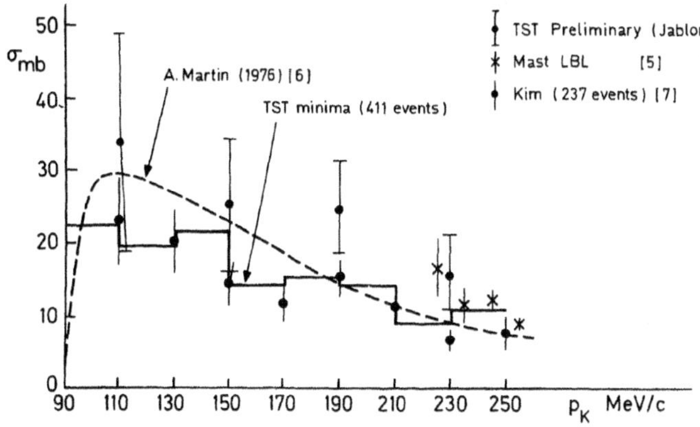

Figure 1 - $K^-p \rightarrow \bar{K}^0n$ cross-sections

It is reassuring that the $\Sigma^+\pi^-$ and $\Sigma^-\pi^+$ cross-sections determined at the same time as the $\bar{K}^0n$ cross-sections (3) are in good agreement with previous workers, so the normalisation is probably correct.

The new result (4) on the R_N ratio (table 1) comes from the only part of the TST analysis where the presence of the plexiglass target has caused significant difficulties. The kinetic energy of the Λ hyperon is used to separate the direct Λ^0 from events with $\Sigma^0 \rightarrow \Lambda\gamma$ but the Λ energy is largely determined by the length of its decay proton. Many of these protons stop in the plexiglass target wall where their precise range cannot be measured. Corrections for this effect had to be made by Monte Carlo and the authors found differences between different programs which led them to increase the errors assigned to the final result. All of their different techniques give values of R_N higher than the two previous experiments but the final shift is only around one standard deviation.

3. Limitations on Low Momentum Experiments

3.1 Beams

Low momentum K^- beams come in two types; bubble-chamber beams with low flux ($\sim$ 20 K^-/burst) and high purity ($\sim$ 90% K^-), or counter beams with much higher flux ($\sim 10^4$-10^5 K/burst) and lower purity (K: $\pi \lesssim 1:3$). For both types of beam it is necessary to balance momentum with separation. Good separation means long separators, giving many decays, and is therefore difficult at low momentum. For experiments below about 550 MeV/c most groups have transported the beam at 550 to 600 MeV/c and then degraded. But energy-loss in a degrader magnifies the momentum-spread in the beam so the experiment has poor momentum resolution unless a precise measurement can be made after the degrader. There have been one or two important exceptions to the general rule. For instance, the K65 LBL experiment (6) used a special high-field small-gap separator which enabled them to transport their beam at around 400 MeV/c, which explains their good resolution around the $\Lambda(1520)$.

3.2 Hydrogen Bubble Chamber

All of the experiments (1, 2, 3, 4, 7, 11) have the same target material of fixed density ≃ 0.06. The residual ranges for low momentum kaons fall-off rapidly as the momentum is reduced (table 2)

Table 2

Momentum	K^- range
250 MeV/c	40 cm
200 MeV/c	19 cm
150 MeV/c	6.5 cm
100 MeV/c	1.5 cm

So a stopping K^- has much more track length at around 200 MeV/c than around 100 MeV/c, and the low energy increase in cross-sections does not compensate much. Figure 2 shows the path-length distribution for part of the TST collaboration data.

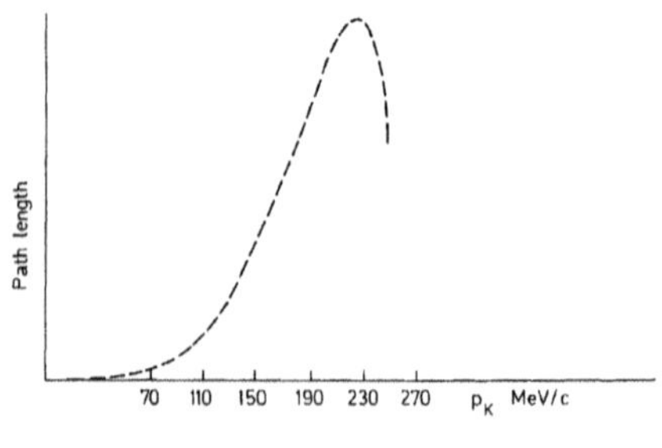

Figure 2 - Path-length distributions for stopping K^- in hydrogen

Since only $\sim$ 15% of all the K^- interact in flight one should imagine a δ function at $p_K = 0$ with approximately seven times the area of the graph. Note that only ∿ 0.2% of K would interact below 80 MeV/c and only 1% below 120 MeV/c, assuming constant cross-sections.

Measurement errors on the incoming K^- beam track after the degrader give around ± 10% accuracy at the mid point, but when this is extrapolated to the end-point of a track the error grows rapidly. Figure 3 is a sketch of the end-point momentum obtained from the beam-track measurements of a typical mixture of in-flight and stopping interactions.

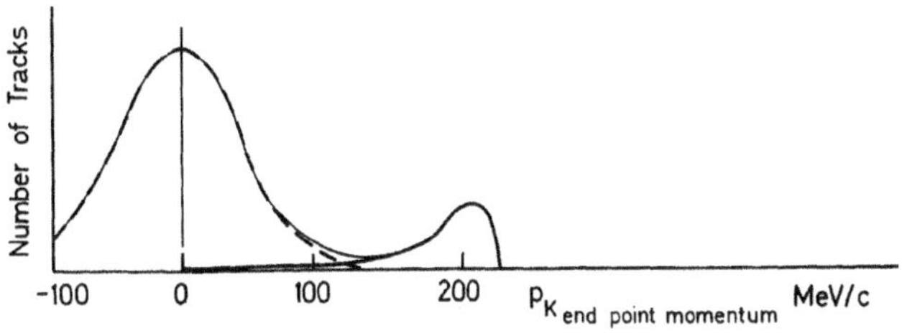

Figure 3 - Sketch of the measured distribution of the end-point momenta

Notice that the δ function at the origin has spread out into a wide distribution which completely swamps the genuine in-flight interactions below ∿ 130 MeV/c. Thus it is only possible to measure cross-sections below this momentum for channels where a kinematic fit can be made. Some channels, such as $\Lambda\pi^o$ or $\Sigma^o\pi^o$, will always have ambiguous fits. Other channels, where fits can be made, need angular cuts to define clear fitable categories of events, and these cuts will automatically produce biases which need to be studied and carefully corrected. Such biases must be present in all low energy bubble chamber experiments.

3.3 Possibilities for Improvement

Higher magnetic fields might improve measurements on the beam track momentum, but not much would be gained and it would be difficult to get the low momentum kaons into a high-field bubble-chamber. There is not much scope for a better conventional bubble-chamber exposure. Some channels would benefit from higher statistics, and there is plenty of film already taken, but the main limitation on statistics is the time taken by physicists to recheck failed events and to accumulate a reliable sample.

In the long term it is possible to suggest that higher beam fluxes and new experimental techniques (H_2 Time Projection Chamber with low concentration of quenching-gas ?) might allow precise experiments around the $\bar{K}^on$ cusp. But the fraction of interactions in this region will still be less than 1% and there would be extra K^- decay background in a gas target. Such experiments may be feasible in 10 years time but they will be expensive and would need a strong theoretical motivation.

References and notes

1. Nowak, R.J., Armstrong, J.L., Davis, D.H., Miller, D.J., Tovee, D.N., Bertrand, D., Goossens, M., Vanhomwegen, G., Wilquet, G., Abdullah, M., Bedford, N.H., Citcrowski, J., Major, J.W., Zakrzewski, J.A., Kielcewska, D., Tymieniecka, T., and Rondio, E. : 1978, Nucl. Phys. B139, 61.

   Marx, R., Natorf, W., and Rondio, E.: 1979, Warsaw University, Institute of Experimental Physics report.

2. Note that the Warsaw group has added another 50% to the statistics for this analysis.

3. Bedford, N.H., Evans, D., Major, J.V., Goossens, M., Wilquet, G., Bartley, J.H., Davis, D.H., Miller, D.J., Tymieniecka, T., Ciborowski, J., Kielcewska, D., Nowak, R.J., Rondio, E., and Zakrzewski, J.A.: 1979, Report presented to the Conference on Hyperfragment and K^- Physics, Jablonna, Warsaw, 11-15 September (to be published in Nukleonika).

4. Goossens, M., Wilquet, G., Armstrong, J.L., Bartley, J.H.; Report submitted to this Workshop.

5. Durham University theses by Fallahi, M.T., and Haman, Y.A. (unpublished).

6. Mast, T.S., Alston-Garnjost, M., Bangerter, R.O., Barbaro-Galtieri, A.S., Solmitz, F.T., and Tripp, R.D.: 1975, Phys. Rev. D11, 3078;

   1976: Phys. Rev. D14, 13.

7. Martin, A.D.: 1977, Nukleonika 22, 857.

   Martin, A.D.: 1976, Phys. Lett. 65B, 346.

8. Humphrey, W.E., and Ross, R.R.: 1962, Phys. Rev. 137, 1305.

   Sakitt, M., Day, T.B., Glasser, R.G., Seeman, N., Friedman, J., Humphrey, W.E., and Ross, R.R.: 1965, Phys. Rev. 139, B719.

   Abrams, G.S., and Sechi-Zorn, B.: 1965, Phys. Rev. 139, B454.

   Kim, J.K.: 1966, Columbia University report, NEVIS 149.

9. Martin, B.R.C., and Sakitt, M.: 1969, Phys. Rev., 183, 1345.

   Martin, A.D., and Ross, G.G.: 1970, Nucl. Phys. B16, 479.

   Kim, J.K. : 1967, Phys. Rev. Lett. 19, 1074.

   Berley, D., Yamin, S.O., Kofler, R.R., Mann, A., Meisner, G.W., Yamamoto, S.S., Thompson, J., and Willis, W.: 1970, Phys. Rev. D1, 1966.

10. Miller, D.J.: 1979, Invited talk at the Conference on Hyperfragment and K^- Physics, Jablonna, Warsaw, 11-15 September; (to be published in Nukleonika).

11. Tovee, D., Davis, D.H., Simonovic, J., Bohm, G., Klabuhn, J.,Wysotski, F., Csejthey-Barth, M., Wickens, J.H., Cantwell, T., NiGhogain, C., Montwill, A., Garbowika-Priewska, K., Priewski, J., Zakrzewski, J.: 1971, Nucl. Phys. B33, 493.

12. Kittel, W., Otter, G., and Wacek, I.: 1966, Phys. Lett. 21, 349.

Also Kim ref. 8.

DISCUSSION

- G. Violini :

Perhaps you remember that at Jablonna Sakitt was suggesting that in order to improve our knowledge of $\bar{K}N$ K-matrix it could be better, instead of increasing statistics, that one looked at some channel like $\Sigma^+\pi^-$, which does not have too flat cross-section as a function of the momenta; actually in that case there is a sort of cusp around 100 MeV/c. Do you have any comment on this proposal ?

- D.J. Miller :

I do not recollect Sakitt's remark and, on reflection, I do not agree with it. With the bubble chamber technique one difficulty is that the error on the beam momentum becomes large at the very lowest momenta, below 100 MeV/c. It is therefore difficult to allocate events to momentum bins without some diffusion into adjacent bins, and this will be especially marked for a rapidly varying cross-section. The channels which I wish to see better measured, and I believe Alan Martin agrees, are the $\Lambda^o\pi^o$ and $\Sigma^o\pi^o$ between 140 and 250 MeV/c.
They are pure I-spin channels, the present fit does not agree very well with the high-precision Berkeley points just below 260 MeV/c, and the errors on the lowest momentum points are enormous. The TST collaboration is working on these channels now.

$\Lambda(1405)$, A CDD POLE, AND THE $\bar{K}N$ INTERACTION AT LOW ENERGIES

K.S. Kumar
Department of Physics - McMaster University, Hamilton

What is described below is a model for the $\bar{K}N$-interaction at low energies. By low energies is meant the energies for which the lab. momentum of the $\bar{K}$ meson is less than about 300 MeV/c, i.e. the energy region in which the $\bar{K}N$ scattering is predominantly S-wave. The model differs from the more conventional ones in the manner in which it treats $\Lambda(1405)$.

Any model for the low energy $\bar{K}N$-interaction must, of course, account for the existence of $\Lambda(1405)$. To be able to do so, we make some assumption, usually tacit, regarding the nature of $\Lambda(1405)$. And the model we construct will depend on this assumption; in other words, the model will depend on how we regard $\Lambda(1405)$. We can of course regard it in one of two ways: we can think of it as a composite particle, an unstable bound state of $\bar{K}$ and N, in which case it is the dynamical consequence of the interaction between $\bar{K}$ and N; or, we can think of it as an "elementary" particle. (Elementary, in the pre-quark sense of the word). Usually we take the former view and regard $\Lambda(1405)$ as an unstable bound state of $\bar{K}$ and N. From this point of view, the existence of $\Lambda(1405)$ is dependent on the interaction between $\bar{K}$ and N; if, say, the interaction did not exist, $\Lambda(1405)$ would not exist. In contrast to this, looking upon it as an "elementary" particle, we can consider $\Lambda(1405)$ from the point of view of the quark model. $\Lambda(1405)$ of course, fits in perfectly well with the quark model. Simply by looking at the different possible combinations of the quarks, wa can predict the existence of a state corresponding to $\Lambda(1405)$. The point here is that the existence of a state having the quantum numbers of $\Lambda(1405)$ is quite independent of the interaction between $\bar{K}$ and N. This would provide an argument against treating $\Lambda(1405)$ simply as a composite $\bar{K}$-N system.

These two points of view (1) - the composite and the quark model - are of course two extremes. In reality one would expect both the direct interaction between $\bar{K}$ and N, <u>and</u> the elementary particle state to contribute to the physically observed state of $\Lambda(1405)$. However, depending on the masses and coupling constants involved, one of these contributions may be negligible. One could think of the conventional view of $\Lambda(1405)$

E. Ferrari and G. Violini (eds.), Low and Intermediate Energy Kaon-Nucleon Physics, 259–265.

as a composite particle as saying just this, that the elementary contribution to $\Lambda(1405)$ is too small to have any observable effects on low-energy $\bar{K}N$-scattering. Here, in contrast, we suppose that there is a significant elementary component, and examine the consequences of this for the $\bar{K}N$-system at low energies. The whole question, as we will show, assumes particular significance in view of the recently reported kaonic H-atom result (2).

Before turning to the model we construct for this purpose, we might mention that the question raised regarding the nature of $\Lambda(1405)$ can be raised for other particles as well - for example, for $\Delta(1236)$ and the πN-interaction. (In a recent paper (3) G. Miller et al. have studied the πN-interaction using an extended version of the Brown-Rho little bag model. What their work does, in essence, is include the elementary or 3 quark component of $\Delta(1236)$ in the πN-interaction). Thus, what we are dealing with here is a specific instance of a general question.

The model Hamiltonian we construct describes only the $I = 0$ channel, for it is in this channel that $\Lambda(1405)$ is formed. We first ignore the inelastic $\pi\Sigma$-channel, and take it into account later by making a coupling constant complex. We also treat the nucleon as if it were static, correcting later for recoil.

The model Hamiltonian H is given by

$$H = H_o + H_I$$

$$H_0 = m_N N^\dagger N + m_o \Lambda_o^\dagger \Lambda_o + \int d\underline{k}\; \omega_k a_k^\dagger a_k \qquad (1)$$

$$H_I = -G \iint d\underline{k} d\underline{k}' u_k u_{k'} a_k^\dagger a_{k'} + g \left\{ \Lambda_o^\dagger N \int d\underline{k}\; u_k a_k + \text{h.c.} \right\}, \qquad (2)$$

where m_o is the mass of Λ_o, μ the kaon mass, $\omega_k = (\mu^2+k^2)^{\frac{1}{2}}$, and $u_k = (2\pi)^{-3/2}(2\omega_k)^{-1/2} v_k$, v_k being the form factor of the interaction source. In accordance with the arguments outlined before, the free part of H contains a field Λ_o - the 'bare' $\Lambda(1405)$- in addition to the $\bar{K}$ and N fields. The interaction Hamiltonian consists of two parts, which are depicted in figure 1.

The first term in (2), corresponding to figure 1a, is a separable interaction for the direct interaction between $\bar{K}$ and N.

This term is meant to represent all contributions to the $\bar{K}N$-interaction arising from processes such as ρ, ω exchanges etc.

The second term in (2), corresponding to figure 1b, represents the coupling of the bare $\Lambda(1405)$ to the $\bar{K}N$-channel.

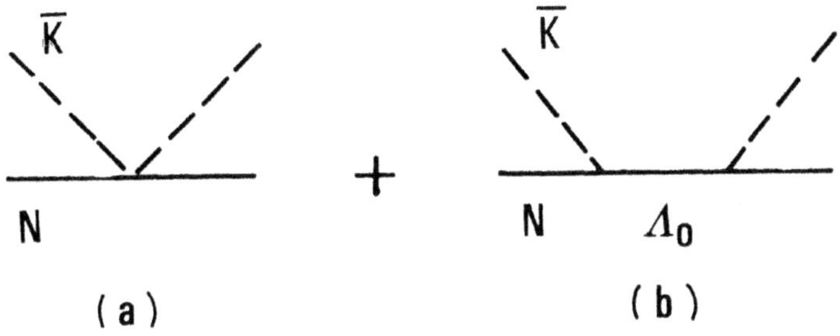

Figure 1 - Diagram of the interactions. Λ_o is the bare $\Lambda(1405)$

There is only S-wave scattering in this model, and it is straightforward to solve the Lippmann-Schwinger equation (4) and determine the scattering amplitude f :

$$f = \frac{e^{i\delta}\sin\delta}{k} = \frac{\lambda_k v_k^2}{1-\lambda_j J_k}, \tag{3}$$

where

$$4\pi\lambda_k = G + \frac{g^2}{\Delta-\omega_k}, \tag{4}$$

$$\Delta = m_o - m_N, \text{ and}$$

$$J_k = \frac{1}{\pi}\int_\mu^\infty \frac{p v_p^2 \, d\omega_p}{\omega_p - \omega_k - i\varepsilon} .$$

With this scattering amplitude we would wish to reproduce the following :

(1) the data on low-energy $\bar{K}$N scattering,

(2) the position and width of $\Lambda(1405)$,

and (3) the kaonic H-atom result.

We will assume that the kaonic H-atom result is sufficiently well-known (2), and only say that the result seems (5) to imply that both the real and imaginary parts of the scattering amplitude are small at threshold - in contradiction of results obtained from previous analyses of low-energy scattering data. Before showing how a small scattering amplitude such as this can be obtained at threshold with this model Hamiltonian, we need to make some general remarks.

If an elementary particle is coupled to a system in such a manner as to leave the discrete spectrum of the total Hamiltonian unchanged, and to preserve certain general features of the system - to be specific, so as to lead to the same Low equation as that of the original system - then, it has the effect of introducing a CDD pole - that is, a zero in the scattering amplitude. In our case, the elementary particle Λ_o is coupled to the $\bar{K}N$-system in just such a manner; hence, we have a CDD pole. The question is, where is this CDD pole located? It is into this question that the question of whether the elementary particle component has a significant effect at low energies resolves itself. As can easily be seen from (3), the position ω_{CDD} of the CDD pole is determined by

$$G + \frac{g^2}{\Delta - \omega_{CDD}} = 0$$

Bearing in mind the kaonic H-atom result, we <u>choose</u> this position of the CDD pole so that it is close to the threshold. This ensures that the scattering amplitude f will be small at threshold, as required. Such a choice also makes the behaviour of f at low energies very different from what has been usually assumed : both the real and imaginary parts of f vary very rapidly in the vicinity of the threshold. So do the K-matrix elements - in constrast to the usually assumed smoothly varying behaviour. Clearly, the effective range expansion will be suspect. In this connection it is interesting to note the work of Weinberg (6) in 1965 where he studied the possibility of there existing an elementary deuteron. The question is much the same as the one studied here for $\Lambda(1405)$. There, as here, a CDD pole appears, as does the question of its location. Weinberg argues that the CDD pole cannot be close to threshold since for the N-N system the effective range expansion works very well. Here the argument runs in the opposite direction : the kaonic H-atom result can be explained if the CDD pole is close to threshold, and this would make the effective range expansion suspect. This, of course, conflicts with the usual assumption of the validity of the effective range expansion.

Introducing a CDD pole close to threshold, though allowing one to explain the kaonic H-atom data, leads also to the following consequence which seems to be in conflict with the scattering data. It can be seen from (3) that if we wish to have a bound state at 1405 MeV, as well as a near-threshold CDD pole, then the real part of the scattering amplitude, Ref, is necessarily positive above the CDD pole. This would mean that Ref > 0 in the low energy region. However, study of Coulomb-nuclear interference seems to indicate that Ref < 0 in this region. Thus, at least in this one-channel treatment of the problem, there seems to be a contradiction.

We will, however, ignore this Coulomb-nuclear interference result, which in any case does not seem to be established beyond all doubt. Thus our assumption is that the sign of Ref above threshsold is not known with certainty, and hence that it could be positive, as required in our model. With this assumption we proceed to fit the data.

We first make G complex, $G = G_1 + iG_2$, to take account of the $\pi\Sigma$ channel. We correct for nucleon recoil by replacing $\bar{\omega}_k$ with $\omega_k = \omega_k + \frac{k^2}{2m_N}$, so that f becomes

$$f = \frac{m_N}{m_n+\omega_k} \quad \frac{\tilde{\lambda}_k v_k^2}{1-\tilde{\lambda}_k \tilde{J}_k} . \tag{5}$$

With this f we attempt to fit the data. We use Chao et al.'s K-matrix parametrization (7) to obtain the experimental scattering amplitude. Here we are faced with difficulties. Firstly, since the scattering amplitude obtained from this parametrization disagrees at threshold with the kaonic H-atom data, we clearly cannot fit it everywhere. We attempt to reproduce Chao et al.'s scattering amplitude f_c only at energies above about 10 MeV in the CM frame, justifying this by observing that the scattering data below this region, where it exists, is very uncertain (8). Secondly, Chao et al.'s K-matrix parametrization gives Ref < 0 above threshold, as do all previous parametrizations. Our model, as mentioned before, requires Ref > 0. We get around this by simply taking the scattering amplitude to be fitted above 10 MeV to be $-f_c$ (8). This procedure is admittedly open to objection.

Figure 2 shows our f for the following set of parameters (in units of μ): g = 0.465, G = 19.06 + 1.42i, Δ = 0.989, and k_c = 2.

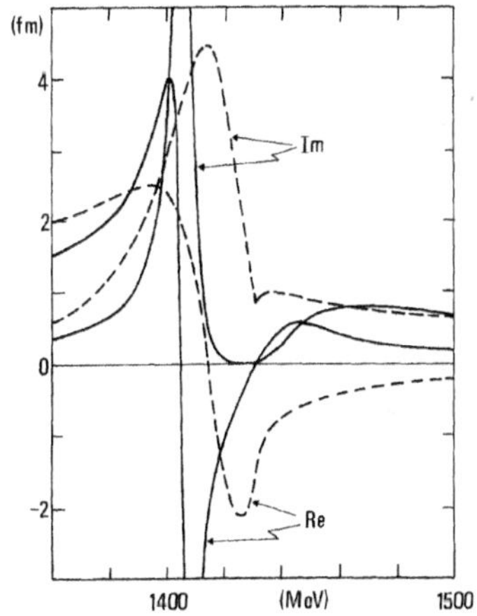

Figure 2 - The scattering amplitude f_0 in fm versus the total CM energy in MeV. The K^-p threshold energy is 1432 MeV. The solid lines represent our f. Imf = 8.6 fm at its peak, and Ref = -4.7 fm at its downward peak. The dashed lines are for Chao et al.'s solution B

Here k_c is a cutoff parameter in v_k which we have taken to be $v_k = \theta(k_c-k)$. In choosing the parameters we imposed the following conditions : (a) the CDD pole condition, $\tilde{f}(k=0) \approx 0$, (b) $\tilde{f} = -f_c$ at $k_{CM} = 0.3\ \mu$, $(\tilde{\omega}-\mu = 33$ MeV), and (c) $\mathrm{Re}(1-\tilde{\lambda}_k\tilde{J}_k) = 0$ at $m_N+\tilde{\omega} = 1405$ MeV. At the K^-p threshold we obtained $f = 0.001 + 0.029i$ (fm). Figure 2 also shows the f_c obtained from Chao et al.'s set B, and this is not very different, in magnitude, from f for $\tilde{\omega}-\mu \gtrsim 10$ MeV. At the threshold if we combine f with Chao et al.'s $a_1 = 0.08 + 0.69i$, we obtain $a(K^-p) = 0.04 + 0.36i$, which is consistent with the K^-p atom result. For the width of $\Lambda(1405)$ we obtain $\Gamma \approx 8$ MeV, which is much too small as compared with the experimental value of $\Gamma \approx 40$ MeV. The width, however, is related to the $\pi\Sigma$ channel, and one would not expect it to be well reproduced unless the $\pi\Sigma$ channel is treated properly. In an extension of the model described here, where we explicitly take account of the $\pi\Sigma$ channel in a coupled channel treatment, we in fact find that the width can be fitted very closely.

The model we have proposed is thus able to fit the K^-p atom results as well as the position of $\Lambda(1405)$. But it also requires that the real part of the scattering amplitude be positive in the low energy region, which is in conflict with what Coulomb nuclear interference results seem to indicate. How serious this conflict is, and whether it can be avoided in the coupled channel calculation we are currently carrying out, remains to be seen.

References and notes

1. Cline, D., et al.: 1971, Phys. Rev. Lett. 26, 194.
2. Davies, J.D. et al.: 1979, Phys. Lett. 83B, 55.

   Also, Batty, C.J., Paper presented at this meeting.
3. Miller, G., et al., Preprint.
4. Alternatively one can solve the Low equation and select the solution with one CDD pole.
5. We assume here that the Coulomb interference is negligible.
6. Weinberg, S.: 1965, Phys. Rev. 137, B672.
7. Chao, Y.A., et al.: 1973, Nucl. Phys. B56, 46.
8. Here we have an underlying assumption that $|\mathrm{Re}f_1| << |\mathrm{Re}f_0|$.

DISCUSSION

- G.C. Oades :

How is inelasticity included in your model and could you extend the model to also fit the $\bar{K}N - \pi\Sigma$ $(I = 0)$ data ?

- K.S. Kumar :

We take account of absorption into the inelastic $\pi\Sigma$ channel by making the separable interaction coupling constant complex. We are currently working on a straightforward extension of the model which does explicitly take account of the $\pi\Sigma$ channel in a coupled channel treatment of the problem. In this extended version we will certainly try to fit the $\bar{K}N - \pi\Sigma$ data.

- A.D. Martin :

There is strong experimental evidence that Re $f(I = 0)$ is negative at threshold. Your amplitudes will not describe the K^-p scattering data and 'at - rest' ratios.

SEARCH FOR STRANGE DIBARYON SYSTEMS (Rome-Saclay-Vanderbilt Collaboration)

G. Auriemma, G. D'Agostini, G. Marini, G. Martellotti, F. Massa, A. Nigro, A. Rambaldi, A. Sciubba
Istituto Nazionale di Fisica Nucleare - Sezione di Roma
Istituto di Fisica - Università di Roma

J.P. De Brion, A. Caillet, J.B. Cheze, J. Derré, G. Marel, E. Pauli, C. Pigot, G. Vrana
DPhPE, Centre d'Etudes Nucléaires de Saclay, Gif-sur-Yvette

J. Marraffino, C. Roos, J. Waters, W. Webster, E. Williams
Vanderbilt University, Nashville

A. Bezaguet, S. Reucroft, K. Schultze
CERN, Geneva

R. Llosa
JEN, Madrid

Presented by G. Marini

In this paper preliminary results of an experiment, which is being run at the CERN Proton Synchrotron (PS) since May 1979, are reported.

The experiment is designed to search for dibaryon systems with strangeness -2, produced in the reaction

$$K^- + d \rightarrow K^+ + MM \qquad (1)$$

at 1.4 GeV/c incident K^- momentum. The missing mass (MM) sensitivity ranges from 2.1 up to 2.4 GeV/c^2, the resolution being $\sim \pm$ 3.5 MeV/c^2.

Several indications on the possible existence of dibaryon systems have been reported from various experiments (1). But, in strange dibaryon systems, the only real indication comes from the results of bubble chamber experiments (2) which have analyzed the reaction $K^- + d \rightarrow \pi^- + (\Lambda p)$, at different K^- beam momenta (.75 and 1.5 GeV/c). A strong and narrow

E. Ferrari and G. Violini (eds.), Low and Intermediate Energy Kaon-Nucleon Physics, 267–275.

enhancement in the (Λp) invariant mass (at M (Λp) = 2.13 GeV/c^2, Γ = 5.4 MeV/c^2) is observed, close to the (ΣN) threshold. Model calculations, using the relativistic deuteron form factor and the phenomenological K-N amplitudes, predict the presence of a "kinematical" cusp at the (ΣN) threshold, due to the deuteron dynamics, even if the strong conversion $\Sigma N \rightarrow \Lambda p$ is smooth. This effect alone does not seem to be adequate to explain the experimental data (3) : the presence of a (ΣN) bound state, named H_1^+, $\sim$ 4 MeV/c^2 below the (ΣN) threshold is hypothesized, with a total cross section of $\sim$ 20-25 μb and a very low momentum transfer t, to the deuteron (t dependence $\sim e^{-29t}$). A tentative assignement of the J^P quantum numbers could be 1^+.

From a theoretical point of view, possible dibaryon states were suggested in 1963 by Oakes (4) in a $\overline{10}$ structure of SU(3), with the deuterium as the non strange component of the antidecuplet.

The interest in the existence of such states was recently strongly enhanced, after the advent of the Quantum-Chromo-Dynamics. In this framework there are in fact several theories, which are able to describe conventional hadrons in terms of confined coloured quarks and gluons, and which, at the same time, foresee a large number of multiquark states (5). As an example the MIT bag model, applied by Jaffe (6) to a six quark system in S-wave, predicts the existence of a stable $J^P = 0^+$ dibaryon flavour singlet at 2.15 GeV/c^2 and of a $J^P = 1^+$ dibaryon octet in the mass range 2.2 - 2.5 GeV/c^2.

It is interesting to notice how these kind of models do not account for weakly bound states of two baryons, like the deuteron. However, a recent theoretical paper (7) suggests that deuterium is, for a small fraction ($\sim$ 5%), a system of six directly interacting quarks, rather than a loosely bound state of two nucleons.

It is generally true that it is difficult to find a clear experimental distinction between a genuine six quark state and a weakly bound baryon-baryon system. The above mentioned H_1^+ enhancement at 2.13 GeV/c^2 in the (Λp) mass, could be interpreted as a I = ½ (ΣN) hypernuclear ground state, decaying, via strong interactions, into Λp (8). In fact the existence of a whole family of Λ hypernuclei is well known; and, although the ΛN interaction does not seem sufficient to bind a Λ to a single nucleon, our present level of knowledge does not exclude the possibility that other hyperons (like Σ or Ξ) join with a single nucleon and form a hypernuclear bound state.

Going back to the reaction (1), we have tried to estimate the order of magnitude of the cross section (σ) for the production of an H_2 (B = 2, S = -2) state (see figure 1a). Following the analysis of ref. (3), we have considered only a production mechanism through the triangular diagram of figure 1b. On the basis of the experimental value of the cross section for the H_1^+ production ($\sim$ 25 μb at 1.5 GeV/c) one can derive a rough estimate of the cross section for the production, via the hypothesized mechanism, of the H_2 state; the resoluting value of σ is of the order of 0.1 μb for an H_2 mass around 2.35 GeV/c^2, and it turns out to be strongly dependent on the mass of the produced H_2 state.

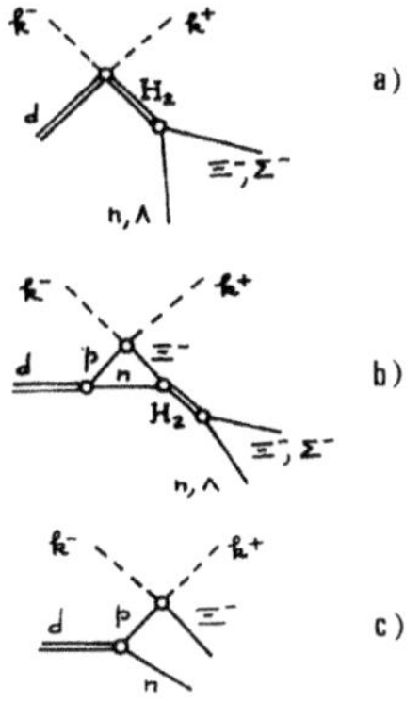

Figure 1 - Diagrams for the production of a dibaryon state Q = -1 S = -2, (a) and (b), and for the physical non resonant background, (c)

The non resonant background is mainly due to the reaction

$$K^- + p + (n) \to K^+ + \Xi^- + (n) \qquad (2)$$

with neutron spectator (figure 1c). The shape of the missing mass spectrum to the K^+, given by this reaction, can be evaluated by a Monte Carlo calculation.

The experimental apparatus consists basically of a double magnetic spectrometer (see figure 2) able to identify both the incident K^- and

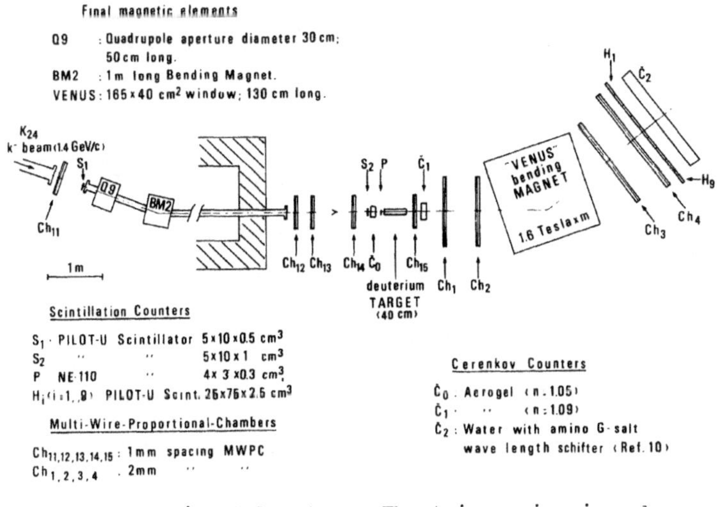

Figure 2 - The experimental set-up. The trigger is given by $S_1 \cdot S_2 \cdot \check{C}_0 \cdot P \cdot \check{C}_1 \cdot H_i$ $(m = i) \cdot \check{C}_2$ (m being the hodoscope counter multiplicity)

the outgoing K^+, and to measure their momenta. The K^- beam is separated from the large π^- contamination ($\sim 8/1$) by an aerogel Čerenkov counter $\check{C}_0$ ($n = 1.05$) and by a time-of-flight (TOF) measurement between two scintillation counters, S_1 and S_2. The hardware selection of the outgoing K^+ is achieved by two Čerenkov counters : $\check{C}_1$ (aerogel, $n = 1.09$) which is in anticoincidence with the produced pions, and $\check{C}_2$ (water) which cuts the slow protons. A clean sample of the produced K^+ is obtained off-line from a TOF measurement between the counter S_2 and one scintillation counter, H_i, in a big hodoscope. The momenta of the incoming and the outgoing particles are accurately measured by a system of nine Multi-Wire-Proportional-Chambers (MWPC).

In figure 3a it is seen that we can unambiguously separate the K^- beam from the pions, while, in figure 3b, the measured momentum vs the squared mass of the outgoing particles is plotted.

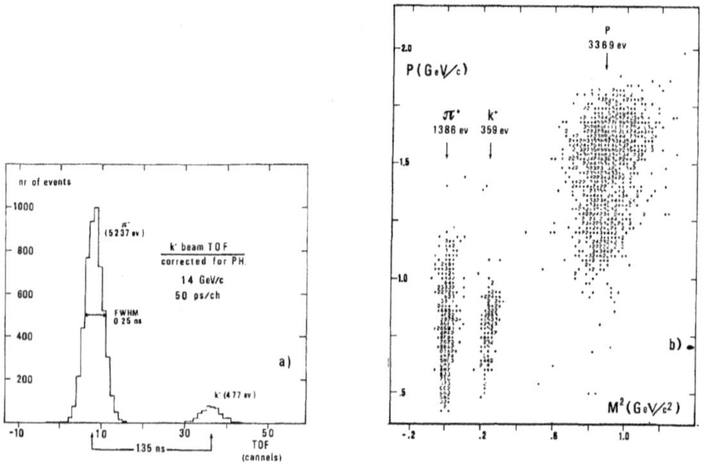

Figure 3 - a) Time-of-flight of the beam particle between S_1 ans S_2 (~ 7 m distance), showing the separation between π^- and K^- (the sample was collected without $\check{C}_0$ in the trigger); b) Plot of the momentum vs the squared mass of the outgoing trigger particle, as reconstructed from the momentum, the time-of-flight and the path length from S_2 to H_i (~ 5 m)

The squared mass is calculated from the momentum, the TOF, and the path length between counters S_2 and H_i. It is apparent that we can easily separate the produced K^+ events from the other reactions.

The acceptance of the experimental apparatus has been evaluated by a Monte Carlo calculation both for the expected background and for the H_2 dibaryon production.

For the H_2, the geometric acceptance is almost constant in the mass range 2.1 - 2.4 GeV/c^2, and its value (times the survival probability of the K^+) turns out to be of the order of .5%, if isotropic production in the center of mass is assumed. (The value of the acceptance increases strongly for processes with a low momentum transfer to the deuteron). For the physical background (reaction (2)) the average geometric acceptance, times the survival probability of the K^+, is increased to 1.13%, at 1.4 GeV/c.

To take into account instrumental inefficiencies and to check the Monte Carlo results, we have calibrated our apparatus by measuring some known reactions on an hydrogen target. The experimental missing mass resolution, as obtained from the calibration reaction $\pi^+ + p \to K^+ + \Sigma^+$, is shown in figure 4 : the measured standard deviation is ± 3.5 MeV/c^2, in very good agreement with the expected value. At the present stage of data analysis, the overall inefficiency is ∿ 50%.

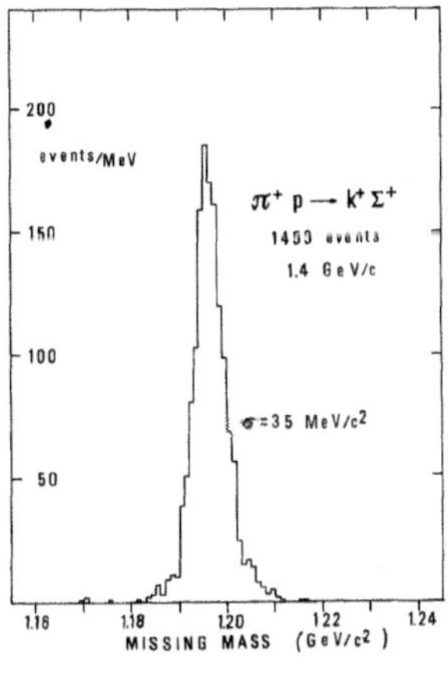

Figure 4 - Missing mass spectrum to the K^+ for the calibration reaction $\pi^+ + p \to K^+ + \Sigma^+$. At this preliminary stage of data analysis, the systematic error on the mass is of the order of 5 MeV/c^2

The features of the data taking for the reaction (1) are the following

- incident beam: ∿ 20.000 K^-/machine burst;

- rate of triggers on a 40 cm long deuterium target: ∿ 20 events/machine burst.

A fast and rather sophisticated on-line analysis is performed on a NORD-10 computer. This allows rejection of accidental triggers, bad tracks, and also a great fraction of the proton triggers, on the basis of a mass reconstruction of the outgoing trigger particle. The final rate of the "accepted events" is ∿ 1 events/machine burst.

A preliminary sample of the data collected on the reaction (1) is shown in figure 5. The 5311 plotted events correspond to an incident flux of 3.1 10^{10} K^-, i.e. a production of 60 events/nb on the target.

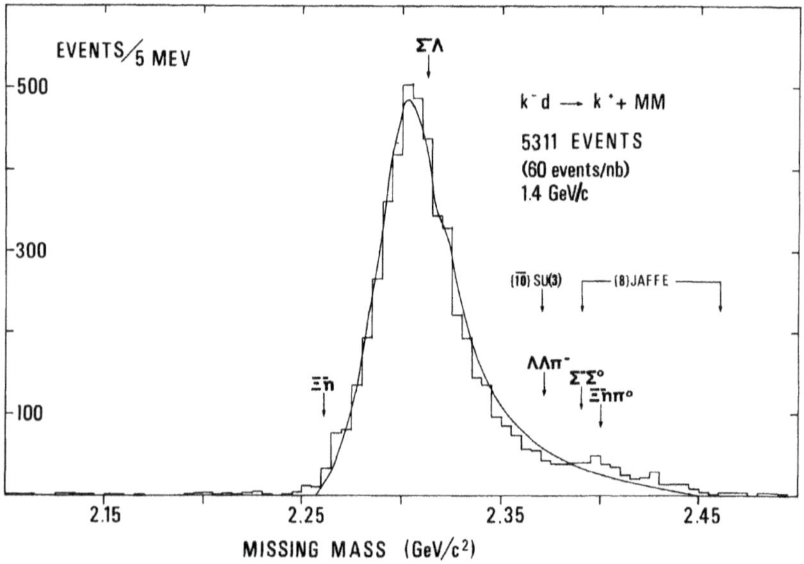

Figure 5 - Missing mass spectrum to the K^+, for the reaction $K^- + d \rightarrow K^+ + MM$. The solid curve is the result of a Monte Carlo calculation which refers to the Ξ^- production on the p, with a neutron spectator. The thresholds for various possible non resonant contributions are shown. Predictions from ref.(4) and (6) are also indicated

The solid line is a Monte Carlo prediction which refers only to the Ξ^- production on the p, with a neutron spectator (reaction (2)). This was computed taking into account the Fermi motion of the nucleons (according to ref. (9)) and the experimental acceptance of the apparatus, and assum ing a flat distribution in the forward direction of the outgoing K^+. The agreement with the experimental data is quite good, confirming that the contribution of the other possible channels is negligible.

Although the present status of the analysis is very preliminary, there is no clear evidence for the presence of narrow structures in the explored mass range, neither above nor below the (Ξ^- n) threshold. The few events below threshold are consistent with other possible physical mechanisms, like K^+ production through two charge exchanges and an intermediate regeneration $\bar{K}^o \rightarrow K^o$:

$$K^- + d \rightarrow \bar{K}^o + n + (n)$$

$$\downarrow \text{by regeneration}$$

$$K^o + d \rightarrow K^+ + n + (n)$$

To set an upper limit to the production cross section of a resonant state H_2, one should take into account the correct mass dependence of the acceptance, and the mass width, Γ, of the considered state. As an example, assuming a $\Gamma \leq 10$ MeV/c^2 and a value of the H_2 mass of 2.35 GeV/c^2, at a confidence level of 99.7%, we get an upper limit of $\frac{d\sigma}{d\Omega}$ $\sim$ 17 nb/sr (for the K^+ in the forward direction). More data are now being collecting on this reaction, and we expect to reach a much lower value ($\sim$ 8 nb/sr) for the upper limit of the cross section.

We have also collected data on the reaction

$$\pi^- + d \rightarrow K^+ + MM \qquad (3)$$

at 1.4 GeV/c incident π^- momentum. Figure 6 shows the Missing Mass spectrum for this process. The preliminary sample of 26 066 events corresponds to an incident beam of 3.17 10^{10} π^- (sensitivity at the production of 62 events/nb). The solid line is the result of a Monte Carlo calculation taking into account only the Σ^- production on the proton, with neutron spectator.

There is a general good agreement with the experimental data up to the opening of the pion threshols, $\Lambda\pi^-$ (n), whose production is expected to be significant.

A more accurate evaluation of the non resonant background is in progress. However, no resonant structures are visible in the spectrum neither above nor below the (Σ^- n) threshold.

As an example, in the region between 2.25 and 2.35 GeV/c^2 (geometric acceptance $\sim$ .57%) the same procedure used in the previous case results in a upper limit for an "H_1^-" production, of $\sim$ 30 nb/sr.

It has to be pointed out that there is no theoretical prediction for the existence of S = 1, Q = -1 dibaryon systems, either in the Oakes scheme or in the framework of the bag model. Vice-versa, a possible signal

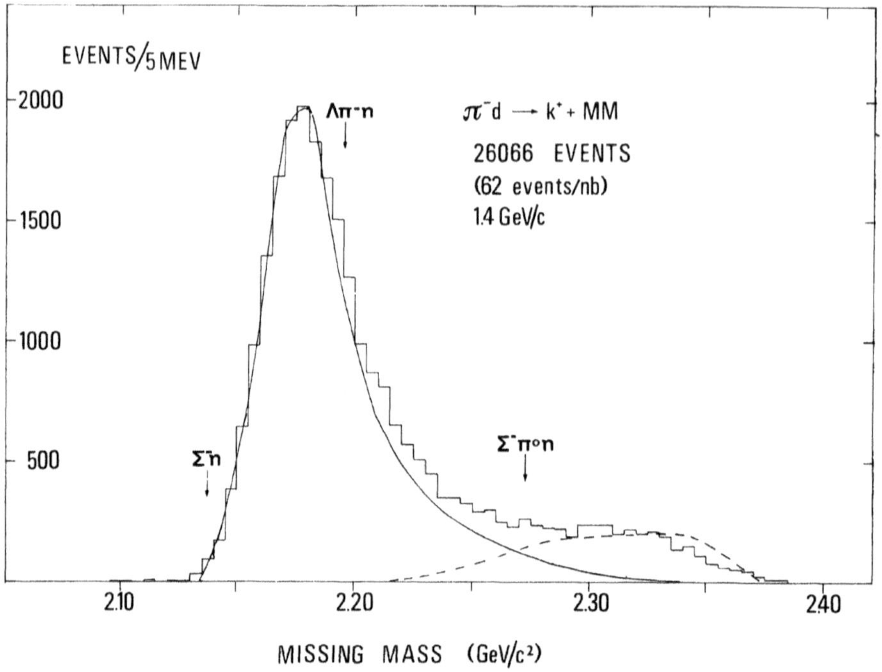

Figure 6 - Missing mass spectrum to the K^+, for the reaction $\pi^- + d \to K^+ + MM$. The solid line is a Monte Carlo prediction for the Σ^- production on the proton, with neutron spectator. The dashed line shows the shape of the $\Lambda\pi^-$ (n) background, as it results from a Monte Carlo calculation. The threshold for the other possible background channels are also indicated

just below the $(\Sigma^- n)$ threshold could have been interpreted as an hypernuclear state I = 3/2, as mentioned before. But it has to be realized that the reaction (3) is certainly not the most appropriate to produce "hypernuclear" states. In fact the $d\sigma/d\Omega$ for Σ^- backward production through the reaction $\pi^- + p \to K^+ + \Sigma^-$ is quite small ($\lesssim$ 10 μb/sr at 1.4 GeV/c) and the kinematics strongly depresses the formation of a loosely bound state near threshold. For this reason we are now taking data also on the line-reversal reaction

$$K^- + d \to \pi^+ + MM$$

at 1.4 GeV/c, which should give us a more conclusive answer on the possible existence of an "H_1^-".

References

1. Miller, D., et al. : 1976, Phys. Rev. Lett. 36, 763.
Auer, I.P., et al. : 1976, Phys. Rev. Lett. 37, 1727.
Auer, I.P., et al. : 1977, Phys. Lett. 67B, 113.
Auer, I.P., et al. : 1977, Phys. Lett. 70B, 475.
Hidaka, H., et al. : 1977, Phys. Lett. 70B, 479.
Kamae, T., et al. : 1977, Phys. Rev. Lett. 38, 468.
Auer, I.P., et al. : 1978, Phys. Rev. Lett. 41, 354.
Auer, I.P., et al. : 1978, Phys. Rev. Lett. 41; 1436.
Mac Gregor, M.H. : 1979, Phys. Rev. Lett. 42, 1274.
Ikeda, H., et al. : 1979, Phys. Rev. Lett. 42 , 1321.

2. Cline, D., et al. : 1968, Phys. Rev. Lett. 20, 1452.
Alexander, G., et al. : 1969, Phys. Rev. Lett. 22, 483.
Tai Ho Tan : 1969, Phys. Rev. Lett. 23, 395.
Eastwood, D., et al. : 1971, Phys. Rev. 3D, 2603.
Brann, O., et al. : 1977, Nucl. Phys. B124, 45.

3. Dosch, H.G., and Hepp, V. : 1978, Phys. Rev. D18, 4071.

4. Oakes, R.J. : 1963, Phys. Rev. 131, 2239.

5. Chodos, A., et al. : 1974, Phys. Rev. D9, 3471.
De Grand, T., et al. : 1975, Phys. Rev. D12, 2060.
Chan Hong-Mo, et al. : 1978, Phys. Lett. 76B, 634.
Mulders, P.J., et al., THEF - NYM - 78-3.
Mulders, P.J., et al., THEF - NYM - 79-19.
Chew, G.F., and Rosenzweig, G. : 1978, Phys. Reports 41, 263.

6. Jaffe, R.L. : 1977, Phys. Rev. Lett. 38, 195.

7. Matveev, V.A., and Sorba, P.: 1977, Nuovo Cimento Lett. 20, 145.

8. Povh, B. : 1978, Ann. Rev. Nucl. Part. Sci. 28, 1.

9. Alberi, G., et al. : 1975, Phys. Rev. Lett. 34, 503.

10. Bezaguet, A., et al. : 1979, Nucl. Instr. & Meth. 158, 303.

DISCUSSION

- D. Miller :

There is still a small tail on the right of your missing-mass spectrum for $K^- \to K^+$ which is not fitted by your $\Xi^- p$ curve. Since you are just above threshold for ϕ meson production, is it possible that you are seeing some forward K^+ from $\phi \to K^+ K^-$ decays ?

- G. Marini :

We have not calculated this.

STUDY OF REACTION INVOLVING FRAGMENT ^{8}Li AND ^{8}B PRODUCTION BY K^{-} MESON CAPTURE IN EMULSIONS

M.K. Jurić
Institute of Physics - Beograd

The study of the mechanism of reactions induced by K^{-}-mesons is rather complex since they include processes : 1) absorption of K^{-} mesons described as a process of K^{-} interaction with one or more nucleons :

$$K^{-} + N \rightarrow Y + \pi \tag{1}$$

$$K^{-} + nN \rightarrow Y + (n-1)N \tag{2}$$

$$Y = \Sigma^{+}, \Sigma^{-}, \Sigma^{0} \text{ or } \Lambda^{0}; \quad n = 2, 3, 4, \ldots$$

2) secondary interactions of reaction products with nucleons of parent nuclei, among them being also the strong interaction of hyperon with nucleons in which Σ hyperon converts to Λ^{0} hyperon :

$$\Sigma + N \rightarrow \Lambda^{0} + N \tag{3}$$

3) deexcitation of the residual nucleus. The specific nature of the target nucleus plays important role in these processes. The presence and the role of various associations in the target nucleus should significantly affect the realization of reaction channels.

In the present work we study the reactions induced by the absorption of stopped K^{-} mesons (RK^{-}) by light nuclear emulsion nuclei, specially the reaction having in the final channel one fragment with $A = 8$. That fragment disintegrates into two alpha particles accompanied by beta emission. These decays may be due to the decay of the following nuclei:

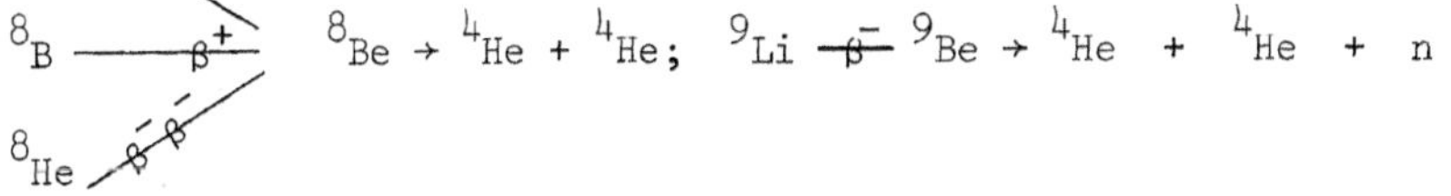

Figure 1

E. Ferrari and G. Violini (eds.), Low and Intermediate Energy Kaon-Nucleon Physics, 277–287.

This fragment because of the collinearity of the momenta of decay products, in nuclear emulsion has a characteristic configuration of the hammer, and is called a hammer fragment (H). A hammer fragment in emulsion is rather simply recognized and so it facilitates the study of reactions in which it occurs.

Up to now the reaction frequency for the production of hammer fragments in nuclear emuslion induced by K^- mesons of different energies has been examined (1-8). Extensive analysis of this reaction with hammer fragment induced by stopped mesons ($R\pi^-$) in nuclear emulsion is also done (9-13).

In the present work we are trying to study the primary interaction of K^--mesons on the basis of the recognition of outgoing particles in the final interaction channel.

If the possible, final channels of RK^- reactions on C, N and O nuclei are analyzed according to the kind of (H) and according to the number and kind of emitted charged particles, then because of the conservation of the baryon number, strangeness and charge, with the requirement that the product be relatively stable nuclei, without hyper nuclei, which leave in the emulsion clearly marked tracks, the possibility of production of individual (H) will be restricted and specific. Table 1 gives data on possible inclusive reaction channels in which only charged particles are involved. Here N_K denotes the total number of charged particles of the reaction channel; H, He and Li stand for one of the corresponding stable isotopes of nucleus and ^{7}Be designates the only possible fragment with $Z = 4$; ± 1 denotes respectively the positively and negatively charged π meson or Σ hyperon.

On table 1 it is seen that only in the case of $N_k = 3$ all three target nuclei take part in the production of (H)-^{8}Li and (H)-^{8}B and that in these stars the negatively charged particle in the final channels is accompanied by (H)-^{8}B, but that the emission of ^{8}He is generally possible in the groups with $N_k \geq 4$.

Experimental

Experimental data on the interaction of K mesons at rest with emulsion nuclei were obtained from a part of Ilford emulsion stack of the European K^- Collaboration (15). During the area scanning, together with the stars induced by stopped K^- mesons, the stars with a hammer-like prong were recorded. As a hammer-like prong we consider a prong having at its end two short ($R < 20$ μm) nearly collinear prongs ($\theta \geq 150°$).

For each star with (H) prong we established the total number of prongs, and each of them was followed till its end in the emulsion, or to the points where it left our stack. The length of each prong in the emulsion is measured and the end of each track was carefully considered with respect to the decay or an interaction.

Table 1

N_k	Target	Inclusive final channel with ^{8}B, *^{8}Li and ^{8}He (Chaerged particles only)		
1	C N O	^{8}B	- - -	- - -
2	C N O	- ^{8}B+1(H)$^{+}$ ^{8}B+He	^{8}Li+He - ^{8}Li+^{7}Be	- - ^{8}He+^{8}B
3	C N O	^{8}B-1+1(H) ^{8}B-1+He ^{8}B-1+Li ^{8}B+1(H)+H	^{8}Li+1(H)+H ^{8}Li+1(H)+He ^{8}Li+He+He ^{8}Li+1(H)+^{6}Li	^{8}He+1+He - - -
4	C N O	- ^{8}B-1+1(H)+H ^{8}B-1+1(H)+He	^{8}Li-1+1(H)+He +^{8}Li+1(H)+H+H +^{8}Li+1(H)+H+He	^{8}He+1(H)+H+H ^{8}He+1(H)+H+He ^{8}He+1(H)+He+He
5	C N O	- - ^{8}B-1+1(H)+H+H -	^{8}Li-1+1(H)+H+H ^{8}Li-1+1(H)+H+He ^{8}Li-1+1(H)+He+He ^{8}Li+1(H)+H+H+H	- ^{8}He+1(H)+H+H+H ^{8}He+1(H)+H+H+He -
6	C N O	- - - -	- ^{8}Li-1+1(H)+H+H+H ^{8}Li-1+1(H)+H+He -	^{8}He-1+1(H)+H+H+H ^{8}He-1+1(H)+H+H+He ^{8}He-1+1(H)+H+He+He ^{8}He+1(H)+H+H+H+H
7	C N O	- - -	- - ^{8}Li-1+1(H)+H+H+H+H	- ^{8}He-1+1(H)+H+H+H+H ^{8}He-1+1(H)+H+H+H+H
8	C N O	- - -	- - -	- - ^{8}He-1+1(H)+H+H+H+H

*^{8}Li comprises also ^{9}Li

$^{+}$1(H) may be a positively charged hyperon, π-meson or the nucleus of an isotope of H

In order to identify mesons among hadrons, and the charge of (H) we made the investigation of the change in ionization in relation to the length of the particle track according to the number of grains (14) or the width of the track (20), respectively. The sign of the charge of $\pi\pm$ mesons was determined only in these cases where they followed to the rest. $\Sigma^{\pm}$ hyperons were identified only in the following cases: a) when their track terminated by a pion decay in flight ($F\Sigma_{\pi}$) or at rest ($R\Sigma_{\pi}$); b) by a proton decay of a Σ^{+} hyperon at rest ($R\Sigma^{+}_{p}$), and in flight only where the other interpretation is not plausible; and c) by the interaction of a Σ^{-} hyperon at rest ($R\Sigma^{-}_{\sigma}$) giving a star with no reasonable doubt as to its nature. So the identification of charged unstable products among baryons was acceptable only if decay or interaction were clearly seen, and it could be excluded that the event was a case of scattering or the interaction of stable particles (16). Each nonidentified Σ hyperon was ascribed to the nucleus of a hydrogen isotope. This specially comprises Σ^{-} hyperons which can be stopped in the emulsion without a visible interaction (Σ_{ρ}). In order to estimate the total amount of Σ^{-} hyperons in the stars with (H) we have separated cases of Σ^{-} hyperons according to the special criterion and multiplied them by experimentally determined coefficient (17).

Results and discussion

i) Among $2.85 \cdot 10^5$ star interactions by stopped K^- mesons (RK) it has been found 455±8 stars with (H) prong. The cited error is due to the star in which the origin of all prongs was not established without ambiguity. The double centered stars with (H) prong in the hyperfragment star is not taken into account (about 20 cases). There are found only two cases in which it was easy to recognize that both one (H) and one hyperfragment are simultaneously emitted from the star.

Beside the RK stars with (H) prong there was found 133±5 stars produced by the interaction of stopped π^- mesons ($R\pi^-$).

ii) In figure 2 the survey is given of the distribution of stars with (H) induced in RK^- and $R\pi^-$ interactions according to the number of charged prongs (N_k) together with the data on general RK^- interaction in the emulsion studied by the K^- Collaboration (16).

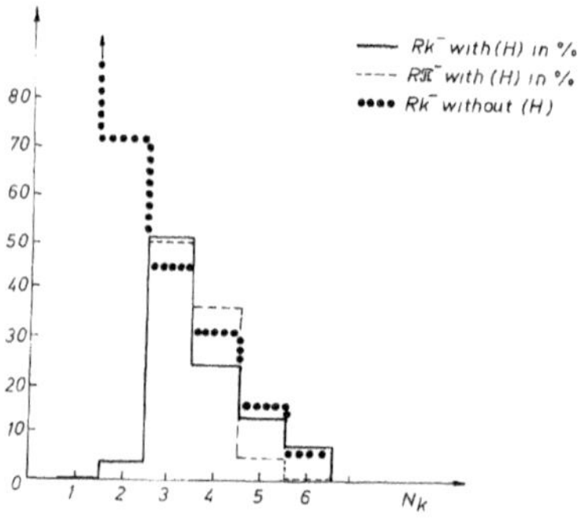

Figure 2 - Distribution of stars with (H) induced in RK^- and $R\pi^-$ interactions according to the number of charged prongs (N_k) together with the data on general RK^- interaction with emulsion (16) nuclei

It can be seen that among the stars with (H) there is a maximum distribution over 50 percent of the events with the number of prongs $N_k = 3$. In general RK^- interaction in the distribution over N_k is no maximum and the number of stars increases with the decrease of N_k. From $N_k = 3$ to $N_k \geq 6$ in all the cases the behavior of the distribution is similar and so the results of the interaction can be compared.

iii) The range distribution of (H) from RK^- and $R\pi^-$ interactions in the emulsion is given in figure 3. Both distributions are normalized to the same number of cases. In the case of RK$^-$ interaction there are more stars with longer range of (H) prong.

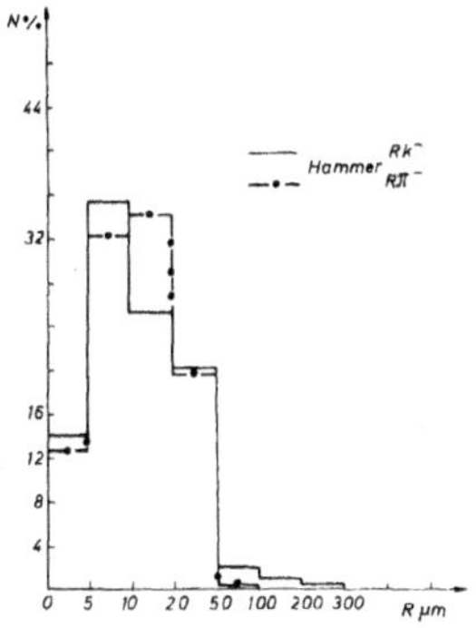

Figure 3 - The distribution of the range hammer-fragments from RK$^-$ and Rπ^- interaction with emulsion nuclei

iv) After a careful investigation whether from (H) is emitted the electron, comparison is made between the relative number of cases when the electron is not emitted in RK$^-$ and Rπ^- interactions. Since in RK$^-$ interactions there was found a greater percentage of (H) not followed by the emission of electrons, this surplus comes probably from hyper-fragments which according to their configuration in the emulsion are similar to (H). There are found about 5% of such fragments among (H) (18).

v) We divided the observed stars with (H) into groups according to directly identified outgoing particle, i.e. into the groups : N_π, N_Σ, N_g and N_b. Table 2 gives a review of the distribution of stars over these groups where N_π and N_Σ group comprise stars in which one outgoing particle was identified as a π meson or hyperon, respectively. N_g and N_b are the groups in which no prong was found to belong to the unstable particle.

Table 2

N_k	N_π	N_Σ	N_g	N_b	Total	% heavy
1	-	-	-	1	1	100
2	-	-	1	18	19	20
3	58±3	25±3	93	58	233	5.2
4	25±2	16±2	36	28	105	13
5	16±2	9±1	25	7	60	12.5
6	9±1	3±1	9	8	29	52
Total	108±8	53±6	164	120	455	12.5%
%	24	12	37	27	100	

In the N_g group a star has at least one prong of the length larger than 3700 μm. (The proton with the track of this range in emulsion corresponds to the energy of 30 MeV and this is unlikely to have been produced in an evaporation process).

This table shows that the most numerous group is the N_g group (37%), while the groups $N_g + N_b$ comprise 64% of all the interactions. The data obtained by the K^- Collaboration for the general case of interaction of stars with $N_k = 3-\geq 6$ show also that 64,8% belong to the stars with stable prongs i.e. to the group $N_g + N_b$, but that the stars of the group N_g make only 14,8%. In our case there are more stars with emitted baryons of higher energies.

vi) The distribution of particles over the ranges of particle tracks in the emulsion is given in figure 4, for π-mesons of known charge, in figure 5a for $\Sigma^{\pm}$-hyperons in the group N_Σ and in figure 5b for the particles with the longest range in the star of the N_g and N_b groups.

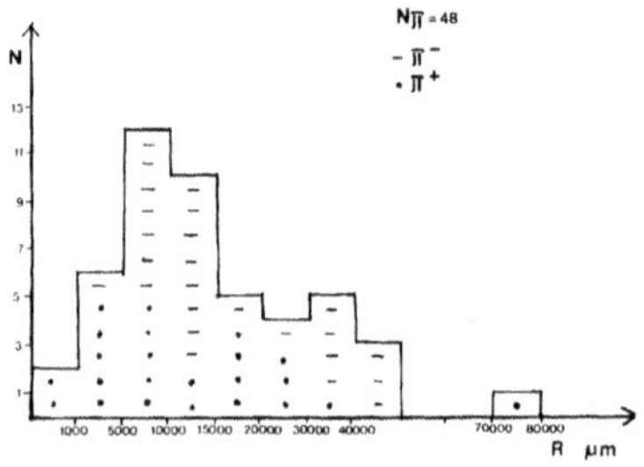

Figure 4 - The range distribution of π-meson of known charge emitted in the RK^- reactions with emulsion nuclei which in the final channel have one (H)

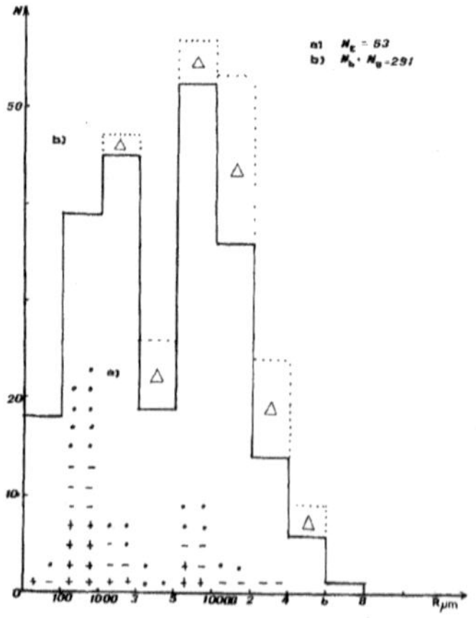

Figure 5 - Range distribution of: a) charged Σ hyperon, and b) fragment of the highest range in group N_b and N_g

It is assumed that the baryons of energies higher than 60 MeV (19) come from the interaction K^-+nN. Supposing that all tracks in figure 5b belong to protons, from figure 5 it may be stated that there are baryons with energy higher than 60 MeV (Σ hyperons $R \geq 10\,000$ μm and protons $R \geq 12\,500$ μm). If we take all these cases with higher energy and if we add to these also the events from the N_Σ group in which $\Sigma^\pm$ of shorter range are accompanied by proton of an energy higher than 50 MeV, we can find the lower limit of the primary RK^- interactions with two or more nucleons. Thus we obtained 19% for all events with (H) and in case of the stars of the group $N_k = 3$, 21. 7%. In the general interaction this lower limit is 9%.

vii) Stars with (H) induced by K^- mesons of higher momenta predominatly are due to the interaction with heavy emulsion nuclei (Ag, Br) (1-8). This is not the case in RK^- and $R\pi^-$ interactions, where the interaction with light nuclei (C, N, O) are dominant. According to the usual criterion (charge, Coulomb barrier, Auger effect of electrons, blob and recoil) it is assumed that RK^- interaction with heavy nuclei occur if the star has $N_k \geq 8$, if there is no prong with $2 < R \leq 30$, and if the blob or electron are emitted. On table 2 there are indicated the percentages of the interactions on heavy nuclei. 12.5% of RK^- interactions with (H) occur on heavy nuclei. In the general RK^- interaction with emulsion nuclei this percentage is 60% (19), and in the case of RK^- interactions followed by the emission of hypernuclei it is 20% (19).

viii) Table 3 gives a review of emission of charged π-mesons where the sign of π is known as well as the emission of π mesons, whose charge

Table 3

N_k	$N_{\pi+}$	$N_{\pi-}$	N_s	Total	$N_{\pi-}/N_{\pi+}$	K^- Collab (16) $N_{\pi-}/N_{\pi+}$
3	13	14	31±3	58±3	1.05	46/9 = 5.1
4	7	5	13±1	25±1	0.71	20/1
5	3	5	7±2	16±2	1.66	12/3 = 4
6	-	5	3±1	8	5/0	5/2
Total	23	29	54±6	106±7	1.3	83/15 = 5.5

could not be determined (N_s) in relation to the number of prongs N_k in the star. The main uncertainties in the determination of the number of meson come from the fast particle of a large dip whose range usually does not end in the emulsion. In such recorded particle we estimated a part which belongs to the mesons and enters in the N_s. On the basis of the experimental data the distribution of the charge of mesons in the group N_s can be estimated.

According to table 1 the existence of (H) fragment of the determined Z in reaction with C, N and O is connected with the distribution of charge in the final reaction channel. In the group $N_k = 3$ the emission of (H)-^{8}B occurs together with the emission of negatively charged hadron.

It is found that the group $N_k = 3$ has at least 12% of reactions in which π^--meson is emitted (20).

For the determination of the number of negatively charged hyperons we used the coefficient (17) by which the experimentally obtained number of Σ^- is multiplied (this being obtained by the use of a special criterion which enables reliable identification of Σ^- hyperons). So it was found that the group $N_k = 3$ has 26% of stars with the emitted negatively charged hadron. It follows that the smallest number of the emitted (H)^{8}B for all RK^- interactions with (H) is 13% (20). This percentage is, for example, much greater than the one obtained in the case of interactions induced by K^- mesons of higher energies (1-8).

For the ^{8}He emission the lower limit was found to be 5%.

ix) The reactions induced by K^- mesons in which (H) is emitted have special characteristics as compared to the general interactions in the stars with the same number of prongs (N_k).

The dependence of the ratio of the number of charged pions $N_{\pi-}/N_{\pi+}$ on N_k in comparison with data of the K^- Collaboration on RK^- interaction for the stars with $N_k = 3$, 4, 5 and ≥ 6, is given on table 3. It can be seen that the differences are great.

The data on the ratio $N_{\pi-}/N_{\pi+}$ in RK^- interactions induced by other targets are given in table 4.

Table 4

Target	H	D	He	Emulsion			
				Total (16)	$N_k = 3 - \geq 6$ (16)	Stars with (H)	
						Total	$N_k = 3$
$N_{\pi-}/N_{\pi+}$	0.45	1.95	5.5	3.9	5.5	1.3	1.1

The ratio of the emitted π^- and π^+ mesons remains practically unchanged in spite of the effect of distorsion which may provoke the nucleus to the outgoing particle. Therefore the ratio $N_{\pi-}/N_{\pi+}$ may give information on the relative frequency of various interactions of K^- meson with one nucleon of the nucleus.

In our case the decrease of the number of π^- mesons indicates to the domination of $K^- + p$ interaction as compared to $K^- + n$ interaction. In the group $N_k = 3$ (table 1) this indicates to the emission of ^{8}Li which when taken for spectator, makes the rest association of nucleons to contain a surplus of protons, and stimulates $K^- + p$ interaction (12,13). In this manner the effect of the probability of cluster formation in nucleus becomes apparent.

On table 5 the review is given of other characteristics of RK^- interactions in which (H) is emitted as well as the general RK^- interactions in which stars with three or more prongs are emitted. The data are taken from the K^- Collaboration (16).

Table 5

	RK^- general (16)		RK^- with (H)	
	$N_k = 3- \geq 6$	$N_k = 3$	$N_k = 3- \geq 6$	$N_k = 3$
$\frac{N_\pi + N_\Sigma}{N_g + N_b}$	0.54	0.62	0.52	0.54
$\frac{N_g + N_b}{N_{all}}$	65%	62%	64%	65%
$N_g/(N_b + N_g)$	23%	27%	58%	65%
N_g/N_{all}	15%	17%	37%	40%
N_π/N_{all}	29%	33%	24%	25%

It can be seen that the ratio of the number of emitted nonstable and stable particles in both interactions are approximately equal, but that the percentage of the emitted unstable baryons is greater. Among stable particles the contribution of fast protons (N_g) is considerable. For example, the percentage of the number of pairs of emitted fast baryons in RK interactions with (H) in groups $N_g + N_b$ (12%) is nearly equal to the percentage of one fast baryon emitted in the general RK^- interaction (14.8%).

x) We tried to learn more about the reaction mechanism on the basis of experimental data by determining the total kinetic energy of the charged reaction products. In figure 6 is given a survey of the total kinetic energy of reaction product assuming that all the (H) fragments are ^8Li, while the rest of the charged outgoing particles are protons if they are not identified as hyperons or mesons. For particles whose tracks do not end in the emulsion, the energy is estimated taking into account the minimal possible value. It is shown specially when $N_k \geq 6$, and 5 that the distribution has two regions separated at 95 MeV. If we take this energy for criterion to distinguish between RK^- interaction with 1N and interactions with nN, then there are 21% events with the interaction of nN. The value obtained in this way does not differ considerably from that obtained in vi). Both indicate to the considerable primary interaction of RK^- mesons with several nucleons.

It is of interest to enquire the size of the cluster of nucleons with which the initial interaction took place.

Some information about this is suggested by the outgoing particles which share the greatest part of the available energy of primary interaction. In determining the total kinetic energy of the outgoing particle

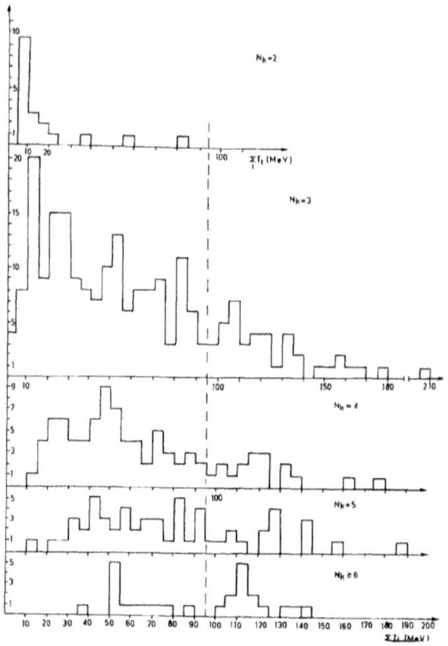

Figure 6 - Distribution of the total kinetic energy of emitted charged particles in reactions RK^- with emulsion nuclei which in final channel have one (H) assuming that all (H) fragments are ^8Li and other outgoing charged particles are protons if they are not identified as hyperons or mesons

it is found that there are cases when the energy of (H) contributes considerably to the energy higher than 95 MeV e.g. on table 6 the case when the energy of (H) is 36 MeV.

Table 6

N°	Range µm	Energy MeV	
32/7	R_H = 132	36	108 MeV
	R_g = 12885	60	
	R_b = 744	12	

It may be concluded that the cluster (H) may be so correlated in nuclei, that it may take part in direct primary interactions.

References

1. Jones, B.D., et al.: 1962, Phys. Rev. 127, 236.
2. European Collaboration: 1963, Proceedings Siena International Conference on Elementary Particles, 323.
3. Harmsen, D.M., et al.: 1964, Phys. Letters 9, 274.
4. Bauman, G., et al.: 1965, Phys. Rev. B138, 350.
5. Lemone, J., et al.: 1966, Nuovo Cimento A41, 235.
6. Gerber, J.P., et al.: 1968, Nucl. Phys. B5, 75.
7. Kurtz, N.: 1969, Thesis No ordre 508, Fac. des Sci. Univ. Strasbourg.
8. Jurić, M., et al.: 1968, Nucl. Phys. B8, 565.
9. Patusov, Yu. A., et al.: 1967, Yad. Fiz. 6, 1151; 1967: Preprint, Dubna, PI-3306; 1967: Yad. Fiz. 6, 1149; 1969: Yad. Fiz. 10, 354.
10. Agabyan, B.S., et al. : 1971, Yad. Fiz. 13, 283; 1972: Preprint, Dubna, PI-6796.
11. Massue, J.P.: 1970, Thesis No ordre 592, Fac. des Sci. Univ. Strasbourg.
12. Zelenskaya, N.S., et al. : 1965, Izv. AN SSSR ser. Fiz. 29, 186; 1968: Yad. Fiz. 7, 515.
13. Golovanova, N.F., et al. : 1968, Nucl. Phys. A113, 1.
14. Grote, C., et al. : 1959, Nuovo Cimento 19, 532.
15. Jurić, M., et al. : 1973, Nucl. Phys. B52, 1.
16. Bhowik, B., et al. : 1959, Nuovo Cimento 13, 690.
17. Jurić, M.K., et al.: 1971, Fizika 5, 135.
18. Jurić, M.K., et al. : 1977, Nucl. Instr. and Meth. 147, 251.
19. Burhop, E.H.S., et al.: 1964, Progress in Nuclear Physics 9, 155.
20. Jurić, M.K. : 1979, Fizika 11(4), 233.

MEASUREMENT OF THE REACTIONS $K^+n \rightarrow K^+\pi^-p$ AND $\pi^+n \rightarrow \pi^+\pi^-p$ AT 6 AND 12 GeV/c ON A TRANSVERSELY POLARIZED DEUTERON TARGET

A. de Lesquen, M. Babou[+], J. Bystricky, G. Cozzika,
T. Dobrowolski, Y. Ducros, M. Fujisaki[++], A. Gaidot,
C.F. Hwang[+++], A. Itano[*], F. Khantine-Langlois, F. Lehar,
L. van Rossum, M. Svec[**]
Département de Physique des Particules Elémentaires -
Centre d'Etudes Nucléaires, Saclay

1. Introduction

The experiment studies the pion production reactions $K^+n \rightarrow K^+\pi^-p$ and $\pi^+n \rightarrow \pi^+\pi^-p$ on neutrons polarized transversely to the beam direction. The events are described by the variables (s, m, t, ψ, θ, ϕ) where s is the center of mass energy squared, t is the four-momentum transfer squared and m is the invariant mass of the ($K\pi$) or ($\pi\pi$) system in the final state. The angle ψ defines the normal to the production plane with respect to the direction of the target polarization. The angles (θ, ϕ) describe the direction of the positive particle in the ($K^+\pi^-$) or ($\pi^+\pi^-$) rest frame. The analysis is carried out in both the s- and t-channel frames of the dimeson. Information on the amplitudes, at given values of (s, t, m), is contained in the cross sections $d^2\sigma/(dtdm)$ and in the angular distributions $d^3\sigma/(d\psi d\theta d\phi)$. At dimeson invariant masses of $m \lesssim 1$ GeV all partial waves with angular momentum $J \geq 2$ are small. The eight complex nucleon transversity amplitudes describing S- and P-wave interaction can be determined - except for the absolute phase and one of the relative phases - from the distribution in (ψ,θ,ϕ) on a transversely polarized target. The missing relative phase can be obtained only by measuring also the recoil nucleon polarization.

In the limit of high energy these amplitudes correspond to exchanges with definite naturality in the t-channel. The t-channel quantum numbers allowed in $K^+n \rightarrow K^+\pi^-p$ are A_2, ρ(natural) and π, A_1, B, Z (unnatural). For $\pi^+n \rightarrow \pi^+\pi^-n$ the allowed exchanges are A_2 and π, A_1.

2. Experiment

The experiment was carried out at the CERN-PS, in an unseparated beam, for reasons of intensity, with a polarized deuteron target. Using $^3He-^4He$ dilution technique for the cryostat the CERN Polarized Target Group had achieved more than 40% deuteron polarization in deuterated

E. Ferrari and G. Violini (eds.), Low and Intermediate Energy Kaon-Nucleon Physics, 289–303.

propanediol (1). The apparatus (2) (figure 1) had been designed primarily for measuring the polarization in $K^+n \to K^o p$ charge exchange (3), the K^o being detected by a $(\pi^+\pi^-)$ pair in the forward spectrometer.

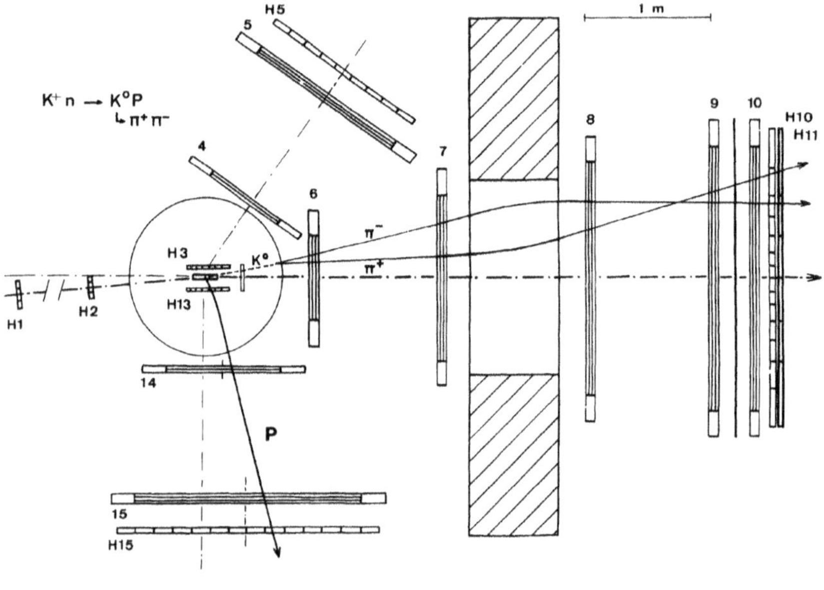

Figure 1 - Schematic lay-out of the apparatus at 6 GeV/c

This trigger could easily be complemented by other triggers (4), in particular for direct production of meson pairs $(K^+\pi^-)$ and $(\pi^+\pi^-)$ for incident kaons and pions, respectively. In all cases the recoil proton is momentum analyzed in one of the two recoil arms, allowing reconstruction of the Fermi momentum of the target neutron, but introducing a cut-off at $|t| \lesssim 0.1$ $(GeV/c)^2$. The aperture of the forward spectrometer limits the acceptance to $|t| \lesssim 1.5$ $(GeV/c)^2$ and to invariant dimeson masses $m \lesssim 1$ GeV. At large invariant masses the acceptance pattern in (θ, ϕ) becomes very complicated and requires careful Monte Carlo studies to extract information on the produced angular distributions for the observed samples. The recoil arms accept only events with the production plane not too far from the horizontal. The direction of target polarization being vertical up or down, the distribution in ψ, measured in the target neutron rest frame, is limited to intervals of $\pm$ 30° around the directions parallel and antiparallel to the target polarization. The average transverse pola rization perpendicular to the production plane, $\langle P_T \cos\psi \rangle$, is therefore large, but the statistical precision on those observables depending on $\langle P_T \sin\psi \rangle$ is poor.

The raw dimeson mass spectra, uncorrected for acceptance, show dominant $K^{o*}(892)$ and $\rho^o(770)$ production, respectively. In $K^+n \rightarrow K^+\pi^- p$ only the mass region $0.8 \leq m(K\pi) \leq 1.0$ GeV contains sufficient numbers of events for data analysis: 11 000 events at 6 GeV/c and 2000 events at 12 GeV/c, within $0.1 \leq |t| < 1.0\ (\text{GeV/c})^2$. For $\pi^+n \rightarrow \pi^+\pi^-p$ the analysis covers the mass range $0.36 \leq m(\pi\pi) < 1.04$ with 57 000 events at 6 GeV/c and 38 000 at 12 GeV/c in the same t-interval. For this reaction the trigger rate at 6 GeV/c actually had to be reduced by a factor 4 to avoid excessive deadtime losses for the other triggers.

From the on-line produced data tapes the events are reconstructed, fitted and selected by cuts on external variables such as vertex positions, target neutron effective mass and Fermi momentum $|P_F|$, and by several other kinematical criteria. The resulting data summary tapes transmit for each event the measured quantities (m, t, ψ,θ,ϕ) and $|P_F|$. The experimental distribution in P_F for events from deuterons and from the heavier nuclei C and O is known from runs with liquid d_2 and carbon targets. From the measured value $|P_F|$ for an event from the polarized target one calculates the probability to be produced on a polarized or on an unpolarized neutron. Determination of the target polarization independent observables involves both classes of events whereas events produced in C or O do not contribute to the polarization dependent terms of the distribution in (ψ, θ,ϕ). The angular distribution is analyzed in terms of normalized spin density matrix elements with the normalization $d^2\sigma/(dmdt) = 1$, thus avoiding the need for absolute calibration of the cross sections.

The distribution in (ψ, θ,ϕ) of events in a fixed region of (m, t) and at fixed s is given by the following expression :

$$
\begin{aligned}
4\pi W(\psi, \theta,\phi) = {} & (\rho_{00} + 2\rho_{11} + \rho_{SS}) \\
& + (\rho_{00} - \rho_{11}) && (3\cos^2\theta - 1) \\
& - \rho_{1-1} && 3\sin^2\theta\cos(2\phi) \\
& - \text{Re}\,\rho_{10} && 3\sqrt{2}\sin(2\theta)\cos\phi \\
& + \text{Re}\,\rho_{OO} && 2\sqrt{3}\cos\theta \\
& - \text{Re}\,\rho_{1S} && 2\sqrt{6}\sin\theta\cos\phi \\
+ P_T\cos\psi & \Big[(\rho^Y_{00} + 2\rho^Y_{11} + \rho^Y_{SS}) \\
& + (\rho^Y_{00} - \rho^Y_{11}) && (3\cos^2\theta - 1) \\
& - \rho^Y_{1-1} && 3\sin^2\theta\cos(2\phi) \\
& - \text{Re}\,\rho^Y_{10} && 3\sqrt{2}\sin(2\theta)\cos\phi \\
& + \text{Re}\,\rho^Y_{OO} && 2\sqrt{3}\cos\theta \\
& - \text{Re}\,\rho^Y_{1S} && 2\sqrt{6}\sin\theta\cos\phi \Big]
\end{aligned}
$$

$$+ P_T \sin\psi \quad \Big[\text{Im}\,\rho^X_{1-1} \quad 3\sin^2\theta \;\sin(2\phi)$$
$$+ \text{Im}\,\rho^X_{10} \quad 3\sqrt{2}\,\sin(2\theta)\,\sin\phi$$
$$+ \text{Im}\,\rho^X_{1S} \quad 2\sqrt{6}\,\sin\theta \;\sin\phi \Big]$$

This expansion in spherical harmonics contains 15 real parameters which are expressed in terms of the dimeson spin density matrix elements $\rho(s, m, t)$. It assumes contribution from partial waves with $J < 2$ only (5, 6). The first six terms are independent of target polarization. The next six are analogous but proportional to $P_T \cos\psi$. The last three terms, only two of which are independent observables, depend on the polarization component $P_T \sin\psi$ in the production plane.

The first parameter $(\rho_{OO} + 2\rho_{11} + \rho_{SS})$ is equal to unity by our choice of normalization. The remaining 14 parameters, representing normalised spin density matrix elements, are determined by a maximum likelihood fit to the sample of events in the particular (m, t) bin.

The effect of non-uniform acceptance of the apparatus in (ψ, θ, ϕ) is not corrected for by weighting individual events. It is taken into account by introducing acceptance integrals into the distribution function in a way which restores the independence of its normalisation on the parameters. The 15 acceptance integrals are calculated for each (m, t) bin from high statistics Monte Carlo samples generated uniformly in (m, t) and isotropically in (ψ, θ, ϕ) but following closely the real event distributions in the external variables.

3. Results

The data for either reaction, at both energies and for both s- and t-channel observables have been analysed in several sets of (m, t) bins. One of the sets covers the entire (m, t) region (figure 2). Other sets display the t-dependence at the resonances $K^{o*}(892)$ and $\rho^o(770)$ or the detailed mass dependence in the t-region of high statistics $0.2 \leq |t| \leq 0.4$. Figure 3 shows the t-dependence of the s-channel parameters for $K^+n \to K^+\pi^-p$ at the K^{o*} mass and figure 4 shows the t-dependence of the t-channel parameters for $\pi^+n \to \pi^+\pi^-p$ at the ρ^o mass.

The five polarization independent parameters are obtained with a very good statistical accuracy but may be subject to systematic errors of the same order. The polarization dependent parameters are obtained with larger statistical errors but are less sensitive to the acceptance of the apparatus. The parameters proportional to $P_T \sin\psi$ have large statistical errors and, especially for $K^+n \to K^+\pi^-p$, will not much constrain the amplitude analysis.

For $K^+n \to K^+\pi^-p$ only the polarization independent matrix elements

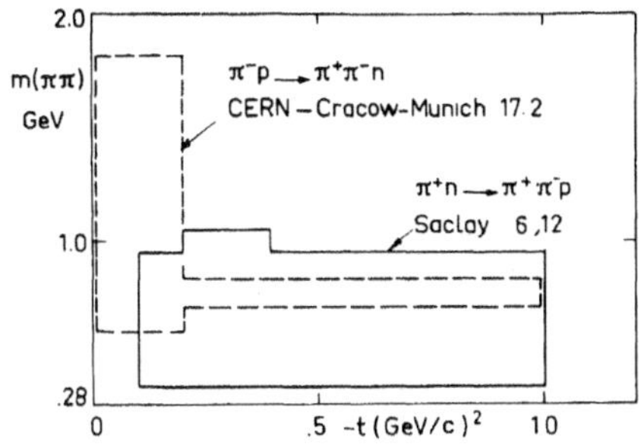

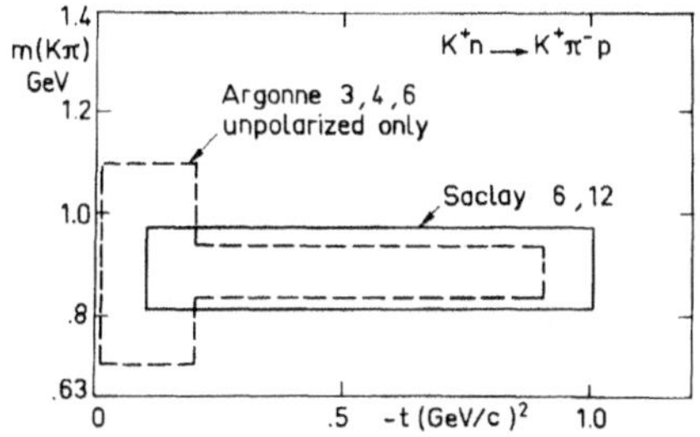

Figure 2 - Kinematic domain of this experiment (solid lines) and of previous experiments (5, 6) (dashed lines)

had been measured before, with the Argonne Effective Mass Spectrometer at 3, 4 and 6 GeV/c (5) and in three Bubble Chamber experiments (7) at 4.6, 8.25 and 12 GeV/c. Most of the events in these experiments are at low four-momentum transfer. Our data at 6 GeV/c considerably improve the statistics in the region of $|t| > 0.1$ $(GeV/c)^2$. Comparison of the t-dependence of the 5 parameters, at the K^{*o} mass, shows good agreement with the Argonne results at the same energy.

The polarization dependent parameters in pion production reactions had been measured so far only in one high statistics experiment for $\pi^-p \to \pi^+\pi^-n$ at 17.2 GeV/c by the CERN-Cracow-Munich group (6).Comparison with our results for $\pi^+n \to \pi^+\pi^-p$ at 6 and 12 GeV/c for the t-dependence of the matrix elements at the ρ^o mass shows smooth energy dependence. Our errors on the $P_T \cos\psi$ terms are typically 2 to 3 times larger. For the $P_T \sin\psi$ terms the difference in precision is even bigger since the CERN experiment, not detecting the recoil neutron, has an isotropic acceptance in ψ. The Argonne experiment also ignores the recoil nucleon. Both experiments therefore extend to very low four-momentum transfer (figure 2).

Without extracting the full information through the amplitude analysis

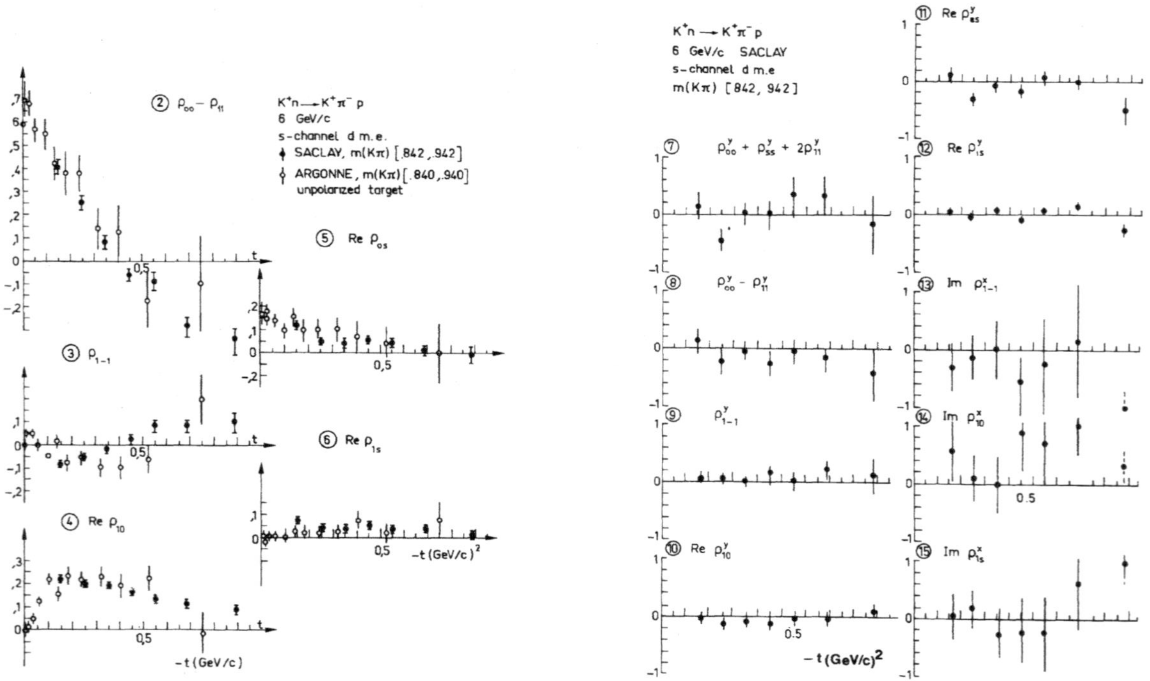

(a) (b)

Figure 3 - Normalised s-channel spin density matrix elements for $K^+n \to K^+\pi^-p$ at 6 GeV/c
a) Polarization independent s.d.m.e. from this experiment and from ref. (5)
b) Polarization dependent s.d.m.e. (this experiment)

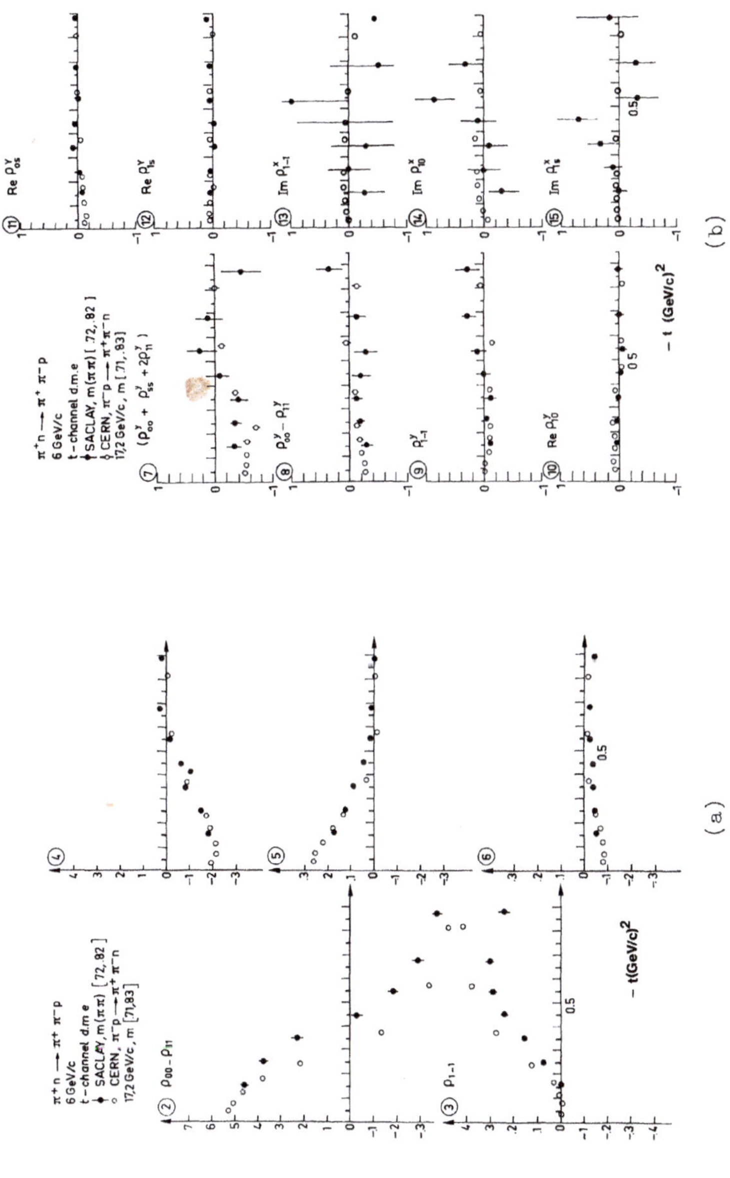

Figure 4 - Normalised t-channel s.d.m.e. for $\pi^+n \rightarrow \pi^+\pi^-p$ at 6 GeV/c (this experiment) and for $\pi^-p \rightarrow \pi^+\pi^-n$ at 17.2 GeV/c (6)
a) Polarization independent s.d.m.e.
b) Polarization dependent s.d.m.e.

it is possible to gain some insight into the partial wave intensities using only the polarization independent quantities. The m- and t- dependence of these observables can be studied in detail using small (m, t) bins containing only a few hundred events each. The first three terms of the expansion provide three relations between the four partial wave intensities and can be combined in the following way:

$$\rho_{00} + \frac{1}{3}\rho_{SS} = \sigma_L + \frac{1}{3}\sigma_S$$

$$\rho_{11} - \rho_{1-1} + \frac{1}{3}\rho_{SS} = \sigma_U + \frac{1}{3}\sigma_S$$

$$\rho_{11} + \rho_{1-1} + \frac{1}{3}\rho_{SS} = \sigma_N + \frac{1}{3}\sigma_S$$

Here σ_S is the normalised S-wave intensity, σ_L is the zero dimeson helicity P-wave intensity and σ_U, σ_N are those combinations of the ± 1 helicity P-wave intensities which, in the high energy limit, correspond to natural (σ_N) and unnatural (σ_U) exchange amplitudes.

Figures 5 and 6 show the t-dependence of the three combinations at the K^{*o} and ρ^o mass for reactions $K^+n \rightarrow K^+\pi^-p$ and $\pi^+n \rightarrow \pi^+\pi^-p$, respectively.

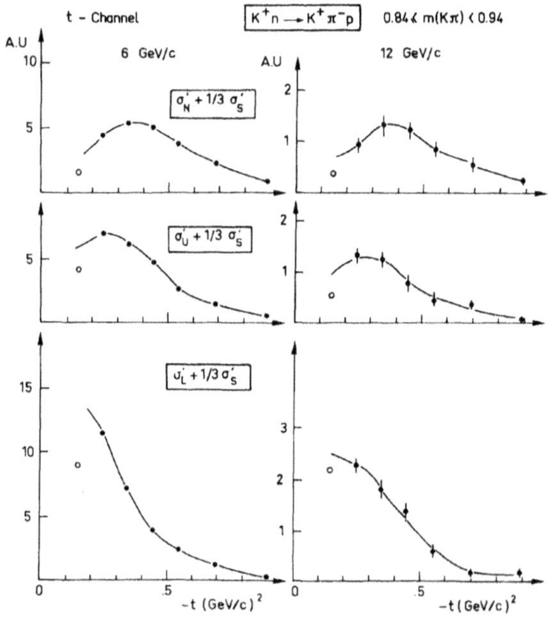

Figure 5 - Combinations of partial wave intensitities for $K^+n \rightarrow K^+\pi^-p$ at 6 and 12 GeV/c. Dependence on t at the K^{*o} mass

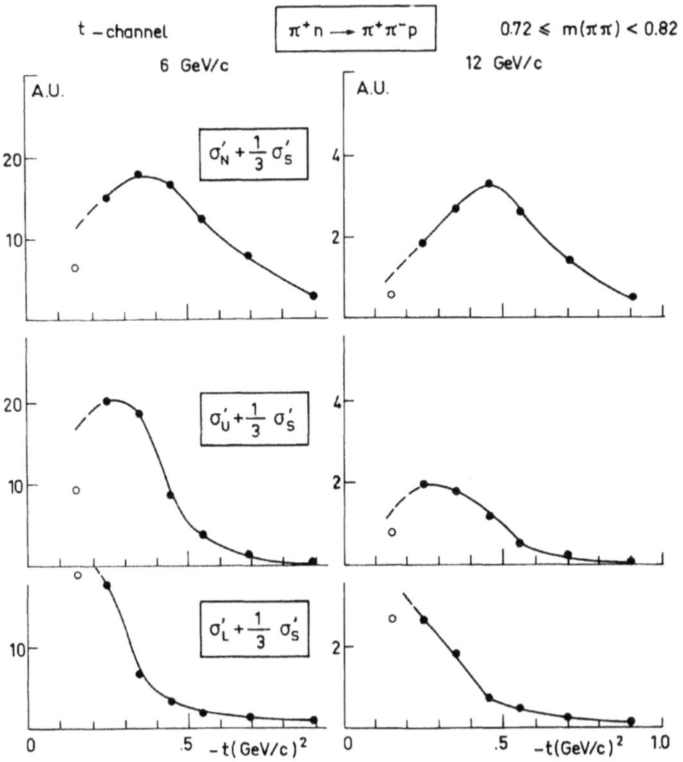

Figure 6 - Combinations of partial wave intensities for $\pi^+n \to \pi^+\pi^-p$ at 6 and 12 GeV/c. Dependence on t at the ρ^0 mass

In order to display more clearly each t-dependence separately the normalised partial wave intensities σ have been multiplied by the number of events N divided by the bin size ($\Delta m \times \Delta t$). The resulting quantities $\sigma' = \sigma N/(\Delta m \Delta t)$ represent absolute cross-sections in arbitrary units and uncorrected for variations of the acceptance. To take into account the acceptance, values in the first t-bin should be multiplied by about 1.7, and in the last two bins by about 1.3.

For either reaction and at both energies the combination $(\sigma_L + \sigma_S/3)$ is the most important contribution at $|t| \leq 0.2$, whereas $(\sigma_N + \sigma_S/3)$ is relatively the most important at $|t| \gtrsim 0.5$.

The mass dependence of the same combinations of partial wave intensities is shown on figure 7 for $\pi^+n \to \pi^+\pi^-p$ at 6 GeV/c, in different regions of momentum transfer.

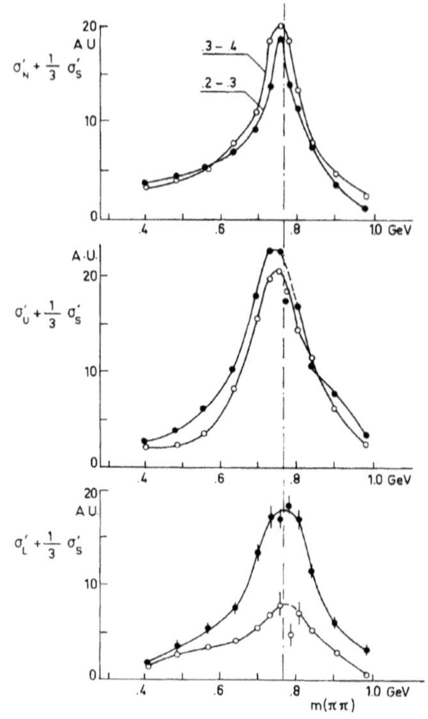

Figure 7 - Combinations of partial wave intensities for $\pi^+ n \to \pi^+\pi^- p$. Dependence on m(ππ) in two regions of t, at 6 GeV/c.

Also in this figure the normalised intensities are multiplied by $N/(\Delta m \Delta t)$ to show the peaking of the cross-sections at the ρ^o-mass. Correcting for acceptance would reduce the low mass tails by factors between two and four. This would not modify, however, the following features of the data : a) the ρ^o peak appears in all three combinations; b) the t-dependence at the ρ^o peak is different for each of the combinations, as already seen in figure 6; c) the width of the peak in $(\sigma'_N + \sigma'_S/3)$, at $-t = 0.25$ for instance, seems to be significantly less than in $(\sigma'_L + \sigma'_S/3)$ at the same t-value.

The figures 8 and 9 show the ratio $R = (\sigma_N + \sigma_S/3)\ (\sigma_L + \sigma_U + 2\sigma_S/3)$ as function of t in different regions of the dimeson mass. This ratio is independent of acceptance and normalisation. At high energy it becomes equal to the ratio of natural to unnatural exchange in the limit of $\sigma_S \ll 3\sigma_N$, and it becomes $R = \frac{1}{2}$ in the limit of pure S-wave.

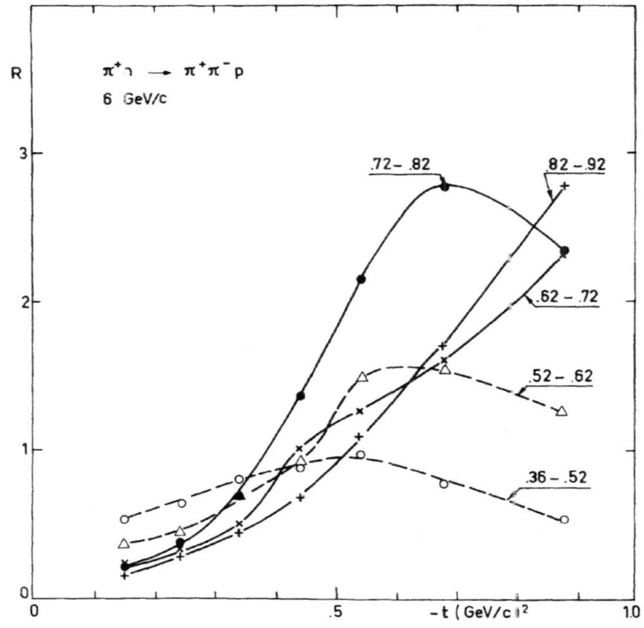

(a)

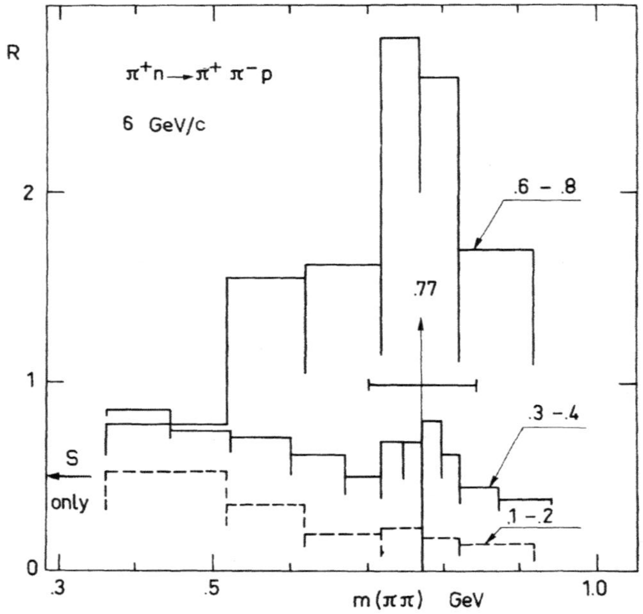

(b)

Figure 8 - Ratio $R = (\sigma_N + \sigma_S/3) / (\sigma_L + \sigma_U + 2\sigma_S/3)$ for $\pi^+n \to \pi^+\pi^-p$ at 6 GeV/c

a) Dependence on t in five regions of $m(\pi\pi)$

b) Dependence on $m(\pi\pi)$ in three regions of t

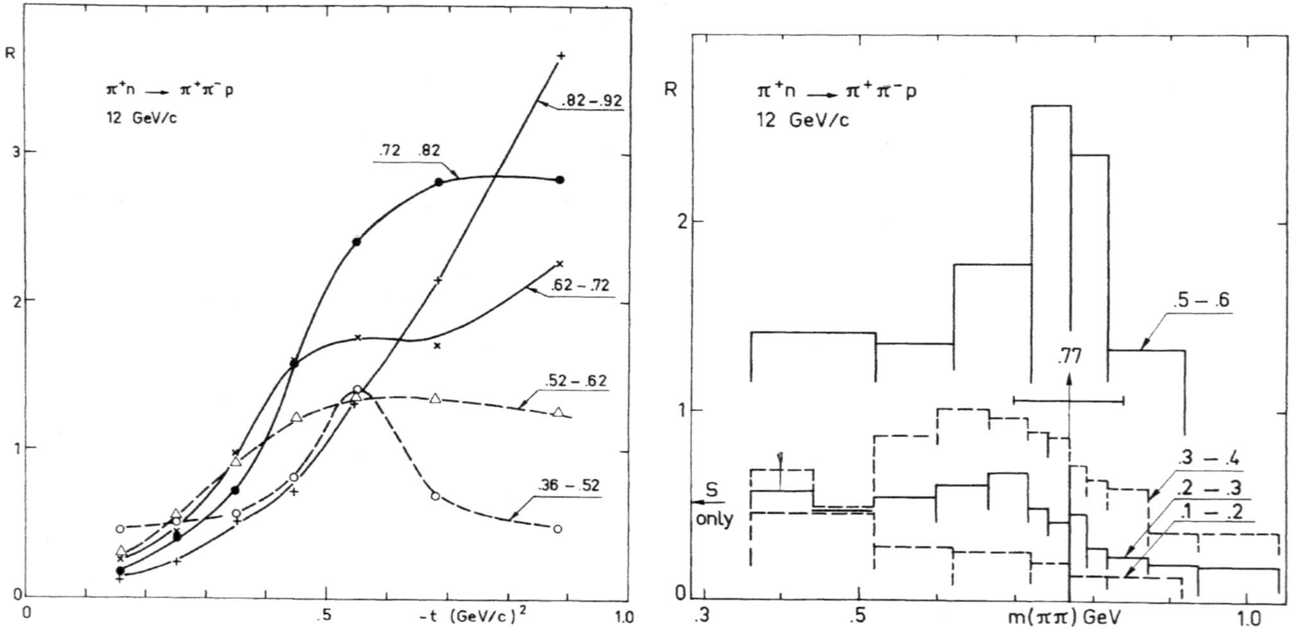

(a) (b)

Figure 9 - Ratio $R = (\sigma_N + \sigma_S/3) / (\sigma_L + \sigma_U + 2\sigma_S/3)$ for $\pi^+n \rightarrow \pi^+\pi^-p$ at 12 GeV/C

a) Dependence on t in five regions of $m(\pi\pi)$

b) Dependence on $m(\pi\pi)$ in four regions of t

In the reaction $\pi^+n \to \pi^+\pi^-p$ at 6 GeV/c (figure 8a) the contribution of $(\sigma_N + \sigma_S/3)$ in the interval $0.72 < m < 0.82$ increases sharply in the region of $-t \simeq 0.5$. In the region $0.4 \leq |t| \leq 0.8$ this combination which includes all of the natural exchange contribution, is significantly larger at the ρ^o mass than in any other mass region. As shown in figure 8b, in the interval $0.6 \leq |t| \leq 0.8$ the ratio R, in a relatively narrow region, (0.77 ± 0.05) GeV, is about 1.6 times larger than in the two adjacent mass bins. The figures 9a, b show that the same observations hold qualitatively also for the data at 12 GeV/c.

The production mechanism for the dimeson systems at the ρ^o mass is clearly changing when going from small four-momentum transfer, $|t| \lesssim 0.2$, to the region $t \gtrsim 0.4$.

For the reaction $K^+n \to K^+\pi^-p$, in the dimeson mass range $0.812 \leq m < 0.972$, we find a ratio $R = (\sigma_N + \sigma_S/3) / (\sigma_L + \sigma_U + 2\sigma_S/3)$ which is about three times less than in $\pi^+n \to \pi^+\pi^-p$. It is also increasing with increasing t, but the mass-dependence shows no significant change of R at the K^{*o} mass, in any of the t-bins (figure 10).

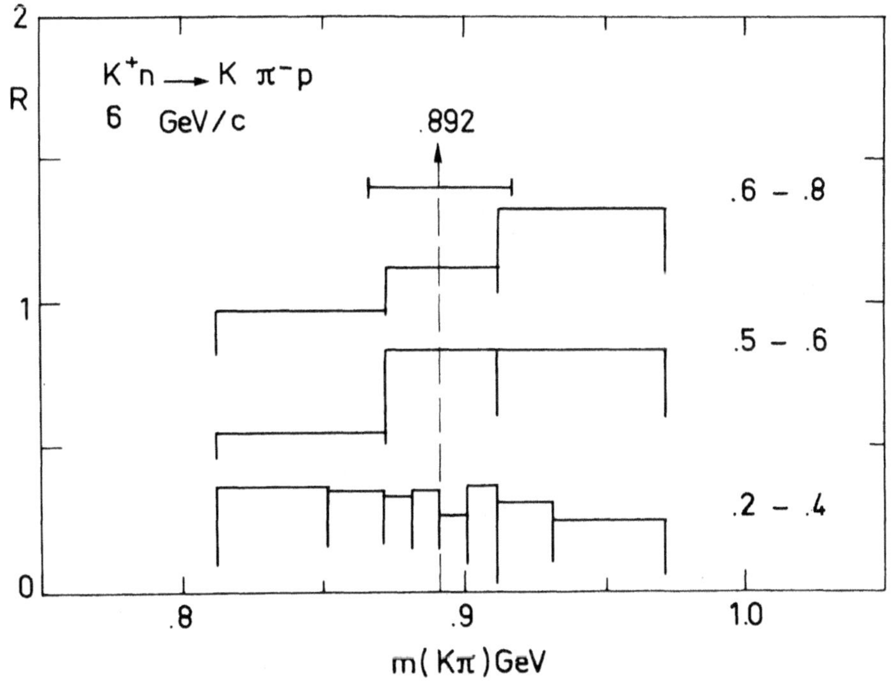

Figure 10 - Ratio $R = (\sigma_N + \sigma_S/3) / (\sigma_L + \sigma_U + 2\sigma_S/3)$ for $K^+n \to K^+\pi^-p$ at 6 GeV/c.
Dependence on $m(K\pi)$ in three regions of t

Making use of all of the measured observables it is possible to separate the four partial wave intensities and also to examine the four partial wave polarizations of the recoil proton. This is one of the results from a model independent amplitude analysis based on this experiment (see contributed paper to this Workshop by M. Svec et al., "Amplitude Analysis in $\pi^+n \to \pi^+\pi^-p$ and $K^+n \to K^+\pi^-p$ reactions at 6 GeV/c") (8).

4. Summary

Measurement of the spin density matrix elements in single pion production reactions on transversely polarized nucleons provide a number of observables almost sufficient to directly determine the S- and P-wave amplitudes describing the reaction. In this paper we present two examples of sets of matrix elements obtained in the CERN-PS experiment with a polarized deuteron target, one for the reaction $K^+n \to K^+\pi^-p$ and the other for $\pi^+n \to \pi^+\pi^-p$.

Without performing an amplitude analysis based on all the data, it is possible to extract interesting information on partial wave intensities by forming simple linear combinations of some of the matrix elements. In both reactions and at all dimeson masses sufficiently above threshold the relative strength of different partial waves is changing considerably as function of t from $-t = 0.1$ to $-t = 1(\mathrm{GeV/c})^2$. Furthermore in reaction $\pi^+n \to \pi^+\pi^-p$, in the region $0.4 \leq |t| \lesssim 0.8$, the production of dipions at the ρ^0 mass involves more contribution from natural exchange amplitudes than at masses immediately below or above, in contrast with the situation at t-values below $|t| \simeq 0.2$. No similarly large effect is observed at the $K^{*0}(892)$ mass in reaction $K^+n \to K^+\pi^-p$. After completing the amplitude analysis, these features of the data will be studied in terms of the moduli of each of the eight amplitudes separately.

+ - Present address : GIXI Ingénierie et Informatique, Paris.
++ - " " : University of Michigan, Ann Arbor, MI.
+++ - " " : Los Alamos Scientific Laboratory, Los Alamos.
* - " " : KEK, National Laboratory for H.E. Physics.
** - Mc Gill University, Montréal.

References

1. Niinikoski, T.O. : 1976, Proc. Symp. on "High Energy Physics with Polarized Beams and Targets", Argonne, p. 458.
   Niinikoski, T.O., and Udo., F.: 1976, Nuclear Instr. Met. 134, 219.
   Guckelsberger, K., and Udo, F. : 1976,Nuclear Instr. Met. 137, 415.

2. Babou, M., et al. : 1979, Nuclear Instr. Met. 160, 1.

3. Fujisaki, M., et al.: 1979, Phys. Letters 80B, 314.
Fujisaki, M., et al.: 1979, Nucl. Phys. B152, 232.

4. Fujisaki, M., et al.: 1979, Nucl. Phys. B151, 206.

5. Wicklund, A.B., et al.: 1978, Phys. Rev. D17, 1197.

6. Becker, H., et al.: 1979, Nucl. Phys. B150, 301.
Becker, H., et al.: 1979, Nucl. Phys. B151, 46.

7. Büchner, K., et al. : 1972, Nucl. Phys. B45, 333.
Hendrickx, K.N., et al. : 1976, Nucl. Phys. B112, 189.
Lissauer, D., et al.: 1972, Phys. Rev. D6, 1852.

8. Svec, M.: 1980, in this Workshop.

DISCUSSION

- C. Dover :

Are you able to extract any information on $\pi\pi \rightarrow \pi\pi$ and $\pi K \rightarrow \pi K$ amplitudes from your data? That is, is the old peripheral model valid for some t-region of your data?

- A. de Lesquen :

No attempt has been made so far to extract information on $K\pi \rightarrow K\pi$ or $\pi\pi \rightarrow \pi\pi$ from our data. In principle, measurement of the polarization dependent spin density matrix elements permits better separation of the different exchange contributions. However, the cut-off in acceptance of our experiment at $|t| \lesssim 0 \cdot 1$ (GeV/C)^2 makes extrapolation to $t=m_\pi^2$ more difficult. Moreover in most of our interval of four-momentum transfer pion exchange is not the dominating contribution. We expect this experiment to provide new information, in particular on the contribution of A_1-exchange to the unnatural amplitudes, and on the natural exchange amplitudes the relative strength of which is rapidly rising in this interval.

AMPLITUDE ANALYSIS IN $\pi^+n \to \pi^+\pi^-p$ and $K^+n \to K^+\pi^-p$ REACTIONS AT 6 GeV/c

M. Svec
Mc Gill University, Montreal

J. Bystricky, M. Fujisaki*, A. Itano**, A. de Lesquen and
L. van Rossum
Département de Physique des Particules Elémentaires - CEN-Saclay

1. Introduction

The construction of amplitudes directly from experimental data (i. e. amplitude analysis) provides model independent information on their t-structure and energy dependence. Scattering experiments with polarized beams and/or targets are required to obtain the necessary set of complete data. The amplitude analysis was first possible for πN two-body scattering at 6 GeV/c (1) and it allowed the isolation of ρ-exchange amplitudes. Later the measurement of K^+n CEX polarization at 6 GeV/c (2) allowed the construction of A_2-exchange amplitudes in KN two-body scattering at 6 GeV/c (3) which showed unexpected features in their t-structure. Similarly motivated experiments in NN scattering (4) revealed large A_1-exchange which had been for long assumed to be a negligible contribution.

Amplitude analysis of reactions with three-body final states provide new information which is not available from analysis of two-body reactions: first, the mass dependence of exchange amplitudes; second, contribution from unnatural amplitudes with t-channel quantum numbers not exchanged in simple two-body meson-nucleon scattering. Also important is the fact that in a single experiment on a trasversely polarized target the measured spin density matrix elements of the produced dimeson state provide enough observables for performing an almost complete amplitude analysis.

The existing data on π-production on polarized targets can be summarized as follows :

$\pi^-p \to \pi^+\pi^-n$	P_{lab} =	17.2 GeV/c	CERN-Cracow-Munich, 1979 (5)
$\pi^+n \to \pi^+\pi^-p$	P_{lab} =	6 and 12 GeV/c	Saclay, 1980 (6)
$K^+n \to K^+\pi^-p$	P_{lab} =	6 and 12 GeV/c	Saclay, 1980 (6)

E. Ferrari and G. Violini (eds.), Low and Intermediate Energy Kaon-Nucleon Physics, 305–319.

Model dependent analyses of $\pi^- p$, $K^- p$ and $K^+ n$ channels measured on unpolarized target only were performed previously by P. Estabrooks and A. Martin (7, 8) and by Wicklund et al. (9). The simplifying assumption of these analyses - the absence of A_1-exchange - has been invalidated by the amplitude analysis based on the CERN-Cracow-Munich data (5).

In this contribution we present preliminary results of an amplitude analysis of $\pi^+ n \to \pi^+\pi^- p$ and $K^+ n \to K^+\pi^- p$ at 6 GeV/c. The final amplitude analysis of the Saclay data at both 6 and 12 GeV/c will be published soon.

2. The amplitudes

The kinematical variables describing the production of a dimeson system are (s, t, m, θ, φ) where s is the c.m.s. energy squared, t is the four-momentum transfer squared and m is the dimeson invariant mass. The angles θ and φ describe the direction of the K^+ or π^+ meson in the dimeson c.m.s. s-channel (or t-channel) helicity frame of reference.

The production process is described by production amplitudes

$$H_{\lambda_p \lambda_n}(s, m, t, \theta, \phi)$$

where λ_p and λ_n are helicities of the proton and neutron, respectively. In the following all helicities are defined in the s-channel helicity frame of reference. The production amplitude can be expanded into partial wave amplitudes :

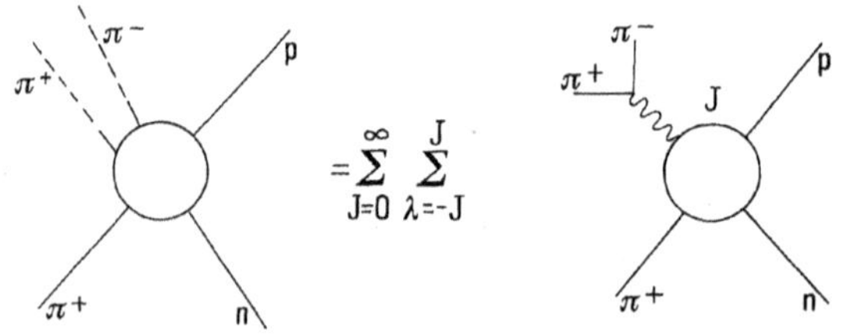

$$H_{\lambda_p \lambda_n}(s, m, t, \theta, \phi) = \sum_{J=0}^{\infty} \sum_{\lambda=-J}^{J} (2J+1)^{\frac{1}{2}} H^{J}_{\lambda\lambda_p, o\lambda_n}(s, t, m) d^{J}_{\lambda o}(\theta) e^{i\lambda\phi}$$

where λ is the helicity of the dimeson state.

For dimeson masses m < 1.0 GeV/c it is possible to assume that only S- and P-waves contribute (5). The contribution of the D-wave is smaller than the combined statistical and systematic errors in the measured data and its neglect will introduce only a small additional systematic error in the amplitude analysis.

In our notation the s-channel helicity amplitudes describing the

S- and P-wave states of the dimeson system are :

S-wave : $0^- \frac{1}{2}^+ \longrightarrow 0^+ \frac{1}{2}^+$

$$H^{o}_{o+,o+} = S_o$$
$$H^{o}_{o+,o-} = S_1 \tag{1}$$

P-wave : $0^- \frac{1}{2}^+ \longrightarrow 1^- \frac{1}{2}^+$

$$H^{1}_{o+,o+} = L_o$$
$$H^{1}_{o+,o-} = L_1 \tag{2}$$
$$H^{1}_{\pm 1+,o+} = \frac{1}{\sqrt{2}} (N_o \pm U_o)$$
$$H^{1}_{\pm 1+,o-} = \frac{1}{\sqrt{2}} (N_1 \pm U_1)$$

The amplitudes S_n, L_n, U_n $(n = 0,1)$ are denominated, at large s, by unnatural exchanges, and the amplitudes N_n $(n = 0,1)$ by natural exchanges. Here $n = |\lambda_n - \lambda_p|$ is the nucleon helicity flip.

If we neglect the ρ-ω interference contributions, the t-channel quantum numbers exchanged in the SU(3) related reactions are summarized in the following table :

	$\pi^+n \to \pi^+\pi^-n$	$K^+n \to K^+\pi^-p$	$K^-p \to K^-\pi^+n$	$\pi^+n \to \pi^+\pi^-\pi^0p$
S_o, L_o, U_o	$J^{PC} = 1^{++}$ A_1	$A_1 - Z$	$A_1 + Z$	2^{--} Z
S_1, L_1, U_1	0^{-+} π	$\pi - B$	$\pi + B$	1^{+-} B
N_o, N_1	2^{+-} A_2	$A_2 - \rho$	$A_2 + \rho$	1^{-+} ρ

The axial vector mesons A_1 and B were recently observed experimentally in production experiments. The Z-exchange was first proposed by Irwing and Michael (10) but no resonance with its quantum numbers has yet been found.

The observables obtained in experiments not measuring recoil nucleon polarizations are most simply related to recoil nucleon transversity amplitudes which are expressed in terms of the s-channel helicity amplitudes as follows :

$$\begin{aligned} S &= \frac{1}{\sqrt{2}}(S_o + i\,S_1) \; ; & \bar{S} &= \frac{1}{\sqrt{2}}(S_o - i\,S_1) \\ L &= \frac{1}{\sqrt{2}}(L_o + i\,L_1) \; ; & \bar{L} &= \frac{1}{\sqrt{2}}(L_o - i\,L_1) \\ U &= \frac{1}{\sqrt{2}}(U_o + i\,U_1) \; ; & \bar{U} &= \frac{1}{\sqrt{2}}(U_o - i\,U_1) \\ \bar{N} &= \frac{1}{\sqrt{2}}(N_o + i\,N_1) \; ; & N &= \frac{1}{\sqrt{2}}(N_o - i\,N_1) \end{aligned} \qquad (3)$$

Asymptotically at high energy the transversity amplitudes S, L, U, $\bar{S}$, $\bar{L}$, $\bar{U}$ are thus dominated by unnatural exchanges while N and $\bar{N}$ are dominated by natural exchanges. The amplitudes S, L, U, N describe production of final states with nucleon spin parallel to the normal to the production plane ("up") and $\bar{S}$, $\bar{L}$, $\bar{U}$, $\bar{N}$ correspond to the production with transversity "down".

For interpretation of results obtained in terms of transversity amplitudes it is important to note that comparison of pairs of moduli of these amplitudes leads to simple conclusions about the corresponding pairs of helicity amplitudes. Let n and f denote the nucleon helicity non flip and flip amplitudes, respectively (e.g. S_o and S_1). Let g and h denote the two corresponding transversity amplitudes (e.g. S and $\bar{S}$). It can be easily demonstrated that

a) if $|g| = |h|$, then (4a)

- either one of the amplitudes n or f vanishes
- or n and f have the same phase;

b) if $|g| \neq |h|$, then (4b)

- none of n and f vanishes
- and n and f must have different phases.

Representing the amplitudes as vectors in a plane, these properties can be illustrated geometrically as shown in the diagram below

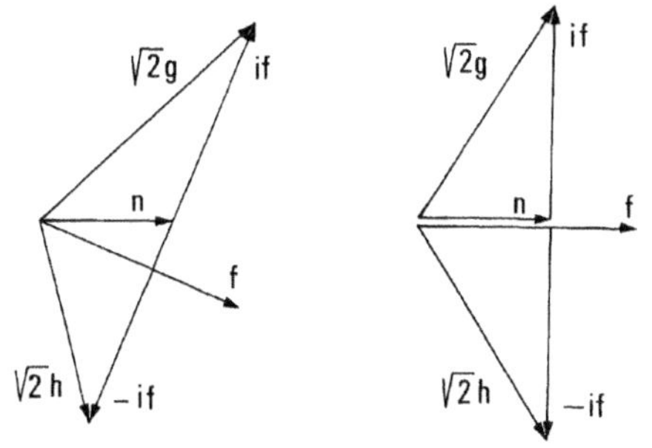

3. The relations between transversity amplitudes and observables

The experiment with unpolarized target measures the differential cross-section $d^2\sigma/dmdt$ and five spin density matrix elements

$$(\rho_{oo} - \rho_{11}),\quad \rho_{1-1},\quad Re\,\rho_{1o},\quad Re\,\rho_{oS},\quad Re\,\rho_{1S} \tag{5}$$

The experiment with transversely polarized target adds information on 9 polarization dependent parameters (11)

$$(\rho^Y_{SS} + \rho^Y_{oo} + 2\rho^Y_{11}),\ (\rho^Y_{oo} - \rho^Y_{11}),\ \rho^Y_{1-1},\ Re\,\rho^Y_{1o},\ Re\,\rho^Y_{oS},\ Re\,\rho^Y_{1S} \tag{6}$$

$$Im\,\rho^X_{1-1},\quad Im\,\rho^X_{1o},\quad Im\,\rho^X_{1S} \tag{7}$$

The parameters (6) depend on the component of target polarization $P_T\cos\psi$ perpendicular to the production plane, and

$$(\rho^Y_{SS} + \rho^Y_{oo} + 2\rho^X_{11})$$

is the total target asymmetry. The parameters (7) depend on the component of polarization $P_T\sin\psi$ in the production plane.

When expressed in terms of the recoil nucleon transversity amplitudes (normalized such that $d^2\sigma/(dmdt) = 1$) the data can be conveniently organized in three separate groups. The first two groups involve only parameters (5) and (6). The first group relates only to the amplitudes S, L, U, $\bar{N}$ and allows determination of their 4 moduli and cosines of two independent relative phases, $\gamma_{1O} = \phi_U - \phi_L$ and $\gamma_{OS} = \phi_L - \phi_S$. Similarly the second group relates only the amplitudes $\bar{S}, \bar{L}, \bar{U}, \bar{N}$. The third group allows the determination (up to sign ambiguities) of the cosines of relative phases $\gamma_N = \phi_N - \phi_U$ and $\gamma_N = \phi_{\bar{N}} - \phi_{\bar{U}}$. These phases are not well determined since the Saclay data (6) have large errors on the parameters (7). The relative phase between the two groups of four amplitudes each would require to measure also the recoil nucleon polarization.

The first group of relations reads as follows :

$$a_1 = \frac{1}{2}\left[(\rho_{SS} + \rho_{00} + 2\rho_{11}) + (\rho^Y_{SS} + \rho^Y_{00} + 2\rho^Y_{11})\right] = |S|^2 + |L|^2 + |U|^2 + |\bar{N}|^2$$

$$a_2 = \frac{1}{2}\left[(\rho_{00} - \rho_{11}) + (\rho^Y_{00} - \rho^Y_{11})\right] = 2|L|^2 - |U|^2 - |\bar{N}|^2$$

$$a_3 = \frac{1}{2}\left(\rho_{1-1} + \rho^Y_{1-1}\right) = |\bar{N}|^2 - |U|^2 \qquad (8)$$

$$a_4 = \frac{1}{2}\left(\mathrm{Re}\,\rho_{10} + \mathrm{Re}\,\rho^Y_{10}\right) = |U||L|\cos(\gamma_{UL})$$

$$a_5 = \frac{1}{2}\left(\mathrm{Re}\,\rho_{0S} + \mathrm{Re}\,\rho^Y_{0S}\right) = |L||S|\cos(\gamma_{LS})$$

$$a_6 = \frac{1}{2}\left(\mathrm{Re}\,\rho_{1S} + \mathrm{Re}\,\rho^Y_{1S}\right) = |U||S|\cos(\gamma_{UL} + \gamma_{LS})$$

The second group of relations is obtained similarly by forming the difference between parameters of the same pairs (5) and (6) :

$$\bar{a}_1 = \frac{1}{2}\left[(\rho_{SS} + \rho_{00} + 2\rho_{11}) - (\rho^Y_{SS} + \rho^Y_{00} + 2\rho^Y_{11})\right] = |\bar{S}|^2 + |\bar{L}|^2 + |\bar{U}|^2 + |N|^2$$

$$\vdots \qquad (8')$$

$$\bar{a}_6 = \frac{1}{2}\left(\mathrm{Re}\,\rho_{1S} - \mathrm{Re}\,\rho^Y_{1S}\right) = |\bar{U}||\bar{S}|\cos(\bar{\gamma}_{UL} + \bar{\gamma}_{LS}),$$

Note that $(\rho_{SS} + \rho_{00} + 2\rho_{11}) = 1$ by definition of the spin density matrix.

The system of equations (8) can be solved in terms of $|L|$:

$$|S|^2 = a_1 + a_2 - 3|L|^2$$

$$|U|^2 = |L|^2 - \frac{1}{2}(a_2 + a_3) \qquad (9)$$

$$|\bar{N}|^2 = |L|^2 - \frac{1}{2}(a_2 - a_3)$$

$$\cos(\gamma_{UL}) = a_4/(|U||L|) \quad ; \quad \cos(\gamma_{LS}) = a_5/(|L||S|)$$

The amplitude squared $|L|^2 \equiv X$ is obtained as a solution of a cubic equation

$$3X^3 - \left[a_1 + \frac{5}{2} a_2 + \frac{3}{2} a_3 \right] X^2 + \left[\frac{1}{2}(a_1 + a_2)(a_2 + a_3) - 3 a_4^2 + a_5^2 + a_6^2 \right] X + (a_1 + a_2)a_4^2 - \frac{1}{2}(a_2 + a_3)a_5^2 - 2 a_4 a_5 a_6 = 0 \qquad (10)$$

The physical solution of the equation (10) must satisfy several independent constraints to ensure the positivity of all moduli and the physical values of the cosines of relative phases :

$$\begin{aligned}
&\text{i)} && 0 \leq X \\
&\text{ii)} && \frac{1}{2}(a_2 + |a_3|) \leq X \leq \frac{1}{3}(a_1 + a_2) \\
&\text{iii)} && a_4^2 \leq X \left[X - \frac{1}{2}(a_2 + a_3) \right] \\
&\text{iv)} && a_5^2 \leq X \left[a_1 + a_2 - 3X \right]
\end{aligned} \qquad (11)$$

The second group (8') leads to a similar cubic equation for $|\bar{L}|^2 \equiv \bar{X}$ and to similar constraints in terms of the observables a_i.

4. The results

$\pi^+n \to \pi^+\pi^-p$

In general we find 3 real solutions of the equations (10), one of which is negative and is rejected. The two positive solutions satisfy the constraints (11) and are reffered to as Solution 1 and Solution 2. For certain (m, t) bins the average values of the data give solutions 1 and 2 such that either the constraints (11) are violated or the solutions become complex. In this case it is necessary to move away from the average values of data, within the experimental errors, to obtain physical solutions. Full treatment of such cases and full account of the error matrix in evaluating the errors of the amplitudes will be given in the final publication.

At the present stage of analysis the figures show those (m, t) bins where the average values lead to physical solutions (dots with error bars) and those cases where the physical solutions are obviously within the range of the experimental errors (bars only).

The t-dependence, at the ρ^0 mass, of the moduli of the normalized recoil transversity amplitudes for $\pi^+n \to \pi^+\pi^-p$ at 6 GeV/c is shown in figure 1 for Solution 1 and in figure 3 for Solution 2. The t-dependence of the cosines of the relatives phases γ_{LS}, $\overline{\gamma}_{LS}$, γ_{UL} and $\overline{\gamma}_{UL}$ is shown in figures 2 and 4 for the 2 solutions.

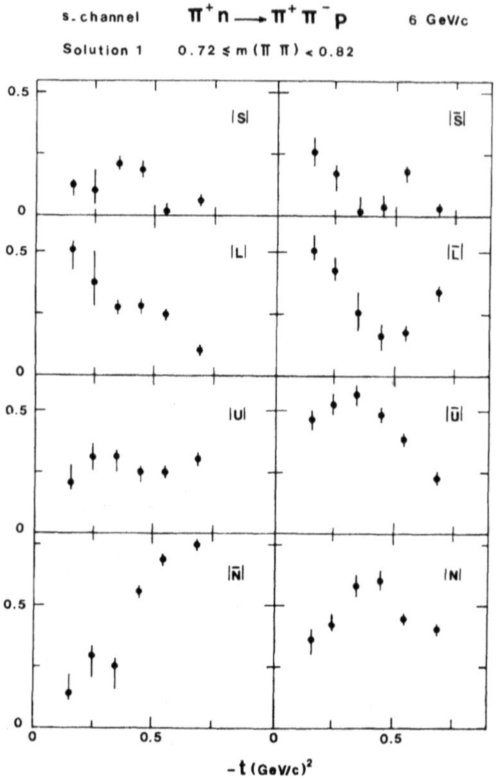

Figure 1 - The reaction $\pi^+n \to \pi^+\pi^-p$ at 6 GeV/c. The figure shows the t-dependence of moduli of the normalised recoil transversity amplitudes, Solution 1

By comparing Solutions 1 and 2 we observe :

i) S-wave amplitudes are small in Solution 1 but are large in Solution 2.

ii) The amplitudes $|L|$ and $|\overline{L}|$ are large and steeply decreasing, with increasing $|t|$, for $|t| \leq 0.3$ in both solutions. This suggests that π- exchange dominance of these amplitudes at small t is a common feature of both solutions.

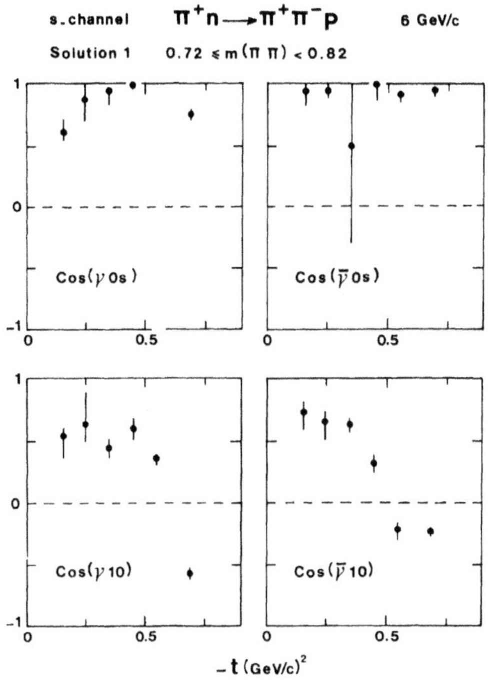

Figure 2 - The reaction $\pi^+n \to \pi^+\pi^-p$ at 6 GeV/c. The figure shows the t-dependence of the cosines of relative phases γ_{OS}, $\overline{\gamma}_{OS}$, γ_{10} and $\overline{\gamma}_{10}$, Solution 1

iii) The amplitudes $|U|$ and $|\overline{U}|$ have similar t-dependence for both solutions but have smaller magnitude in Solution 2. The amplitude $|\overline{U}|$ is decreasing rapidly for $|t| \gtrsim 0.35$.

iv) The amplitudes $|\overline{N}|$ and $|N|$ have also similar t-dependence for both solutions. The contribution of the amplitude $|\overline{N}|$ is increasing rapidly for $|t| \gtrsim 0.45$ while the contribution of $|N|$ is decreasing for the same t.

v) In the Solution 1 the cos (γ_{LS}) and cos $(\overline{\gamma}_{LS})$ are close to +1. Consequently, the amplitudes S and L and the amplitudes $\overline{S}$ and $\overline{L}$ have approximately the same phase. The relative phases between unnatural amplitudes L and U, and between amplitudes $\overline{L}$ and $\overline{U}$, are however different from zero. The situation is approximately reversed in the Solution 2.

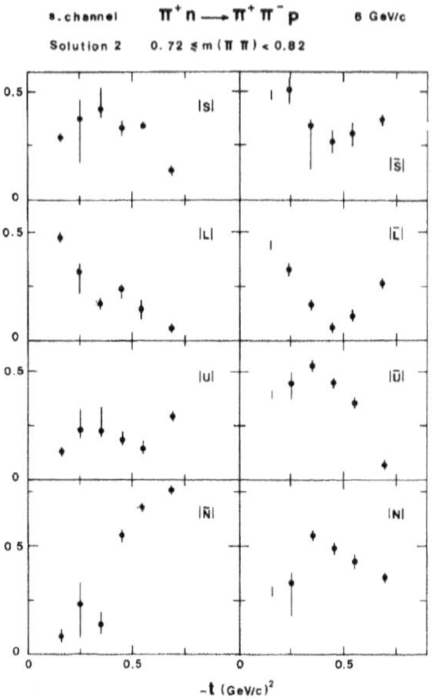

Figure 3 - Same as figure 1, Solution 2

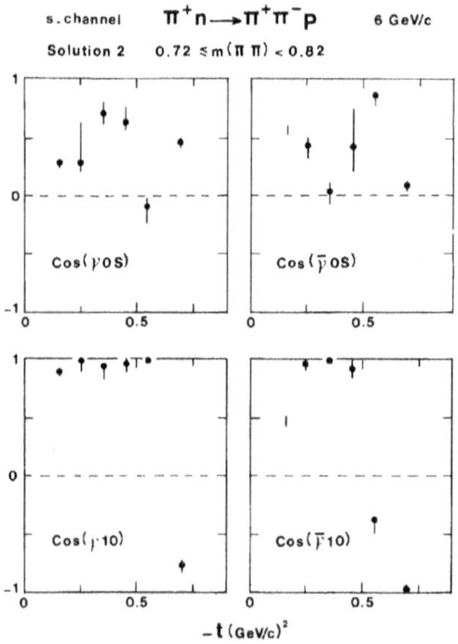

Figure 4 - Same as figure 2, Solution 2

vi) We observe large differences in the moduli of the pairs of amplitudes $|S|$ and $|\bar{S}|$, $|U|$ and $|\bar{U}|$, and $|N|$ and $|\bar{N}|$. This observation holds for both Solutions 1 and 2 and can be explained (see equations 4) by the presence of large s-channel helicity amplitudes S_o, U_o and N_o in both solutions. Also, the amplitudes S_o and S_1, U_o and U_1 and N_o and N_1 must have different phases. We recall that amplitudes S_o and U_o are (asymptotically) dominated by A_1-exchange while the amplitude N_o is dominated by A_2-exchange.

vii) The amplitudes $|L|$ and $|\bar{L}|$ have approximately equal moduli for $|t| \le 0.3$ but not for large t. This suggests that for $|t| \le 0.3$ the amplitude L_o is either small or is in phase with the amplitude L_1.

Using the calculated moduli of the amplitudes it is now possible to separate the normalised partial wave intensities

$$\sigma_S = |S|^2 + |\bar{S}|^2 \qquad \sigma_U = |U|^2 + |\bar{U}|^2$$

$$\sigma_L = |L|^2 + |\bar{L}|^2 \qquad \sigma_N = |N|^2 + |\bar{N}|^2$$

The two solutions for each group lead to four possible values of σ. From the figures 3 and 4 it appears that associating the same type of solution yields extreme values. In the following we discuss only these two cases.

In figure 5 we show the mass dependence, at $0.2 \le |t| \le 0.3$ of the partial wave intensities for the reaction $\pi^+n \to \pi^+\pi^-p$. Clearly the contribution σ_S of the S-wave is small for Solution 1. At $m(\pi\pi) \approx 0.76$, near the ρ^o resonance, σ_S drops to about 7% of the cross-section in Solution 1. In Solution 2 σ_S is larger and near $m = 0.76$ all partial waves are approximately equal. In this solution we observe a broad enhancement of σ_S below the ρ_o mass.

$K^+n \to K^+\pi^-p$

The analysis of reaction $K^+n \to K^+\pi^-p$ also gives two sets of solutions of the same analytical type of Solutions 1 and 2 in the $\pi^+n \to \pi^+\pi^-p$ reaction.

In figures 6 and 7 we present the t-dependence, at the K^{*o} mass, of the Solution 1 for the reaction $K^+n \to K^+\pi^-p$.

The difference in the moduli of pairs of amplitudes with opposite transversities demonstrate the presence of both nucleon helicity flip and non flip amplitudes. When comparing the moduli in $K^+n \to K^+\pi^-p$ with the corresponding moduli of Solution 1 for $\pi^+n \to \pi^+\pi^-p$, the observed differences confirm the presence of B- and possibly Z- exchange in the

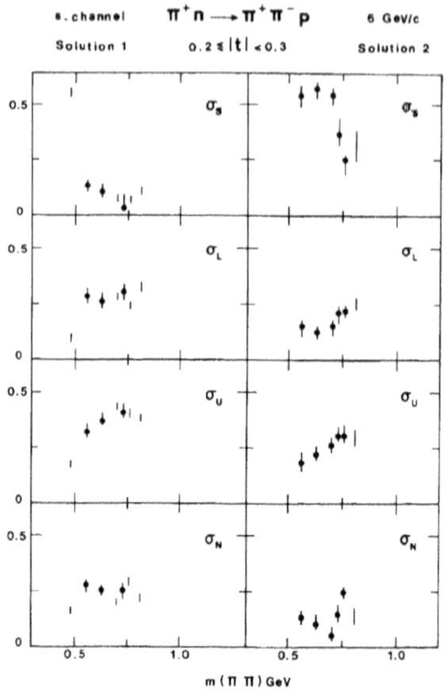

Figure 5 - The reaction $\pi^+n \rightarrow \pi^+\pi^-p$ at 6 GeV/c. The figure shows the mass dependence of normalised partial wave intensities σ_S, σ_L, σ_U and σ_N at $0.2 \leq |t| < 0.3$ $(GeV/c)^2$, Solutions 1 and 2

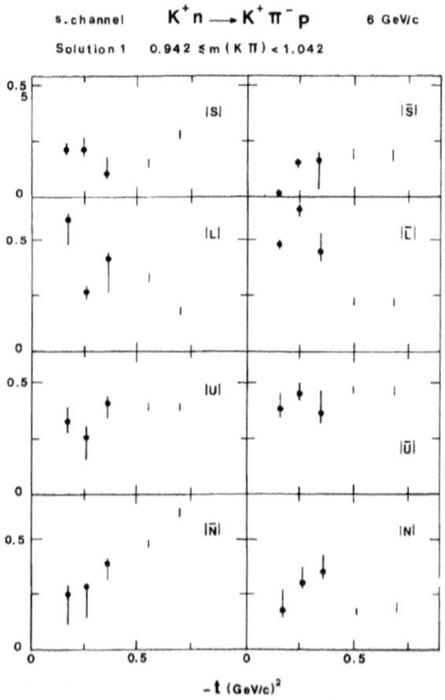

Figure 6 - The reaction $K^+n \rightarrow K^+\pi^-p$ at 6 GeV/c. The figure shows the t-dependence of moduli of the normalised recoil transversity amplitudes, Solution 1

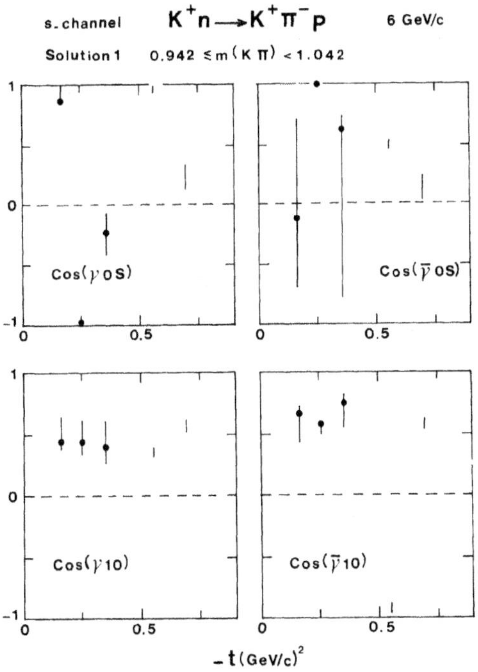

Figure 7 - The reaction $K^+n \rightarrow K^+\pi^-p$ at 6 GeV/c. The figure shows the mass-dependence of the cosines of relative phases γ_{OS}, $\bar{\gamma}_{OS}$, γ_{10} and $\bar{\gamma}_{10}$, Solution 1

unnatural amplitudes, and of ρ-exchange in the natural amplitudes. We also note that there is little evidence for the phase coherence of amplitudes S and L, or of amplitudes L and U, in contrast with the $\pi^+n \rightarrow \pi^+\pi^-p$ reaction

5. Summary

Preliminary results of the amplitude analysis of the reactions $\pi^+n \rightarrow \pi^+\pi^-p$ and $K^+n \rightarrow K^+\pi^-p$ using the Saclay data has been presented. This data on transversely polarised target give two sets of solutions for 8 moduli of the recoil nucleon transversity amplitudes and 4 relative phases.

Experiments which measure the recoil nucleon polarisation are necessary to select a unique solution.

The Solution 1 is characterised by a small S-wave contribution while this contribution is large in the Solution 2. Both solutions provide clear evidence for the presence of nucleon helicity non flip amplitudes with phases different from those of nucleon helicity flip amplitudes.

In addition we find evidence for presence of B- and possibly Z- exchange in the reaction $K^+n \to K^+\pi^-p$.

Arguments based on $\pi^o\pi^o$ phase shift analyses previously used to reject the Solution 2 assumed small A_1-exchange. The production experiments on polarized target do not however support this assumption.

The final analysis will include the Saclay results at 12 GeV/c. Together with the CERN-Cracow-Munich data at 17.2 GeV/c, this will allow to examine also the energy dependence of the amplitudes.

*. Present address : University of Michigan, Ann Arbor, USA.

**. Present address : KEK, National Laboratory for High Energy Physics, Japan.

References

1. Halzen, F., and Michael, C. :1971, Phys. Letters 36B, 367.
Kelly, R.L. : 1972, Phys. Letters 39B, 635.
Cozzika, G., et al.: 1972, Phys. Letters 40B, 281.
2. Fujisaki, M., et al.: 1979, Phys. Letters 80B, 314.
3. Fujisaki, M., et al.: 1979, Nucl. Phys. B152, 232.
4. Johnson, P.W., and Thomas, G.H.: 1977, Phys. Rev. D16, 2783.
5. Becker, H., et al.: 1979, Nucl. Phys. B150, 301.
Becker, H., et al.: 1979, Nucl. Phys. B151, 46.
6. de Lesquen, A., et al., Contributed paper to this Workshop.
7. Estabrooks, P., and Martin, A.D. : 1972, Phys. Letters 41B, 350.
Estabrooks, P., and Martin, A.D.: 1974, Nucl. Phys. B79, 301.
8. Estabrooks, P., et al.: 1976, Phys. Letters 60B, 473.
Estabrooks, P., et al.: 1978, Nucl. Phys. B133, 490.
Estabrooks, P., et al. : 1978, Phys. Rev. D17, 658.
9. Wicklund, A.B., et al. : 1978, Phys. Rev. D17, 1197.
10. Irwing, A.C., and Michael, C.: 1974, Nucl. Phys. B82, 282.
11. Kimel, J.D., and Owens, J.F.: 1977, Nucl. Phys. B122, 464.

DISCUSSION

- C. Dover :

Have you considered a reaction like

$$K^- p \rightarrow K^- K^+ \Lambda \qquad (\text{or } \Sigma^0)$$

in order to get some information on the KK → KK amplitudes ?

- M. Svec :

No. I do not think these reactions have been yet measured on polarized target despite the obvious interest in obtaining the KK → KK amplitudes by extrapolation.

- L. van Rossum :

Recently the CERN-Cracow-Munich group reported their results on partial wave analysis of $\pi^- p \rightarrow K^- K^+ n$ reaction at 18 GeV/c (1). This measurement on polarized proton target could provide some information on $\pi\pi \rightarrow K\bar{K}$ amplitudes.

1. Görlich, L., et al., 'A model independent partial wave analysis of the reaction $\pi^- p \rightarrow K^+ K^- n$ at approximately 18 GeV/c'. CERN preprint, CERN-EP/79-139.

DETERMINATION OF $g_{\pi\Lambda\Sigma}$ AND $g_{\pi\Sigma\Sigma}$ COUPLINGS FROM ADLER CONDITIONS

R. Prasad and C.P. Singh
Department of Physics - Banaras Hindu University, India

Presented by C.P. Singh

1. Introduction

Recently we have derived (1) the PCAC Adler consistency conditions on the invariant amplitudes of the elastic as well as inelastic pion-charmed baryons processes $\pi C_0 \to \pi C_0$, $\pi C_1 \to \pi C_1$, $\pi C_0 \to \pi C_1$ and $\pi C_0 \to \pi C_1^*$. Numerical evaluations of these conditions with the low-lying intermediate states together with the assumption of unsubtracted form of dispersion relations for the invariant amplitudes yield the various hadronic couplings of charmed baryons. The values of the coupling constants are found as follows :

$$\frac{g^2_{\pi c_0 c_1}}{4\pi} = 58.3, \quad \frac{g^2_{\pi c_1 c_1}}{4\pi} = 28.6, \quad \frac{g^2_{\pi c_1 c_1^*}}{4\pi} = 0.034, \quad \frac{g^2_{\pi c_1^* c_1^*}}{4\pi} = 1.43,$$

if we use

$$\frac{g^2_{\pi c_0 c_1^*}}{4\pi} = 0.15$$

(i.e. $\Gamma c_1^* = 20$ MeV) as input value in the sum rules. In a later investigation (2), we used superconvergence and finite-energy sum rules to determine these coupling constants and now we find;

$$\frac{g^2_{\pi c_0 c_1}}{4\pi} = 51.5, \quad \frac{g^2_{\pi c_1 c_1}}{4\pi} = 24.5, \quad \frac{g^2_{\pi c_1 c_1^*}}{4\pi} = 0.024, \quad \frac{g^2_{\pi c_1^* c_1^*}}{4\pi} = 16.4, \quad \frac{g'^2_{\pi c_1^* c_1^*}}{4\pi} = 6.33$$

using again the same input value of $\frac{g^2_{\pi c_0 c_1^*}}{4\pi}$.

The purpose of this investigation is to use these techniques separately and to determine the values of the various coupling constants for the parallel cases of pion-hyperon scatterings so that we can get a check

E. Ferrari and G. Violini (eds.), Low and Intermediate Energy Kaon-Nucleon Physics, 321–330.

whether the above inconsistency in the values of some coupling constants is a more general one. Moreover, the experimental results (3) for the couplings $g_{\pi\Lambda\Sigma}$ and $g_{\pi\Sigma\Sigma}$ do not satisfy either the SU(3) or the Goldberger-Treiman relations (4). Consequently these coupling constants have been determined by using many different methods (5-8). However, all these values still do not resolve the controversy. It is, therefore, worthwhile problem to estimate the values of these coupling constants by some alternative methods. We will further test the assumption of writing unsubtracted dispersion relation obtained in the numerical evaluation of the Adler constraints on the invariant amplitude and we use finite-energy sum rule for this purpose. Lastly we will test the assumption of saturation of the dispersion relation with the low-lying states by taking all the known resonances for the cases of $\bar{K}N \to \Lambda\pi$ and $\Lambda\pi \to \Sigma\pi$.

2. Method of Calculation

The derivation of the Adler consistency conditions for the processes

$$\pi + \Lambda \to \pi + \Lambda \tag{1}$$

$$\pi + \Sigma \to \pi + \Sigma \tag{2}$$

has been done by many authors (5, 9). If we decompose the matrix elements for the meson-baryon scattering $M_1(K) + B_1(p_i) \to M_2(q) + B_2(p_f)$ into the usual invariant amplitude :

$$\bar{u}(p_f)\, Mu\,(p_i) = \bar{u}(p_f)\left[-A(s,t,K^2) + \hat{K}\,B(s,t,k^2)\right]u(p_i) \tag{3}$$

where

$$\nu = -K\cdot(p_i + p_f)/(m_{B_1} + m_{B_2})$$

$$\nu_B = K\cdot q/(m_{B_1} + m_{B_2})$$

$$s = -(K + p_i)^2$$

$$t = -(K - q)^2$$

Here we have shown K^2 explicitly in A and B, since we want to explore the consequences when the incident pion is taken off its mass-shell. The Adler constraints on the invariant amplitude A for the processes (1) and (2) are found as follows :

$$A\,(\nu = o,\ \nu_B = o,\ K^2 = o) = o \tag{4}$$

$$A\,(\nu = o,\ \nu_B = o,\ K^2 = o) = \frac{16\pi}{m_\Sigma}\ \frac{g^2_{\pi\Sigma\Sigma}}{4\pi}\ K_{\Sigma\Sigma\pi}(o) \tag{5}$$

Here $K_{\Sigma\Sigma\pi}$ is the pion form factor for the $\pi\Sigma\Sigma$ vertex evaluated at $K^2 = o$. This form factor is normalized as $K_{\Sigma\Sigma\pi}\ (K^2 = m^2_\pi) = 1$. Expanding

$K_{\Sigma\Sigma\pi}$ at $K^2 = m_\pi^2$ and assuming the deviation at $K^2 = o$ to be small, we can take $K_{\Sigma\Sigma\pi}(o) \approx 1$. The numerical result of Adler in the πN case support this conclusion.

The PCAC consistency condition for the strong interaction inelastic process :

$$\pi + \Lambda \rightarrow \pi + \Sigma \tag{6}$$

can be derived (1) and found as :

$$A\,(\nu = o,\ \nu_B = o,\ K^2 = o) = \frac{g^2_{\pi\Lambda\Sigma} \quad K_{\pi\Lambda\Sigma(o)}}{m_\Lambda + m_\Sigma} \tag{7}$$

Similarly for the process

$$\pi + \Lambda \rightarrow \pi + Y_1^* \tag{8}$$

We get four invariant amplitudes because Y_1^* has spin $\frac{3}{2}$ and the general structure is

$$\bar{\psi}_\alpha(p_f)\, M_\alpha\, \gamma_5 u(p_i) = \bar{\psi}_\alpha(p_f) \left[a_1 K_\alpha + a_2 q_\alpha + \hat{K}\, b_1 K_\alpha + \hat{K}\, b_2 q_\alpha\right]\gamma_5\, u(p_i) \tag{9}$$

We get the consistency condition on only one amplitude a_2 :

$$a_2\,(\nu = o,\ \nu_B = o,\ K^2 = o) = o \tag{10}$$

Similarly for the inelastic process

$$\bar{K} + N \rightarrow \Lambda + \pi \tag{11}$$

We get the consistency condition as follows :

$$A\,(\nu = o,\ \nu_B = o,\ K^2 = o) = \frac{g_{KN\Lambda} \quad g_{\pi\Lambda\Sigma}\, K_{\pi\Lambda\Sigma}\ (o)}{m_\Lambda + m_\Sigma} \tag{12}$$

3. Numerical Estimation

For numerical estimation of these conditions, we write the unsubtracted dispersion relation for the amplitude A as follows :

$$A(o,\, t) = (\text{Born terms}) + \frac{2}{\pi}\int_0^\infty \frac{\mathrm{Im}\ A(\nu' t')}{\nu'}\, d\,\nu' \tag{13}$$

To find the resonance contribution, we relate A with the amplitude f_1 and f_2 as follows :

$$\frac{A}{4\pi} = \left[\frac{\left(W + \frac{m_{B_1} + m_{B_2}}{2}\right)}{\left\{(E_i + m_{B_1})\,(E_f + m_{B_1})\right\}^{\frac{1}{2}}} \, f_1 - \frac{\left(W - \frac{m_{B_1} + m_{B_2}}{2}\right) f_2}{\left\{(E_i - m_{B_1})\,(E_f - m_{B_2})\right\}^{\frac{1}{2}}} \right] \qquad (14)$$

where $W = \sqrt{s}$ is the total c.m. energy and

$$E_i = \frac{W^2 + m_{B_1}^2}{2W}, \quad E_f = \frac{W^2 + m_{B_2}^2 - m_\pi^2}{2W}, \qquad (15)$$

for all the processes except $\bar{K}N \to \Lambda\pi$ for which

$$E_i = \frac{W^2 + m_N^2 - m_K^2}{2W}, \quad E_f = \frac{W^2 + m_\Lambda^2}{2W} \qquad (16)$$

The amplitudes f_1 and f_2 are given by

$$f_1 = \sum_{\ell=o}^{\infty} \left[f_{\ell+}\, P'_{\ell+1}\,(x) - f_{\ell-}\, P'_{\ell-1}\,(x) \right], \quad f_2 = \sum_{\ell=1}^{\infty} (f_{\ell-} - f_{\ell+})\, P'_\ell\,(x) \qquad (17)$$

where

$$x = \cos\Theta = 1 + \frac{m_M^2}{E_i^2 - m_{B_1}^2}\,.$$

For off-mass shell nature of pion, we multiply (17) by a factor of

$$\left(\frac{q_{off}}{q_{on}}\right)^{\ell}$$

where q_{off} (q_{on}) is the magnitude of the centre-of-mass momentum of pion in the off mass (on-mass) shell limit. We evaluate the contributions in the narrow-width-approximation :

$$I_m\, f_{\ell\pm} = \frac{1}{(|\vec{P}_i|\;|\vec{P}_f|)^{\frac{1}{2}}}\;\frac{\pi M_s\;\phi\;(\Gamma_1\;\Gamma_2)^{\frac{1}{2}}}{(M_s^2 - s)} \qquad (18)$$

where M_s is the mass of the resonance, Γ_1, Γ_2 are the partial decay widths of the resonance and ϕ is the relative phase of the resonant amplitudes.

4. Superconvergence Sum Rules

We write superconvergence sum rules for the processes

$$\pi^+ + \Sigma^- \rightarrow \pi^- + \Sigma^+ \quad (19)$$

$$\pi^+ + \Sigma^- \rightarrow \pi^- + Y_1^{*+} \quad (20)$$

$$\pi \ + \Lambda \rightarrow \pi \ + \Lambda \quad (21)$$

$$\pi \ + \Lambda \rightarrow \pi \ + Y_1^{*} \quad (22)$$

Processes (19) - (20) are 'exotic' exchange processes. Thus by assuming that the possible t-channel Regge trajectories with I = 2 for such processes have negative intercept at zero momentum transfer, we can write

$$\int_{-\infty}^{\infty} \mathrm{Im} \ B^{I=2} \ (\nu, \ t = o) \ \ d\nu = o \quad (23)$$

$$\int_{-\infty}^{\infty} \mathrm{Im} \ a_2^{I=2} \ (\nu, \ o) \ \ d\nu = o \quad (24)$$

$$\int_{-\infty}^{\infty} \mathrm{Im} \ b_1^{I=2} \ (\nu, \ o) \ \ d\nu = o \quad (25)$$

Similarly we get for the process (22), the following superconvergence relation :

$$\int_{-\infty}^{\infty} \mathrm{Im} \ b_2^{I=1} \ (\nu, \ o) \ \ d\nu = o \quad (26)$$

where $\nu = (s-u)/2 \ (m_{B1} + m_{B2})$. For the process (21), we use the finite-energy sum rule :

$$\int_{o}^{\omega} \mathrm{Im} \ B^{(I=o)} \ (\nu, \ t) \ d\nu = \sum_i \frac{\beta_i \ (t) \ \ \omega^{\alpha_i +1}}{\alpha_i + 1} \quad (27)$$

However, for (21) Renner and Zerwas (8) have found that the contributions of f and ε trajectories are small and can be ignored. So we finally get :

$$\int_{o}^{\omega} \mathrm{Im} \ B^{(o)} \ \ (\nu, \ o) \ d\nu = o \quad (28)$$

5. Results and Discussions

The coupling constants obtained from the Adler consistency conditions for the processes (1), (2), (6) and (8) by saturating with the low-lying states Λ, Σ, Y_1^*, are only

$$\frac{g^2_{\pi\Lambda\Sigma}}{4\pi} = 13.2, \quad \frac{g^2_{\pi\Sigma\Sigma}}{4\pi} = 12.2, \quad \frac{g^2_{\pi Y_1^* Y_1^*}}{4\pi} = 1.7$$

We have used $g_{\pi\Lambda Y_1^*}$, and $g_{\pi\Sigma Y_1^*}$ as the input experimental values as

$$\frac{g^2_{\pi\Lambda Y_1^*}}{4\pi} = 6.7\, m_\pi^2, \quad \frac{g^2_{\pi\Sigma Y_1^*}}{4\pi} = 4.2\, m_\pi^2.$$

We have ignored p-wave coupling $g'_{\pi Y_1^* Y_1^*}$ in the low-energy approximation.

Similarly the coupling constants obtained from the superconvergence relations (23), (24), (25, (26) and (27) when the above experimental values for $g_{\pi\Lambda Y_1^*}$ and $g_{\pi\Sigma Y_1^*}$ are again used as input values are :

$$\frac{g^2_{\pi\Lambda\Sigma}}{4\pi} = 11.02, \quad \frac{g^2_{\pi\Sigma\Sigma}}{4\pi} = 5.0, \quad \frac{g^2_{\pi Y_1^* Y_1^*}}{4\pi} = 19.3, \quad \frac{g'^2_{\pi Y_1^* Y_1^*}}{4\pi} = 0.06.$$

Thus we find that the inconsistency reported in our papers (1, 2) is found again for the parallel processes in pion-hyperon scattering. Thus we can safely say that the assumptions of unsubtracted dispersion relations and the saturation with low-lying states are acceptable only to a limited extent.

In order to explore these questions further. We saturate the condition (12) for the process $\bar{K}N \rightarrow \Lambda\pi$ and (7) for $\pi\Lambda \rightarrow \pi\Sigma$ with all the known poles and resonances. The coupling constants used are given (4, 10, 11) as follows :

$$\begin{aligned}
&g(N(1470) \rightarrow \pi p) = 4.5 && g(N(1470) \rightarrow K\Lambda) = -4.6 \\
&g(N(1550) \rightarrow \pi p) = 0.4 && g(N(1550) \rightarrow K\Lambda) = 0.2 \\
&\frac{1}{m_\pi}\, g(N(1518) \rightarrow \pi p) = 10.8 && \frac{1}{m_\pi}\, g(N(1518) \rightarrow K\Lambda) = -11.6 \qquad (29) \\
&\frac{1}{m_\pi}\, g(\Sigma(1385) \rightarrow \pi\Lambda) = 9.1 && \frac{1}{m_\pi}\, g(\Sigma(1385) \quad Kn) = -\,7.4 \\
&g_{\bar{K}N\Lambda} = -\,11.2 && g_{\bar{K}N\Sigma} = 4.5
\end{aligned}$$

The contribution of various pole term and resonance contributions are tabulated in table (1) and (2). Thus finally we get

Table 1 - Pole term and resonance contributions for KN $\longrightarrow$ $\Lambda\pi$

State J^P	Γ_{tot}(GeV)	$10^2 x(\Gamma_1\Gamma_2)^{\frac{1}{2}}$	Contributions	
			s-channel	u-channel
$\Sigma(1197)\ \frac{1}{2}^+$	...	...	$\frac{g_{\pi\Lambda\Sigma}\ g_{\pi\Sigma\Sigma}}{m_\Lambda + m_\Sigma}$	0
$\Sigma(1385)\ \frac{3}{2}^+$	0.036	0.85	36.62	44.83
$\Sigma(1670)\ \frac{3}{2}^-$	0.05	2	15.85	17.90
$\Sigma(1750)\ \frac{1}{2}^-$	0.08	0	0	0
$\Sigma(1765)\ \frac{5}{2}^-$	0.10	0.39	2.64	2.92
$\Sigma(1915)\ \frac{5}{2}^+$	0.05	0.07	-0.45	-0.49
$\Sigma(2030)\ \frac{7}{2}^+$	0.12	1.59	-9.40	-10.12

Table 2 - Pole term and resonance contributions for $\pi\Lambda \longrightarrow \pi\Sigma$

State J^P	Γ_{tot}(GeV)	$10^2 x(\Gamma_1\Gamma_2)^{\frac{1}{2}}$	ϕ	Contributions
		s-channel		
$\Sigma(1197)\ \frac{1}{2}^+$	...	...	...	$-1.94\ g_{\pi\Lambda\Sigma}$
$\Sigma(1385)\ \frac{3}{2}^+$	...	...	...	-15.704
$\Sigma(1670)\ \frac{3}{2}^-$	0.05	0.5	- 1	5.80
$\Sigma(1750)\ \frac{1}{2}^-$	0.08	2.0	+ 1	0.62
$\Sigma(1765)\ \frac{5}{2}^-$	0.10	2.72	+ 1	28.25
$\Sigma(1915)\ \frac{5}{2}^+$	0.07	0.49	+ 1	-4.89
$\Sigma(1920)\ \frac{1}{2}^+$	0.17	2.38	+ 1	-1.92
$\Sigma(1940)\ \frac{3}{2}^-$	0.28	3.92	+ 1	-12.23
$\Sigma(2030)\ \frac{7}{2}^+$	0.17	3.4	- 1	-33.48
$\Sigma(2080)\ \frac{3}{2}^+$	0.25	2.25	+ 1	2.44
		t-channel		
$N(938)\ \frac{1}{2}^+$	...	...	...	0
$N(1470)\ \frac{1}{2}^+$	...	...	...	8.4
$N(1518)\ \frac{3}{2}^-$	...	...	...	-0.13
$N(1550)\ \frac{1}{2}^-$	...	...	...	-4.77
$N(1675)\ \frac{5}{2}^-$	0.17	0.18	- 1	-8.27
$N(1688)\ \frac{5}{2}^+$	0.13	0.10	- 1	12.52
$N(1700)\ \frac{1}{2}^-$	0.30	3.4	- 1	-1.18
$N(1780)\ \frac{1}{2}^+$	0.227	1.1	+ 1	-1.64
$N(1860)\ \frac{3}{2}^+$	0.296	1.48	- 1	-2.94

$$\frac{g^2_{\pi\Lambda\Sigma}}{4\pi} = 8.02, \qquad \frac{g^2_{\pi\Sigma\Sigma}}{4\pi} = 13.6 \tag{30}$$

The use of unsubtracted dispersion relation can always be questioned. In order to see the effect of this assumption we write the finite energy sum rule for $\bar{K}N \rightarrow \Lambda\pi$,

$$A^+(o, t) = (\text{Born terms}) + \frac{2}{\pi} \int_{\nu_o}^{\nu_1} \frac{d\nu}{\nu} \, \text{Im } A^+(\nu, t) - \frac{2}{\pi} \sum_i \frac{\beta_i(t)}{\alpha_i(t)} \left(\frac{\nu_i}{\nu_o} \right)^{\alpha_i(t)} \tag{31}$$

where

$$A^+(\nu, t) = \frac{1}{2} \left[A(\nu, t) + A(-\nu, t) \right]$$

The last term on the right-hand side is the high energy contribution arising due to the exchange degenerate K^*, K^{**} Regge trajectories and the parameters used (10) are :

$$\beta_i(t) = 1.57 \, (\text{GeV})^{-1}, \qquad \alpha_i(t) = 0.15 + t$$

$$\nu_1(t) = \nu_1(o) + \frac{t}{2(m_N + m_\Lambda)}, \qquad \nu_o = \frac{s_o}{(m_N + m_\Lambda)}$$

with $s_o = 1 \, (\text{GeV})^2$ and $\nu_1(o) = 1.8$ GeV. We get from (31),

$$\frac{g^2_{\pi\Lambda\Sigma}}{4\pi} = 10.5$$

which is about 25% different than the value given in equation (30). However, this difference matches with the uncertainty involved in the values of the coupling constants arising due to widths of the resonances and, therefore, justifies our assumption regarding unsubtracted dispersion relation.

References

1. Prasad, R., and Singh, C.P.: 1979, Phys. Rev. D20, 256.
2. Prasad, R., and Singh, C.P.: in press, Phys. Rev.
3. Chan, C.H., and Meiere, F.T.: 1968, Phys. Lett. 20, 568.
4. Nagels, M.M. et al.: 1976, Nucl. Phys. B109, 1.
5. Chan, C.H., and Smalley, L.L.: 1970, Phys. Rev., D2, 2635.

6. Singh, C.P.: 1971, Phys. Rev. D4, 270.
7. Engels, J., and Pilkuhn, H.: 1971, Nucl. Phys. B31, 531.
8. Renner, B., and Zerwas, P.: 1971, Nucl. Phys. B35, 397.
9. Martin, B.R.: 1966, Nuovo Cimento 43 A, 629.
10. Field, R.D., and Jackson, J.D.: 1971, Phys. Rev. D4, 693.
11. Dubricka, S. et al.: 1974, Nucl. Phys. B88, 477.

DISCUSSION

- J. E. Bowcock :

I wanted to query the assumption of saturating the dispersion integrals with resonance poles. One knows that in general the non-resonance contributions to integrals can be large and could certainly amount to the 20-30% effects which you have considered.

- C. P. Singh :

Yes, you are right. The background contribution is usually large. However, in the saturation of superconvergence relations and FESR, we neglect these contributions.

But in the derivation of Adler consistency conditions we use the fact that the background contributions vanish in the required low-energy limit.

- C. Dover :

Where did you obtain the $C_1^* \to C_1\pi$ width that you quoted? Your values for $g^2_{C_1 C_0\pi}$ and $g^2_{C_1 C_1\pi}$ look very large. Presumably your coupling constants deviate greatly from SU(4) symmetry.

- C. P. Singh :

We use $C_1^* \to C_0\pi$ width of 20 MeV as input value. We do not have any experimental information on the widths of any charmed baryon resonances.

Regarding your second question, there is a very large difference between the values of $g_{\pi\Lambda\Sigma}$ and $g_{\pi C_0 C_1}$ (the charmed baryon analogue of strange baryons). Therefore, you are right when you say that SU(4) symmetry is very badly broken.

- G. Violini :

I think that the case you mentioned at the beginning is not so strong. Actually the values of the couplings you showed are in reasonable agreement except perhaps one. However, going to your results I have a few comments.

First why do you assume SU(3) for so many couplings; this obviously affects the $g^2_{\pi\Lambda\Sigma}$ value (look for instance at the relation which gives $g_{K\Lambda N}\ g_{\pi\Lambda\Sigma}$). Indeed a reasonable change in NΛK coupling would change by about 20% the πΛΣ one. My opinion is that this would not be so bad because it is hard to see how approximate saturation may give a better value.

There is another point; I wonder if the present status of K matrix multichannel analysis both at low and intermediate energy would not better suggest to write dispersion relations directly for πΛ or πΣ scattering. Now obviously the latter will have the problem of the I=2 multichannel analysis both at low and intermediate energy would not better suggest to write dispersion relations directly for πΛ or πΣ scattering. Now obviously the latter will have the problem of the I=2 part coming from crossing, but probably the former can be used to obtain $g^2_{\pi\Lambda\Sigma}$, with suitable assumptions for the high-energy.

- C. P. Singh :

Yes, so far as the value of $g_{\pi\Lambda\Sigma}$ is concerned, most of the determinations are close to 12 except for the values reported by Chan and Smalby and Chan and Meiere which are about 20 and 18 respectively.

Regarding the use of SU(3) values of the couplings for the vertex, I would like to say that we have used SU(3) values of the couplings for vertices like $N(1470)\rightarrow\bar{K}\Lambda$, $N(1550)\rightarrow\bar{K}\Lambda$, $N(1518)\rightarrow\bar{K}\Lambda$ etc., for which we cannot get information from any other source. So in order to become consistent, I have used SU(3) values for other couplings also. Moreover, Jackson and Feld have successfully used these couplings to evaluate the FESR for this reaction.

CHIRAL SYMMETRY BREAKING : KAON-NUCLEON SYSTEM

A.M. Rodriguez-Vargas*
International Centre for Theoretical Physics, Trieste
and
Departamento de Fisica, Universidad de Los Andes, Bogotà.

1. Introduction

During the last year the KN sigma term has been recalculated by several authors (1-4). As we know the sigma term is a quantity which gives an indication of the chiral symmetry breaking terms in the strong interaction Hamiltonian and this evaluation requires that limit of zero mass pseudoscalar meson is taken, and therefore approximations are required in order to use experimental data obtained from physical particle processes.

The KN sigma term seems to be now between 400 MeV and 600 MeV (central value) but unfortunately the errors are still of the same order of magnitude.

The large error is a reflection of our poor knowledge of low-energy scattering length and of the coupling constant of the Y_o^* and Y_1^*.

2. Theoretical comments on the Sigma Term

We assume that the hadronic Hamiltonian can be decomposed into a part H_o, which is invariant under SU(3) ⊗ SU(3) transformations and an additional part $\varepsilon H'$, which describes the symmetry breaking contributions

$$H = H_o + \varepsilon H' \tag{1}$$

We assume also that the symmetry breaking piece of the total Hamiltonian conserves parity and strangeness and does not admit operators of isospin or hypercharge 2; then the only representation allowed for the components of H' is the $(3, \bar{3}) \times (\bar{3}, 3)$ representation of SU(3) ⊗ SU(3). The most general form for $\varepsilon H'$ is :

$$\varepsilon H' = u_o + c\, u_8$$

which has been suggested by Gell-Mann, Oakes and Renner (GMOR) (5) and

E. Ferrari and G. Violini (eds.), Low and Intermediate Energy Kaon-Nucleon Physics, 331–338.

where c is the only parameter of the model to be determined.

The sigma term is, in some way, a measure of this symmetry breaking and is defined by (6) :

$$\sigma_{NN}^{ba} = \langle N \left| \left[F_b^5(o), \left[F_a^5(o), \varepsilon H'(o) \right] \right] \right| N \rangle \qquad (2)$$

where

$$F_a(x_o) = \int d^3x\, V_a^o(x)$$

$$F_a^5(x_o) = \int d^3x\, A_a^o(x)$$

generate the Lie-Algebra of SU(3) ⊗ SU(3) and $V_a^\mu(x)$ $(A_a^\mu(x))$ are eight SU(3) vector (axial-vector) currents.

After a little algebra the sigma term takes for KN case the form

$$\sigma_{NN}^{KK} = \frac{1}{3}\left(\sqrt{2} - \frac{c}{2}\right) \langle N \mid \sqrt{2}\, u_o + \sqrt{3}\, u_3 - \frac{1}{2} u_8 \mid N \rangle \qquad (3)$$

Using the Wigner-Eckart theorem

$$\langle B_a \mid u_b \mid B_c \rangle = i f_{abc} F + d_{abc} D$$

$$\langle B_a \mid u_o \mid B_c \rangle = \alpha\, \delta_{ac}$$

we obtain (6,7)

$$\langle p \mid u_3 \mid p \rangle = (2\sqrt{3}\, c)^{-1} (2M_N + M_\Sigma - 3M_\Lambda)$$

$$\langle n \mid u_3 \mid n \rangle = -(2\sqrt{3}\, c)^{-1} (2M_N + M_\Sigma - 3M_\Lambda)$$

$$\langle p \mid u_8 \mid p \rangle = (2c)^{-1} (2M_N - M_\Sigma - M_\Lambda)$$

$$\langle n \mid u_8 \mid n \rangle = (2c)^{-1} (2M_N - M_\Sigma - M_\Lambda)$$

Since u_o breaks SU(3) ⊗ SU(3) and u_8 breaks SU(3) as well as SU(3) ⊗ SU(3), we cannot allow $\langle N | u_o | N \rangle$ to be very different from $\langle N | u_8 | N \rangle$.

Therefore, one obtain the following estimates for the sigma term

$$\sigma_{pp}^{KK} \sim 200 \text{ MeV}$$

$$\sigma_{nn}^{KK} < \sigma_{pp}^{KK}$$

because of the (-) sign of the equation.

Most of the experimental results for σ_{NN}^{KK} yield values larger (by about a factor of 2) than the theoretical estimates of the GMOR.

3. The Calculation of the Sigma Term

We consider the scattering amplitude for the process

$$K(q_1) \;+\; N(p_1) \longrightarrow K(q_2) \;+\; N(p_2)$$

Since we are interested in the smooth behavior of the scattering amplitude in the neighborhood of the Weinberg (8) point $\nu = t = q_1^2 = q_2^2 = 0$, we have to construct a smoothly varying amplitude in this region. Thus one has to subtract from T^+ the Λ pole, the Σ pole and the resonances $Y_o^*(1405)$ and $Y_1^*(1385)$.

$$F(\nu, t, q_1^2, q_2^2) = \mathrm{Re}\, T^+ - T_B^+ - T_{Y_o^*}^+ - T_{Y_1^*}^+ \tag{4}$$

Following Weinberg's (8) original suggestion and taking into account the symmetry properties of F we can write

$$F(\nu, t, q_1^2, q_2^2) = Am_K^2 + Bt + C(q_1^2 + q_2^2) + D\nu^2 + R(\nu, t, q_1^2, q_2^2) \tag{5}$$

where R measures the deviation of F from linearity.

For the four expansion parameters A, B, C, D we have to find four equations :

a) at the physical threshold we get

$$F(m_K, 0, m_K^2, m_K^2) = m_K^2(A + 2C + D) + R(m_K, 0, m_K^2, m_K^2) \tag{6}$$

b) at the unphysical point $\nu = t = 0$, but on shell :

$$F(0, 0, m_K^2, m_K^2) = m_K^2(A + 2C) + R(0, 0, m_K^2, m_K^2) \tag{7}$$

c) at $\nu = q_2^2 = 0$; $t = q_1^2 = m_K^2$ (The Adler consistency condition) we have :

$$F(0, m_K^2, m_K^2, 0) = m_K^2(A + B + C) + R(0, m_K^2, m_K^2, 0) = 0 \tag{8}$$

d) expanding F in power of $k^2 \cos\phi$ around, the physical threshold and comparing the coefficients of $k^2 \cos\phi$, we obtain :

$$\frac{1}{4\pi}\left(2B + \frac{m_K}{M_N}D\right) + C_K + \frac{1}{4\pi}\frac{m_K}{M_N}\frac{\partial R}{\partial \nu^2}(m_K, 0, m_K^2, m_K^2) + \frac{1}{2\pi}\frac{\partial R}{\partial t}(m_K, 0, m_K^2, m_K^2) - X \tag{9}$$

Where

$$C_K = \frac{1}{4\pi} \sum_{\Lambda,\Sigma,Y_o^*,Y_1^*} \left.\frac{\partial T_i^+}{\partial (K^2 \cos\theta)}\right| K^2\cos\theta = 0$$

and

$$X=(16M_N^2)^{-1}(2a_1+b_o+b_1)+\frac{1}{4}(2a_{11}+b_{01}+b_{11})+\left[\frac{1}{2}+\frac{3}{4}\left(\frac{m_K}{M_N}\right)\right](2a_{13}+b_{03}+b_{13}) \quad (10)$$

As we know from the low energy theorem

$$\sigma_{NN}^{ba} = - f_a f_b T_{ba}^+ (0, 0, 0, 0) \quad (11)$$

in order to calculate the value of the sigma term, one must know the value of $D_p(o)$ ($D_n(o)$), the real part of the amplitude for the unphysical value of the energy $w = 0$. This quantity can be determined by dispersion relations.

Using equations 4 to 11 it is straightforward to calculate F (0, 0, 0, 0), once we know T^+ (0, 0)

a) we obtain for the $K^{\pm}p$ scattering

$$-m_K^2 A = D_p(0) - 0.5263\, I + 0.2756\, g_\Lambda^2 + 0.0757\, g_\Sigma^2 - 4.5609\, g_{Y_o^*}^2 + 0.0594\, g_{Y_1^*}^2 + 4\pi m_K^2 (X - C_K) + R_\Delta$$

where

$$X = \frac{1}{16M_N^2}(2a_{K^+p} + a_{K^-N}^{I=0} + a_{K^-N}^{I=1}) + \frac{1}{4}(2b_{K^+p} + b_{K^-N}^{I=0} + b_{K^-N}^{I=1}) + \left(\frac{1}{2}+\frac{3}{4}\frac{m_K}{M_N}\right)(2c_{K^+p} + c_{K^-N}^{I=0} + c_{K^-N}^{I=1}) \quad (12)$$

In this equation we use

$$\frac{g_{Y_o^*}^2}{4\pi} = 0.67 \pm 0.14 \; ; \quad \frac{g_{Y_1^*}^2}{4\pi} = 3.95 \pm 1.87$$

b) We obtain for the $K^{\pm}n$ scattering

$$-m_K^2 A = D_n(0) - 0.5263 I + 0.1514\, g_{K\Sigma N}^2 - 4\pi\, m_K^2\, C_K + 4\pi\, m_K^2\, X$$

where

$$X = \frac{1}{16M_N^2} (a_{KN}^{I=1} + a_{KN}^{I=0} + 2a_{\bar{K}N}^{I=1}) + \frac{1}{4} (b_{KN}^{I=1} + b_{KN}^{I=0} + b_{\bar{K}N}^{I=1}) +$$

$$(\frac{1}{2} + \frac{3}{4} \frac{m_K}{M_N}) (c_{KN}^{I=1} + c_{KN}^{I=0} + 2c_{\bar{K}N}^{I=1})$$

we obtain for the sigma term

a) $\sigma_{pp}^{KK} = 493 \pm 717$ MeV $\left(\frac{g_\Sigma^2}{4\pi} = 2 \right)$

b) $\sigma_{nn}^{KK} = 638 \pm 438$ MeV

4. Discussion

Both $K^\pm p$ and $K^\pm n$ have been recently determined by another method (by the use of finite contour fixed t dispersion relations) and one obtains for σ_{NN}^{KK} a value (4)

$$375 \text{ MeV} < \sigma_{pp}^{KK} < 550 \text{ MeV}$$

$$300 \text{ MeV} < \sigma_{nn}^{KK} < 650 \text{ MeV}$$

Our error for σ_{NN}^{KK} is much larger than the current values given in the literature (tables 1 and 2) for a similar approach.

Table 1 - Estimates of the Kp sigma term

Authors	σ_{pp}^{KK} (MeV)
Gensini (1973) (10)	410^{+22}_{-28}
Kopp et al. (1972) (10)	345^{+270}_{-550}
Nasrallah and Schilcher (1973) (10)	480
Reya (1972) (10)	540^{+160}_{-160}
Sastry and Tiwari (1975)	550±50
Oades (1980) (4)	$375 < \sigma < 550$
Di Claudio et al. (1979) (1)	493±717
Gensini (1980) (12)	134^{+93}_{-11}
Reya (1973) (10)	480±110

Table 2 - Estimates of the Kn sigma term

Authors	σ_{nn}^{KK} (MeV)
Harnois (1975) (11)	1160
Oades (1980) (4)	$300 < \sigma < 650$
Rodriguez-Vargas and Violini (1980) (2)	638 ± 438
Gensini (1980) (12)	40^{+115}_{-14}

It is still an open question whether other approaches allow one to determine σ_{NN}^{KK} more reliably.

This situation could suggest to look at the $K^{\pm}n$, which could reduce the size of the errors; however it is necessary to be able to discriminate between the alternative solutions of the $a_{p3/2}$ scattering lengths repor_ ted in the literature.

Acknowledgments

We express our gratitude to Banco de la República de Colombia and COLCIENCIAS for the financial support.

We acknowledge the Centro de Cómputo de la Universidad de los Andes for providing us with the facilities necessary to make our calculations.

We also thank Professor G. Violini for many useful discussions.

*. Permanent address : Departamento de Fisica, Universidad de Los Andes, Bogota.

References

1. Di Claudio B., Rodriguez-Vargas, A.M., and Violini, G.: 1979, Zeit. Physik C3, 75.
2. Di Claudio, B., Rodriguez-Vargas, A.M., and Violini, G. : 1979, Lett. Nuovo Cimento 26, 555.
3. Rodriguez-Vargas, A.M., and Violini, G. : 1980, Zeit.Physik C4, 135.
4. Oades, G.C., "The Low Energy fixed t KN and $\bar{K}N$ Amplitudes", Talk at this Workshop.
5. Gellmann, M., Oakes, R.J., Renner, B. : 1968, Phys. Rev. 175, 2195.
6. Reya, E. : 1974, Rev. Mod. Phys. 46, 545.
7. Gensini, P., Private Communication.
8. Weinberg, S.: 1966, Phys. Rev. Lett. 16, 879.

9. Altarelli, G., Cabibbo, N., and Maiani, L. : 1971, Phys. Lett. 35B, 415; 1971, Nucl. Phys. B34, 621.

10. Gensini, P. : 1973, Nuovo Cimento 17A, 557.
Kopp, G., Walsh, T.F., and Zerwas, P.: 1972, Nucl. Phys. B42, 109.
Nasrallah, N.F., and Schilcher, K. : 1973, Phys. Rev. D7, 810.
Reya, E.: 1973, Phys. Rev. D7, 3472; 1972, Phys. Rev. D6, 200.
Sastry, C.V., and Tiwari, T.N.: 1975, Phys. Rev. D11, 1.

11. Harnois, J.Y.: 1975, Phys. Lett. B56, 379.

12. Gensini, P., "Mesic Atoms and the Sigma-Terms", Talk at this Workshop.

DISCUSSION

- G. Oades :

The RLIC and LBL phase shifts when extrapolated to threshold both favour a negative I=1 $p_{3/2}$ scattering length. A similar result is found when this partial wave is projected out from the fixed-t amplitudes described in my contribution. Finally in the $\bar{K}N$-$\bar{K}N$ partial wave dispersion relation the Y_1^* behaves like an unstable bound state and so a negative scattering length is favoured provided the coupling of the Y_1^* to the $\bar{K}N$ channel is not small.

- G. Violini :

I would like to add a comment on this $p_{3/2}$ scattering length matter. The situation is a little unpleasant because not only different experiments give different sign for it, but also analyses from the same group do give this, even if for energy ranges which are not the same. So this is really a complicated matter. Probably the MAST group data or the TST collaboration may shed some light on this matter. From a theoretical point of view partial wave dispersion relations may help to fix this sign.

- G. Oades :

What does theory have to say about the difference between σ_{pp}^{KK} and σ_{nn}^{KK}.

- G. Violini :

One can answer the question whether the σ_{pp}^{KK} and σ_{nn}^{KK} are supposed to have some relation noticing that the sigma term for Kp and that for Kn should differ for the following reason. They are given by a matrix element in p(or n) states of a sum of operators transforming like u_0, u_3 and u_8. Now whereas u_0 and u_8 have the same value for p and n states, Clebsch Gordan tell us that u_3 changes sign. Moreover that a difference should be present can be inferred from the fact that the sigma term is calculated from the crossing-even amplitude. Now for Kp this is $\frac{1}{2}(I=0+I=1)_{\bar{K}N}+(I=1)_{KN}$ whereas for Kn it is given by $(I=1)_{\bar{K}N}+$ $+\frac{1}{2}(I=0+I=1)_{KN}$.

MESIC ATOMS AND THE SIGMA-TERMS*

P.M. Gensini
Istituto di Fisica - Università di Lecce

1. Introductory remarks

This talk is intended to be a short round trip through the domains of atomic, nuclear and particle physics. As such, it will be very far from the degree of completeness many of us could wish.

But since I am not to give a sort of crash course in modern physics here, I feel we can blame the organizers for my being too much informative. I strove to make up for any lost detail with a wide list of references. Here I owe my apologies to many colleagues whose contributions I may have overlooked. Since this talk is intended to be an introduction, not a review, I hope to be easily forgiven.

1.1 Meson-baryon amplitudes and broken chiral SU(3) ⊗ SU(3)

The emergence of an SU(3) ⊗ SU(3) algebra for the time-components of the electroweak hadronic currents suggested soon the approximate symmetry of the hadrons under the chiral group SU(3) ⊗ SU(3), generated by the corresponding charges (1-3). The pattern, in which the symmetry was broken, generated another, related, pattern for the current divergences, whose matrix elements at low momentum transfers should be dominated by the lowest-lying poles or continuum resonances (all the above conventionally labeled by the acronym PCAC (2-4)).

The complex of SU(3) ⊗ SU(3) current and charge algebras plus PCAC yields a family of zero-momentum theorems for hadronic amplitudes. These are, first, Adler's consistency conditions (2, 5): the amplitude vanish when one of the external, pseudoscalar mesons has zero four-momentum, unless the zero is accidentally canceled by a Born-term pole; second, the soft-meson limits (2, 4, 6-8): for two vanishing meson four-momenta the amplitude (or its derivative with respect to an energy variable) is proportional to a doubly integrated, current-divergence (or current-current) commutator, apart from the already mentioned Born terms.

E. Ferrari and G. Violini (eds.), Low and Intermediate Energy Kaon-Nucleon Physics, 339–362.

A whole generation of papers has been devoted to the extrapolation of these rigorous but unphysical results onto the real world of massive mesons (9-14). Excellent reviews are available, so I shall refer to these for the details (2, 3, 15).

Three methods emerge as the more popular and promising ones. In chronological order, as well as in order of decreasing model-dependence and increasing use of experimental data, they are (see also the review in Reya (16)) :

a) the mass-and-energy, doubly dispersive technique by Fubini, Furlan, de Alfaro and Rossetti (10-13), a.k.a. Fubini-Furlan (or FF) sum rules;

b) the expansion by Altarelli, Cabibbo and Maiani (17) in terms of on-mass-shell, low-energy parameters, related also to the use of Ward identities (18);

c) the linear extrapolation of Cheng and Dashen (19), evaluating the zero-energy amplitude at a momentum transfer equal to the sum of the meson masses squared.

But a test of the chiral-symmetry breaking pattern through the chain outlined below :

symmetry breaking pattern

$\Downarrow$

current divergences

$\Downarrow$

current-divergence (sigma) commutators

$\Downarrow$

meson-baryon amplitudes

should involve, as the history of SU(3) symmetry has taught us, an investigation over the largest number of meson-baryon combinations, involving then not only pion but also kaon beams.

1.2 Mesic atoms : what for ?

The previous idea soon led to attempts at a "global" determination of the sigma commutators

$$\langle B | \left[Q_j^5(x_o), \dot{Q}_i^5(x_o)^\dagger \right] | A \rangle = -i(2\pi)^3 \; \delta^3(\vec{P}_A - \vec{P}_B)(4E_A E_B)^{\frac{1}{2}} \; \Sigma_{AB}^{ij} \qquad (1.1)$$

using the FF sum rules (20-22). These early attempts were doomed to impotence both by the poor quality of meson-baryon data at that time (bar a few) and by theoretical uncertainties (even the renormalizability of strong interactions was seriously questioned, if not plainly disbelieved).

The attack to the problem continued along the other two lines, and has at least led to consensus on a value for the πN sigma commutator close to 65 MeV (23); it has however failed to produce nothing more than very large upper limits, of about 1 GeV (24), for both the KN sigma commutators.

A look at the FF sum rules (13), as we shall see later on, shows that their main variables are the isospin and the hypercharge of the target. Variations in these parameters, even larger than using the whole baryon octet, can be obtained with nuclear targets.

In this case it is hard to obtain all the information needed for methods b and c, but reliable measurements of the scattering lengths could be enough if we understood the dependences of the terms in the FF sum rules on the target quantum numbers. Ideally, we could think of replacing the missing information with a separation of the different, systematic dependences on mass-number and isospin of the target.

Scattering measurements on nuclei are affected, on a dramatically expanded scale, by the same problems we have to face in the $\bar{K}N$ system; but the nuclear Coulomb field binds the negative mesons at energies less than a MeV below the elastic threshold. Moreover the Coulomb field offers a "natural" scale against which we can easily assess the strength of the meson-nucleus strong interaction, from which a "purely nuclear" scattering length can then be computed.

Even having to deal with nuclear absorption and relativistic equations, the theoretical complexity of the problem does not exceed that of the hydrogen atom, since the bound meson "moves" real close to the atomic nucleus.

Once we have a scattering length, we have to saturate the corresponding FF sum rules (or in our case that combination which contains the negative meson amplitude at the threshold) to translate it into a sigma commutator.

The aim of this talk is to show that very simple ideas about nuclei can lead to adequate approximations. Such a "soft technology" approach can of course be backed up by more complex, microscopic treatments (25-28). We have taken this naive stance, in the hope that it will exhibit more clearly the relevant phenomena determining the size of every piece in the sum rules (29-31).

2. The atomic facet (a detour)

Let us write an effective Lagrangian for a (pseudo) scalar meson field ψ in some (external) fields S, T, U_μ :

$$L = - (\partial_\mu\psi^*\partial^\mu\psi + \mu^2\psi^*\psi) - S\psi^*\psi + T\partial_\mu\psi^*\partial^\mu\psi + + iU_\mu\left[(\partial^\mu\psi^*)\psi - \psi^*(\partial^\mu\psi)\right] - U_\mu U^\mu\psi^*\psi \tag{2.1}$$

leading to the one-particle evolution equation

$$(\partial_\mu - iU_\mu)^2\psi = (\mu^2 + S + \partial_\mu T\partial^\mu)\psi . \qquad (2.2)$$

If the boson has charge e and mass m and we take the external fields as static fields we shall have, instead (32),

$$(i\partial_o - e\phi - u)^2 \psi = \left[(\vec{p} - e\vec{A})^2 + m^2 + 2ms + 2m\vec{\nabla}\cdot(t\,\vec{\nabla})\right]\psi . \qquad (2.3)$$

Such an equation can be interpreted in a two-component theory (33), writing $\psi=\phi+\chi$, with ϕ (or χ) the positive (or negative) energy wave-function. We shall then have

$$i\partial_o\Psi = H\Psi \qquad (2.4)$$

where Ψ is the column vector $\begin{pmatrix}\phi\\ \chi\end{pmatrix}$ and

$$H = \Sigma\left[\frac{1}{2m}(\vec{p} - e\vec{A})^2 + s + \vec{\nabla}\cdot(t\,\vec{\nabla})\right] + m\sigma_3 + e\phi + u, \qquad (2.5)$$

with $\Sigma = \sigma_3 + i\sigma_2$, obeys the Hermiticity condition $H = \sigma_3 H^\dagger \sigma_3$ if the external potentials are real. Complex potentials emerge as a consequence of absorption in the one-particle equation (2.3), for which we have the continuity equation, where $\bar{\Psi} = \psi^\dagger \sigma_3$,

$$\partial_o(\bar{\Psi}\Psi) - \vec{\nabla}\cdot\vec{j} = S, \qquad (2.6)$$

where the current $\vec{j}$ is given by

$$\vec{j} = i\left[(\vec{\nabla}\bar{\Psi})\ \Sigma(\frac{1}{2m} - \mathrm{Re}\,t)\Psi - \bar{\Psi}\,\Sigma(\frac{1}{2m} - \mathrm{Re}\,t)\vec{\nabla}\Psi\right] + \frac{e}{m}\left[\bar{\Psi}\,\Sigma\,\vec{A}\,\Psi\right] \qquad (2.7)$$

and we have, then, the surface and volume absorption contribution

$$S = 2\bar{\Psi}\left[\Sigma\,\mathrm{Im}\,s + \mathrm{Im}\,u\right]\Psi + \vec{\nabla}\cdot\left[(\vec{\nabla}\bar{\Psi})\ \Sigma\,\mathrm{Im}\,t\,\Psi + \bar{\Psi}\,\Sigma\,\mathrm{Im}\,t\,\vec{\nabla}\Psi\right] - 2(\vec{\nabla}\bar{\Psi})\cdot(\Sigma\,\mathrm{Im}\,t\,\vec{\nabla}\Psi) . \qquad (2.8)$$

Further restrictions on the potentials may come from the charge-conjugation properties of the field ψ. For neutral pions (of positive C-parity) we shall then have $u = 0$ (recall that the photon has negative C-parity), but we should in principle work with all three nuclear potentials for charged particles. It is then a matter of simplicity which leads to reduce the "static" equation to the usual form

$$(E - e\phi)^2 \psi = \left[-\nabla^2 + m^2 + 2m\,(s + \vec{\nabla}t\vec{\nabla})\right]\psi, \tag{2.9}$$

for the energy eigenvalues of the mesic atoms, where the Coulomb potential ϕ will include also terms due to the vacuum polarization and the very small electron screening and meson polarizability effects.

This amounts to assume dominance of positive-C exchanges, such as two pions in a S-wave, over negative-C exchanges, such as the vector mesons, at threshold; in our opinion this is not a wholly satisfactory situation, though it is fully consistent with today's phenomenology.

Meson-nucleus scattering lengths are then derived as follows: given the complex eigenvalues of a mesic atom energy spectrum, one can find, for given shapes of the nuclear potentials, complex strength parameters to fit the experimental data. This is possible as long as these potentials act, in the portion of space where the meson wavefunction is nonnegligible, as small perturbations with respect to the long-range Coulomb potential, which gives indeed most of the binding.

From the nuclear potentials so derived one computes the meson-nucleus scattering length and volume, using the low-momentum limit of H in eq. (2.5) with $e = 0$. Such a scattering length will then be a purely nuclear one.

The procedure has been shown to be relatively stable with respect to variations in the potentials' shapes, since the scattering lengths, as well as the energy shifts, depend essentially on the "active" portions of the potentials wells, and are therefore strongly correlated to each other and only weakly correlated to other aspects of the interactions (34-36).

Two facts are also playing important roles. In pion-nucleus interactions, the strong gradient potential induced by the $\Delta(1230)$ makes the 1s level shift the only one strongly correlated to the scattering length, since all others are then strongly correlated to the scattering volume. This in turn limits the range of measurements useful to our analysis to $Z = 12$, since for nuclei heavier than ^{24}Mg the 1s level is completely absorbed inside the nucleus.

In kaon-nucleus interactions instead the gradient term is much weaker than the scalar one, as a consequence of the weaker coupling to the $\bar{K}N$ channels of the $Y_1^*(1385)$ with respect to the $Y_0^*(1405)$, close to the elastic threshold. The second piece then strongly affects also the higher $n = \ell + 1$ levels, circumstance extremely fortunate, since the lowest observed transitions, but for Hydrogen, are $3d \to 2p$ ones, and for nuclei not heavier than ^{35}Cl we still cannot observe but $4f \to 3d$ ones.

In this second case, the non-linearity of eq. (2.9) is necessary in order to reconcile strong attraction and absorption in the K^--nucleus interaction with the repulsive real part of the amplitude at threshold. The net effect of the two features of K^--nucleus interaction is indeed an effective repulsion, generated by the strong absorption of the K^- wave function inside the nucleus (36, 37).

The relevant feature which we like to point out again is the strong, generally non-linear correlation existing between the complex energy shifts and the complex meson-nucleus scattering lengths. This confirms the naive expectation based on the observation that mesons and nuclei interact in mesic atoms at energies much closer to the elastic threshold than any scattering experiment will ever make them to do.

We can then be confident that the FF sum rules may have here a possible ground for a significant test of the broken chiral symmetry. Enough good enough data may be already at hand, as witnessed by the many good reviews on the subject of mesic atoms (38-46) and the recent Karlsruhe compilation (47) of experimental data.

3. Sum rules: the nuclear and particle facet

Defining the correlation function of two axial divergences

$$\rho_{iA}(\nu, q^2) = \frac{1}{2} \lim_{\vec{q}\to 0} \int d^4x\, e^{iqx} \langle Ap| \left[\partial_\mu A_i^\mu(x), \partial_\nu A_i^\nu(0)^\dagger\right] |Ap\rangle \tag{3.1}$$

and the integrated commutators

$$C_{iA}^{(+)} = \int d^3\vec{x}\ \langle Ap| \left[\partial_\mu A_i^\mu(\vec{x}, 0), \partial_\nu A_i^\nu(0)^\dagger\right] |Ap\rangle \tag{3.2}$$

$$C_{iA}^{(-)} = i\int d^3\vec{x}\ \langle Ap| \left[\partial_\mu A_i^\mu(\vec{x}, 0), \partial_0(\partial_\nu A_i(0)^\dagger)\right] |Ap\rangle \tag{3.2'}$$

one can derive (10-13) the two sum rules

$$\frac{1}{2\pi} P \int d\nu\, dq^2\, \rho_{iA}(\nu, q^2)\, \delta(q^2-\nu^2/p_o^2)\, (1-q^2/m_i^2)/\nu^2 = 2Q_i - C_{iA}^{(-)}/2m_i^2 p_o \tag{3.3}$$

$$\frac{1}{2\pi p_o} P \int d\nu\, dq^2\, \rho_{iA}(\nu, q^2)\, \delta(q^2-\nu^2/p_o^2)\, (1-q^2/m_i^2)/\nu = -\Sigma(iA) - C_{iA}^{(+)}/2m_i^2 p_o, \tag{3.3'}$$

where

$i = \pi^+$, K^+ and $2Q_\pi = 2T_3$, $2Q_K = \frac{3}{2}Y + T_3$.

Here q and p are the meson and target four-momenta, and $\nu = p\cdot q$, so that the c.m. squared energy s is given by $s = M^2 + q^2 + 2\nu$. The sum rules are nothing more than a pair of double dispersion relations computed at $q^2 = \nu = 0$, where the soft-meson theorems hold.

We can then separate the most singular part at $q^2 = m_i^2$ in the correlation function :

$$\rho_{iA}(\nu, q^2) = 2\pi\delta(q^2 - m_i^2)\,\frac{f_i^2\, m_i^4}{m_i^2 - q^2}\Big[\theta(q_o)\,D_{\bar{i}A}(p_o) - \theta(-q_o)\,D_{iA}(p_o)\Big] + \tilde{\rho}_{iA}(\nu, q^2) \qquad (3.4)$$

PCAC ensures that $\tilde{\rho}_{iA}$ will have at most a double pole at $q^2 = m_i^2$, whose residue is now the absorptive part of the off-mass-shell meson-nucleus forward amplitude

$$\tilde{\rho}_{iA}(\nu, q^2) = \frac{\pi f_i^2 m_i^4\left[F_{iA}(s, q^2) + \partial_{iA}(s, q^2)\right]}{(m_i^2 - q^2)^2} \qquad (3.4')$$

where $\partial_{iA}(s, q^2)$, the PCAC correction, is expected to be slowly varying and small at values of $|q^2| = O(m_i^2)$, with a double zero at $q^2 = m_i^2$.

A little algebra will then turn the sum rules into a dispersion relation in the "collinearity" x, defined as $x = q_o/p_o$ in the "collinear frame" introduced by Fubini and collaborators (11), with two subtractions at $x = 0$. In this frame the set of states coupled to the initial and final meson-target systems is reduced to only s- and p-wave excitations, contributing respectively via direct and z-graphs. The second ones are furthermore depressed with respect to the first ones by a factor $(m_i/p_o)^3$. We shall then have

$$D_{\bar{i}A}(p_o) = \frac{m_i p_o}{f_i^2}\left[2Q_i - \frac{\Sigma(iA)}{m_i} - \frac{C_{iA}^{(+)} + m_i C_{iA}^{(-)}}{2p_o\, m_i^2}\right] + \frac{1}{2}\left(\frac{m_i}{p_o}\right)^2 P\int_{-\infty}^{+\infty} dx\, \frac{F_{iA}(x) + \partial_{iA}(x)}{x^2(x - m_i/p_o)} \qquad (3.5)$$

The dispersive integral can now be broken into three pieces; a continuum one, which we expect to present the same dominant features as in the meson-nucleon case, deformed by the nuclear medium; a second, similar piece corresponding to the meson-nucleon unphysical region interaction, absent for pion scattering, where again we expect analogous deformations from the nuclear medium. Finally we shall have a third, purely nuclear piece, coming from the absorption of the meson and the subsequent excitation of nuclear, or hypernuclear, degrees of freedom.

This last piece can be derived directly from the commutators, making repeated use of the formal identity (11, 13)

$$<A|Q_i^5(0)|B> = -i(2\pi)^3\delta^3(\vec{p}_A - \vec{p}_B)\frac{<A|\partial_\mu A_i^\mu(0)|B>}{E_A - E_B + i\varepsilon} \tag{3.6}$$

and of the impulse approximation for the nuclear axial charges. Under the assumption of a single, dominant isovector or strange state of parity opposite to that of the nucleus or hypernucleus ground state (29-31), we can obtain

$$\frac{1}{2}\left(\frac{m_\pi}{p_o}\right)^2 P\int_{\text{levels}} dx \frac{F_{\pi A} + \partial_{\pi A}}{x^2(x - m_\pi/p_o)} = 8\left(\frac{g_A}{g_V}\right)_N^2 \frac{K_{av}}{m_N}\left(1 - \frac{\varepsilon^2}{m_\pi^2}\right)\frac{m_\pi p_o}{f_\pi^2}\left[2T_3 + A\frac{\varepsilon}{m_\pi}\right] \tag{3.7}$$

and

$$\frac{1}{2}\left(\frac{m}{p_o}\right)^2 P\int_{\text{levels}} dx \frac{F_{KA} + \partial_{KA}}{x^2(x - m/p_o)} = 3\frac{K_{av}}{m_N}\left[3\left(\frac{g_A}{g_V}\right)_\Lambda^2\left(1 - \frac{\varepsilon_\Lambda^2}{m_K^2}\right)\cdot\right.$$

$$\left.\cdot\left(1 + \frac{\varepsilon_\Lambda}{m_K}\right)(A + 2T_3) + \left(\frac{g_A}{g_V}\right)_\Sigma^2\left(1 - \frac{\varepsilon_\Sigma^2}{m_K^2}\right)\left(1 + \frac{\varepsilon_\Sigma}{m_K}\right)(3A - 2T_3)\frac{m_K p_o}{f_K^2}\right]. \tag{3.7'}$$

Both assumptions seem to be upheld by more detailed, microscopic analyses (27, 28, 48, 49). Here ε, ε_Λ and ε_Σ are the excitation energies of the dominant states with respect to the ground state of the target, while K_{av} is the average kinetic energy of the nucleus in its ground state.

Since excitation energies are all measurable quantities, the only model-dependent input is K_{av}, which in the impulse approximation fixes the scale of the nuclear contribution. A Fermi gas contained in a sphere of radius $R \simeq 1.2\ \text{fm}\, A^{1/3}$, value which we shall use throughout this paper, gives a value of 20 MeV, which may perhaps be an underestimate of the true value, justifiable if we think the above radius to be larger than the purely nuclear one due to Coulomb repulsion between the protons.

The continuum contributions, both above and below the elastic threshold, will be evaluated in a narrow-resonance approach, using eq.(3.4') with $\partial_{iA} = 0$ for the mass extrapolation along lines of constant s (for the direct graphs) or u (for the z-graphs).

This will introduce typical "PCAC deviations" in the continuum contributions, which, rather than try to estimate in some model of "extended" PCAC, we prefer to describe as our "theoretical uncertainty".

Collecting all commutators in the terms

$$\begin{aligned} \Delta(\pi A) &= \Sigma(\pi A) + \left[C^{(+)}_{\pi A} + m_\pi C^{(-)}_{\pi A}\right]/2m_\pi^2 M \\ \Delta(KA) &= \Sigma(KA) + \left[C^{(+)}_{KA} + m_K C^{(-)}_{KA}\right]/2m_K^2 M \end{aligned} \tag{3.8}$$

and evaluating the sum rule (3.5) at the elastic threshold $p_o = M$, we obtain then

$$\begin{aligned} L_\pi(A, T_3) &= (1 + m_\pi/M)\mathrm{Re}\, a(\pi^- A) = \frac{m_\pi}{4\pi f_\pi^2}\left[2T_3 - \Delta(\pi A)/m_\pi - \right. \\ &\left. - 8\frac{K_{av}}{m_N}\left(\frac{g_A}{g_V}\right)^2_N \left(1 - \frac{\varepsilon^2}{m_\pi^2}\right)\left(2T_3 + A\frac{\varepsilon}{m_\pi}\right)\right] + \\ &+ \frac{1}{8\pi M}\left(\frac{m_\pi}{M}\right)^2 P\int_{\text{continuum}} dx\, \frac{F_{\pi A}(x)}{x^2\,(x - m_\pi/M)}, \end{aligned} \tag{3.9}$$

and

$$\begin{aligned} L_K(A, T_3) &= (1 + m_K/M)\,\mathrm{Re}\, a(K^- A) = \frac{m_K}{4\pi f_K^2}\left\{\frac{3}{2}A + T_3 - \Delta(KA)/m_K - \right. \\ &- 3\frac{K_{av}}{m_N}\left[3(A + 2T_3)\left(\frac{g_A}{g_V}\right)^2_\Lambda \left(1 + \frac{\varepsilon_\Lambda}{m_K}\right)\left(1 - \frac{\varepsilon_\Lambda^2}{m_K^2}\right) + \right. \\ &\left.\left. + (3A - 2T_3)\left(\frac{g_A}{g_V}\right)^2_\Sigma \left(1 + \frac{\varepsilon_\Sigma}{m_K}\right)\left(1 - \frac{\varepsilon_\Sigma^2}{m_K^2}\right)\right]\right\} + \frac{1}{8\pi M}\left(\frac{m_K}{M}\right)^2 \cdot \\ &\cdot P\int_{\text{continuum}} dx\, \frac{F_{Ka}(x)}{x^2\,(x - m_K/M)}. \end{aligned} \tag{3.9'}$$

The coupling of the narrow resonances will be reduced, with respect to the coherent sum over the nucleons in the target, by the finite density of diffusing centres inside the nucleus; semi-classically we can evaluate this screening effect as for a sphere of uniform density (50), yielding a factor

$$S(a) = 3\left[a^2 + 2(1+a)e^{-a} - 2\right]/2a^3, \quad a = 2\rho\sigma R \qquad (3.10)$$

where ρ is the diffusing centres' density of the target and σ the cross-section at the resonance. For the Born terms, included in the z-graphs, we use the simple coherent sum over the couplings allowed by the Pauli principle, consistently with results from analytic extrapolations (51, 52).

A problem emerges in the kaon sum rule (3.9') : the $Y_o^*(1405)$ contribution, though strongly screened still comes out as large as the current algebra term, proportional to Q_K. We need therefore a strong constraint on its coupling to avoid errors of the order of 100% on K-nucleus sigma-term determinations. Such a constraint is offered by the "odd" sum rule (3.3), which, in the absence of K^+ -nucleus data, we can evaluate only for hydrogen. It implies a negligible octet admixture in the $Y_o^*(1405)$ (53), much less than allowed by some SU(3) fits (54,55). The correlation between its coupling and those of the higher $\frac{1}{2}^-$ hyperon states is positive, so that PCAC errors are also correlated and thus reduced in size by this additional sum rule.

We can thus evaluate all pieces in relations (3.9-9') using data from the latest compilations (55-58), which provide us with a substantially improved input with respect to previous attempts (30, 31), mainly because of

a) extended measurements on Λ-(56), and now even Σ-hypernuclei (57);

b) much better values of axial weak couplings (55, 58);

c) more extended, statistically significant analyses of πN and $\bar{K}N$ at intermediate energies (55).

Our only addition to these compilations is the completion of SU(3) octets and decuplets seen in πN partial waves with the missing Y_o^* and Y_1^* states for $J^P=\frac{1}{2}^-$, whose masses and couplings have been guessed from an "average" over the F/D patterns of $J^P=\frac{1}{2}^{\pm}$ multiplets, which indeed do show some regularities in masses and couplings.

All terms bar $\Delta(\pi A)$ and $\Delta(KA)$, summed together, are displayed as full dots in figures 1 through 4, with the experimental data both from mesic atoms (35, 36, 41, 59-61) and from pion and kaon scattering of hydrogen, deuterium and ^{4}He (58, 62, 63). Large discrepancies are evident both in L_π and L_K between these sums and the data, and they are sure indicators of sizeable contributions from the commutators of broken, chiral SU(3) $\otimes$ SU(3) symmetry to the scattering amplitudes.

This is not a new result (26, 30, 31, 64); but since FF sum rules yielded at their best a $\Sigma(\pi N)$ of 46 MeV, much lower than its on-mass-shell value of about 65 MeV, it is interesting to see if the new inputs did

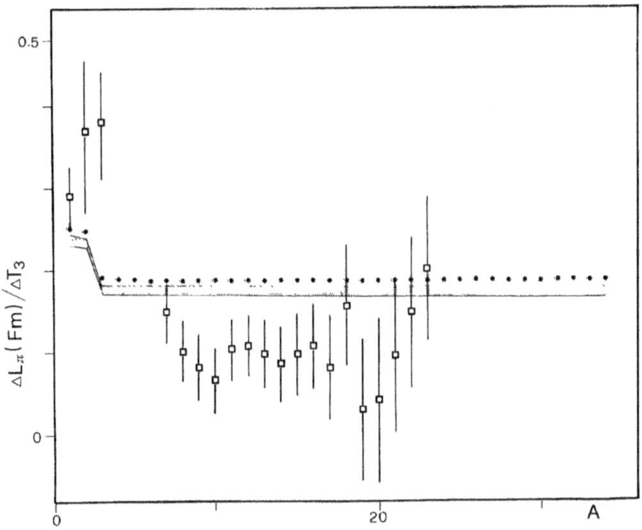

Figure 1 - Isospin dependence of the pion-nucleus scattering lengths : the open squares are the data, while the full dots and the shaded band represent the sum rule predictions without and with the effects of the chiral symmetry breaking

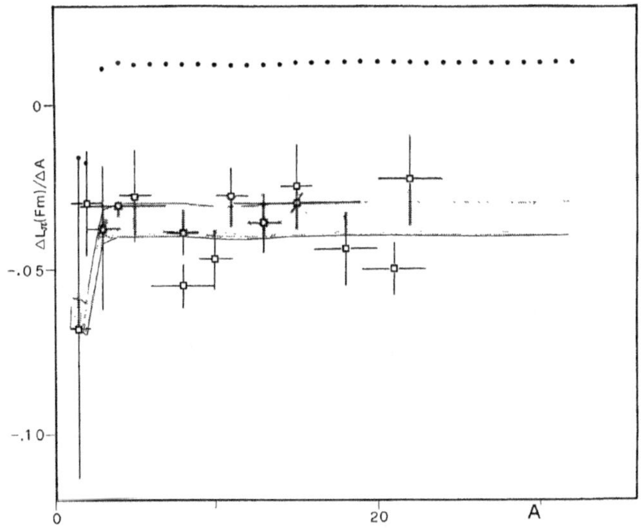

Figure 2 - Mass-number dependence of the pion-nucleus scattering lengths: notations are the same as in figure 1

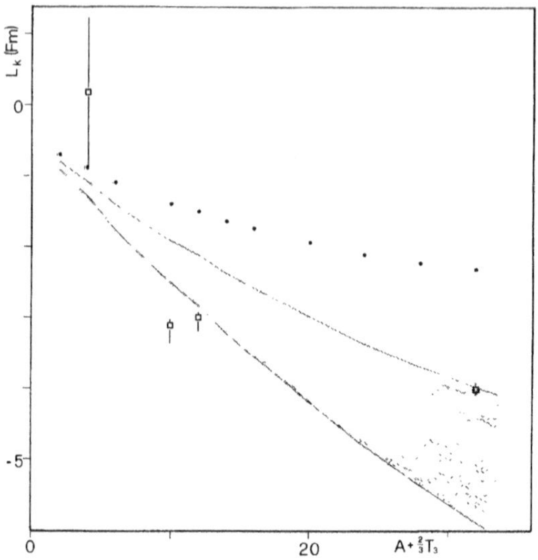

Figure 3 - Kaon scattering lengths on nuclei with $T_3 = 0$: notations are again the same as in figure 1

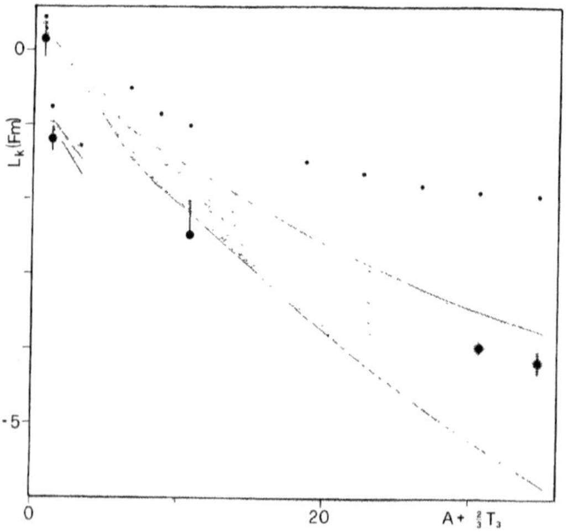

Figure 4 - Kaon scattering lengths on nuclei with $T_3 = +\frac{1}{2}$ (below) and $-\frac{1}{2}$ (above): data are here represented by the large full dots, while the smaller dots and the shaded bands still retain the same meaning as in the previous figures

change things and in which direction. This we intend to do next, after inflicting you a small theoretical digression.

4. The fundamental facet

4.1 How much freedom in models, now ?

To the dismay of group representation athletes, the most likely answer is, today, "close to none". The scene of current algebra itself has changed very little, but it has been entered by, you guess it, gauge theory. Since the early days of abelian gauges (65, 66) we know such a theory to embody current algebra and PCAC in a set of Feynman rules in a renormalizable way, and now, in its coloured, non-abelian version (67, 68), hadron spectroscopy and asymptotic freedom, i.e. both low- and high-momenta phenomenology. Leaving aside sales pitches and some opponents' plain slanders, we can fairly say that even if a "meaningful test" of QCD has yet to come and even if we are able to compute little more than lowest orders in perturbations theory, QCD is the only theory of strong interactions which is both renormalizable and asymptotically free. This, plus the fact that its phenomenology has yet to be really challenged up to momenta of several hundred GeV (68), explain why it is the most popular theory on the market, despite the fanaticism of recent converts and the staunch opposition of few heretics.

The Lagrangian we shall have to work with is then

$$L_{QCD} = \bar{q}(i\not{\partial} + g\not{G} - \underline{M})q - \frac{1}{4} F_{\mu\nu} F^{\mu\nu} \tag{4.1}$$

where all colour and flavour indices have not been written but should be summed over, and $F_{\mu\nu}$ takes the non-abelian form for the colour group SU(3)

$$F^a_{\mu\nu} = \partial_\mu G^a_\nu - \partial_\nu G^a_\mu + g f^{abc} G^b_\mu G^c_\nu , \tag{4.2}$$

Things could be slightly different if L, as in eq. (2.1), were an effective Lagrangian in the presence of some external medium, as for quarks immersed in the nuclear matter. We believe such a case to be highly improbable, since, as in QED, we expect the Van der Waals forces induced by (4.1) to describe hadron-hadron forces; but, to have an estimate of the possible effects of such a remote possibility, we shall introduce it anyway. The symmetry breaking piece of the Lagrangian is then written as

$$L_{8B} = -\varepsilon_o S_o - \varepsilon_3 S_3 - \varepsilon_8 S_8 - \frac{1}{3}\lambda\,(\rho, g^2)\,[S_{11} + S_{22} + S_{33}] , \tag{4.3}$$

where the first three are the usual mass terms, written as the members of a $(3,3^*) + (3^*,3)$ representation of $SU(3) \otimes SU(3)$, allowing for

$m_d \neq m_u$, and the fourth piece belongs to an (8,8) representation, as it suits an OBE piece, with its form dictated by the Zweig rule, assuming no strange quark content in the nucleon (69, 70). Its scale λ is furthermore expected to vanish when either ρ (the medium density) or g^2 vanish: this will turn out to be useful in computing the $C^{(\pm)}$ terms, where this piece in L_{S8} would otherwise generate terms highly singular on the light cone.

We could express the breaking parameters ε_o, ε_8 and ε_3 through the quark masses m_u, m_d and m_s

$$m_u = \frac{1}{\sqrt{6}}\varepsilon_o + \frac{1}{2\sqrt{3}}\varepsilon_8 + \frac{1}{2}\varepsilon_3 \tag{4.4}$$

$$m_d = \frac{1}{\sqrt{6}}\varepsilon_o + \frac{1}{2\sqrt{3}}\varepsilon_8 - \frac{1}{2}\varepsilon_3 \tag{4.4'}$$

$$m_s = \frac{1}{\sqrt{6}}\varepsilon_o - \frac{1}{\sqrt{3}}\varepsilon_8 \tag{4.4''}$$

we shall however use the customary parametrization

$$c = \varepsilon_8/\varepsilon_o \quad , \qquad \varepsilon = \varepsilon_3/\varepsilon_o \tag{4.5}$$

where we fix $\varepsilon \simeq \sqrt{3}\alpha$ from a <u>linear</u> fit to the baryon octet masses, yielding as well

$$F = [m_n - m_{\Xi^o}] \;/\; \sqrt{3}\,(1-\alpha) \;\simeq\; -218.3 \text{ MeV}/c^2 \tag{4.6}$$

$$D = \sqrt{3}\,[m_{\Sigma^o} - m_\Lambda]/2 \;\simeq\; 66.6 \text{ MeV}/c^2 \tag{4.6'}$$

Form the Lagrangian (4.3) we can derive the axial divergences

$$\partial_\alpha A^\alpha_k = \varepsilon_o \left[d_{Ok\ell} + c(d_{8k\ell} + \sqrt{3}\, d_{3k\ell})\right] P_\ell + \frac{2}{3}\lambda f_{kI\ell} S^A_{I\ell} \tag{4.7}$$

and then the sigma commutators, defined as

$$\delta(x_o - y_o)\,[A_i(x), \partial_\alpha A^\alpha_j(y)^\dagger] = \delta^4(x-y)\,\sigma_{ij}(x). \tag{4.8}$$

We shall here write the only two of interest, for charged pion and kaons, namely

$$\sigma_{\pi\pi}(x,\lambda) = \frac{\sqrt{2}+c}{3}\varepsilon_o\,[\sqrt{2}S_o + S_8] + \frac{2}{3}\lambda\,[S_{11} + S_{22} + S_{33}] \tag{4.9}$$

and

$$\sigma_{KK}(x,\lambda) = \frac{\sqrt{2}-\frac{1}{2}c(1-3\alpha)}{3}\,[\sqrt{2}S_o - \tfrac{1}{2}S_8 + \frac{\sqrt{3}}{2}S_3] + \frac{1}{6}\lambda\,[S_{11} + S_{22} + S_{33} + S_{66} + S_{77} + \sqrt{3}S_{38} + \tfrac{1}{2}(S_{44} + S_{55})] \tag{4.10}$$

The estimate of the terms $C^{(\pm)}_{iA}$, which give the coefficients for the most singular terms in the correlation functions in the Bjorken limit (13, 71, 72), is simplified by the fact that they can be evaluated in the same limit; then, as asymptotic freedom dictates g^2, and hence λ, to vanish in the limit, we shall have, using the definitions

$$c^{(-)}_{ij}\ (x) = \int d^3y\ \left[\partial_\alpha A^\alpha_i(x),\quad \partial_\beta A^\beta_j(y)^\dagger\right]_{x_o = y_o} \qquad (4.11)$$

$$c^{(+)}_{ij}\ (x) = -i\int d^3y\ \left[\partial_\alpha A^\alpha_i(x),\ \partial_o\ (\partial_\beta A^\beta_j(y))^\dagger\right]_{x_o = y_o} \qquad (4.11')$$

the two relations

$$c^{(-)}_{ij}\ (x) = i\varepsilon^2_o C^k_i C^\ell_j f_{k\ell m} V^o_m\ (x) + O(\lambda) \qquad (4.12)$$

and

$$c^{(+)}_{ij}\ (x) = -\varepsilon^2_o C^k_i C^\ell_j \sigma_{k\ell}\ (x,\ \lambda=0) + O(\lambda) \qquad (4.13)$$

where

$$C^k_i = d_{Oik} + c\,(d_{8ik} + \sqrt{3}\alpha d_{3ik}). \qquad (4.14)$$

After summing over the indices in the three expressions we expect the following structures for the Δ-terms

$$\Delta(\pi A) = A\Sigma(\pi N)\ (1-\delta_\pi) + 2T_3\cdot\delta_\pi m_\pi + A\sigma(\pi N) = \Delta_o(\pi A) + A\cdot\sigma(\pi N) \qquad (4.15)$$

and

$$\Delta(KA) = (A+\tfrac{2}{3}T_3)\left[\frac{\Sigma(Kp)+\Sigma(Kn)}{2}(1-\delta_k) + \frac{3}{2}\delta_k m_k\right] + 2T_3\ \cdot$$
$$\cdot\ \frac{\Sigma(Kp)-2\Sigma(Kn)}{3}(1-\delta_k) + N\sigma(Kn) + Z\sigma(Kp) = \Delta_o(KA) + N\sigma(KN) + Z\sigma(Kp). \qquad (4.16)$$

Here Σ and σ indicate respectively the conventional, "fundamental" sigma-term and the contribution coming from the hypothetical chiral symmetry breaking in the nuclear interaction.

With <u>linear</u> mass formulae we shall have then

$$\Sigma(\pi N) = \frac{(\sqrt{2}+c)\ (3F-D)}{2\sqrt{3}c}\ (1 + \tfrac{2}{3}\Delta_{ZW}) \qquad (4.17)$$

$$\Sigma(KN) = \frac{\sqrt{2}-\frac{1}{2}c\ (1-3\alpha)}{4\sqrt{3}c}\left[(3F-D)\ (1+\tfrac{4}{3}\Delta_{ZW}) + 2t_3\ (F+D)\right] \qquad (4.18)$$

where t_3 is the nucleon isospin and Δ_{ZW} measures the deviation from the prediction of Zweig rule (69, 70):

$$\langle N|S_0|N\rangle = \sqrt{2}\,\langle N|S_8|N\rangle\,(1+\Delta_{ZW}). \qquad (4.19)$$

We also defined

$$\delta_\pi = \frac{4}{9}\left|\frac{\sqrt{2}/c+1}{1+\alpha}\right|^2\left(\frac{m_s-m_u}{m_\pi}\right)^2 \qquad (4.20)$$

$$\delta_K = \frac{4}{9}\left[\frac{\sqrt{2}/c-\frac{1}{2}(1-3\alpha)}{1+\alpha}\right]^2\left(\frac{m_s-m_u}{m_K}\right)^2 \qquad (4.20')$$

where ε_0 has been related, via the ratio $c=\varepsilon_8/\varepsilon_0$, to the mass difference m_s-m_u, for which we shall use the linear estimate from the decuplet masses

$$m_s-m_u \simeq \frac{1}{3}(m_{\Omega-}-m_{\Delta++}) \simeq 153.7\ \mathrm{MeV}/c^2\ . \qquad (4.21)$$

For the nuclear contributions σ, not to be confused with the Σ-terms, we have taken two extreme alternatives. The first, which we prefer on the bases expounded before, is to ignore them altogether; in this "orthodox" framework, where QCD gives the whole of subnuclear and nuclear forces, we are indeed measuring, apart from the δ_i's brought in by the asymptotic terms, the chiral symmetry breaking due to the renormalized quark masses.

The opposite stance is to blame on the "nuclear", symmetry breaking part the whole of the mass-shifts inside the nuclear medium, i.e. the whole binding energies. To simplify this picture further let our terms to have "octet enhancements" as well, as in weak non-leptonic decays, and then suppress the contributions from the 27-representation of SU(3): we shall then estimate

$$\sigma(\pi N) = \frac{8}{3}b_N \qquad (4.22)$$

$$\sigma(Kp) = \frac{1}{3}(2b_N + b_\Lambda + 3b_\Sigma) \qquad (4.22')$$

$$\sigma(Kn) = \frac{1}{3}(b_N + 2b_\Lambda + 3b_\Sigma) \qquad (4.22'')$$

These terms are not negligible, since they amount to several MeVs per nucleon already for ^{4}He, and increase to a few tens of MeVs per nucleon in ^{35}Cl, as much as some early determination of $\Sigma(\pi N)$. Despite their size, they do not change dramatically the parameters of the chiral symmetry breaking, apart from an unreasonable increase in the central value for $\Sigma(\pi N)$, which corroborates our prejudice about their not having to be there.

4.2 Analysis of the preliminary data

From eqs. (3.9) and (3.9') and the experimental data mentioned before (35, 36, 41, 43, 58-63) we can derive determinations of $\Delta(\pi A)$ and $\Delta(KA)$, where we shall add a 6% "PCAC error" to the experimental errors on the data, "world averaged" whenever possible. From these we subtract, for the second of our two "extreme" attitudes toward nuclear effects, the contributions (4.22-4.22") coming from the hypothetical "nuclear" symmetry breaking.

We shall then fit $\Delta_o(\pi A)$ as a linear function of A, forgetting the small, and still undetectable, dependence on T_3, and $\Delta_o(KA)$ as an equally linear function of $A+T_3/3$ only.

Indeed we have

$$\frac{\Sigma(Kp) - 2\Sigma(Kn)}{3} = \frac{\sqrt{2} - \frac{1}{2}c\,(1 - 3\alpha)}{3\sqrt{3c}} \left[D(1 + \frac{1}{3}\Delta_{ZW}) - F\Delta_{ZW}\right] \qquad (4.23)$$

which we expect to be rather small, particularly with respect to the very large errors on $\Delta_o(KA)$. Better data could make even these small isotopic effect detectable, particularly the one for kaons, making the number of tests double at once.

With our preliminary data the discrepancies Δ_o are fitted by the slopes

$$\partial\Delta_o(\pi A)/\partial A = 54.14 \pm 1.27 \text{ MeV}$$
$$\partial\Delta_o(KA)/\partial(A + \frac{2}{3}T_3) = 175.7 \pm 64.4 \text{ MeV} \qquad (4.24)$$

in the "orthodox" scheme (no σ's) and by the slopes

$$\partial\Delta_o(\pi A)/\partial A = 73.43 \pm 1.54 \text{ MeV}$$
$$\partial\Delta_o(KA)/\partial(A + \frac{2}{3}T_3) = 207.0 \pm 61.6 \text{ MeV} \qquad (4.25)$$

in the extreme case of a fully symmetry-breaking binding term.

Of course, both sets correspond to the same set of discrepancies and then do correspond to the same shaded bands in figures from 1 through 4.

The situation for the two parameters c and Δ_{ZW} implied by values (4.24) and (4.25) are represented in figures 5 and 6 respectively, where the positivity bound (73)

$$\Delta_{ZW} \geq - \frac{3(F-D)}{2(3F-D)} \simeq -0.592 \qquad (4.26)$$

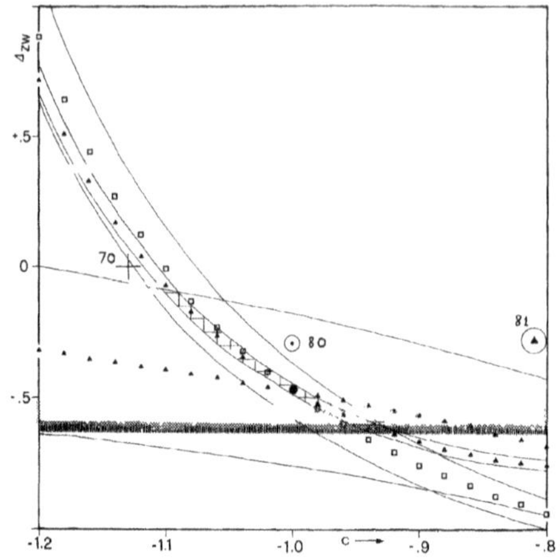

Figure 5 - Solutions for the parameters c and Δ_{ZW} from $\Delta(\pi A)$ and $\Delta(KA)$ are shown as full triangles for the "orthodox" case $\sigma = 0$. Open squares correspond to the dispersion relation result (4.29), corrected for extrapolation effects, and the lower bound (4.26) is marked by the horizontal shaded band. The solution is indicated by the cross-hatched area, with its "central" value marked by a large full dot. Other phenomenological estimates in the range covered here also displayed with numbers indicating the relative references

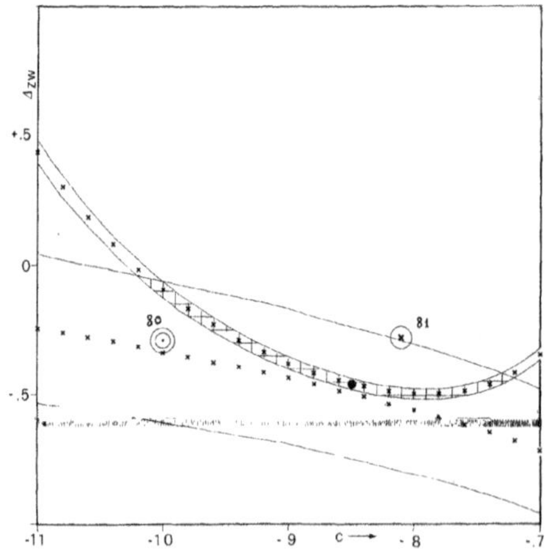

Figure 6 - Crosses indicate the solutions for c and Δ_{ZW} from the slopes for our second choice for the σ's. The remaining notations are the same as in the previous figure

has also been indicated. In figure 5 we have also indicated the dispersion relation result at the Cheng-Dashen point (19), $t = 2m_\pi^2$ and $s = u = m_p^2$, derived by the Karlsruhe-Helsinki group (23)

$$\Sigma(\pi N)_{CD} = 65 \pm 6 \text{ MeV}. \tag{4.27}$$

To relate this to our $\Sigma(\pi N)$ one has to account for the t-channel unitarity and mass-extrapolation effects, of $O(m_\pi^4 \ell n\, m_\pi)$ (see refs. 15 and 74-77); we shall write the phase representation for the amplitude at $q_1^2 = q_2^2 = t/2$ and $s = u = m^2$, obtaining

$$\Sigma(\pi N)_{CD} = \Sigma(\pi N)\, \frac{2m_\pi^2 - t_o}{t_o}\, \exp\left[\frac{2m_\pi^2}{\pi}\int_{4m_\pi^2}^{\infty} dt\, \frac{\phi(t)}{t(t-2m_\pi^2)}\right] \tag{4.28}$$

in the hypothesis of a single, low-lying zero. A reasonable estimate can be obtained with $t_o \simeq m_\pi^2$ and $\phi(t) = \delta_o(t)_{\pi\pi}$ at least up to $t \simeq 18m_\pi^2$, and yields a 10% reduction from value (4.27) to

$$\Sigma(\pi N) = 59 \pm 8 \text{ MeV}; \tag{4.29}$$

the small increase in the error reflects our optimism on the uncertainties in $\phi(t)$ and t_o. Note that eq. (4.28) actually defines both quantities, which are then restricted by imposing PCAC, off-mass-shell unitarity and s-wave dominance in the t-channel.

From figure 5 we can extract the parameters of the symmetry breaking in the "orthodox" version

$$c = -0.994^{+0.058}_{-0.110} \quad ; \quad \Delta = -0.479^{+0.395}_{-0.113} \tag{4.30}$$

corresponding to the Σ-terms

$$\Sigma(\pi N) = 59.8 \pm 4.6 \text{ MeV};\ \Sigma(Kp) = 114^{+88}_{-26} \text{ MeV};\ \Sigma(Kn) = 30^{+95}_{-30} \text{ MeV} \tag{4.31}$$

in remarkable agreement both with eq. (4.29) and the phenomenological estimates by other authors (79-81), and corresponding to a mass ratio for the quarks $m_s/m_d = 8.5^{+4.1}_{-1.4}$, definitely smaller than the one invoked from earlier estimates in broken SU(3) ⊗ SU(3) (82-84). Note the strong correlation between c and Δ_{ZW}, which makes our "determination" still marginally consistent with $c \simeq -1$ and $\Delta_{ZW} \simeq 0$, the naive expectations from spectral sum rules (79-81, 85) and QCD folklore (68).

The extent by which these parameters can be influenced by nuclear effects is well typified by the extreme choice (4.22-22") for the σ's, which leads to the parameters (see figure 6)

$$c = -0.85^{+0.13}_{-0.17} \qquad (m_s/m_d = 5.3^{+3.0}_{-1.3})$$

$$\Delta_{ZW} = -0.459^{+0.424}_{-0.059} \tag{4.32}$$

corresponding to the terms

$$\Sigma(\pi N) = 96^{+16}_{-17}\ \text{MeV} \ ; \ \Sigma(Kp) = 134^{+93}_{-11}\ \text{MeV} \ ; \ \Sigma(Kn) = 40^{+115}_{-14}\ \text{MeV} \tag{4.33}$$

The large value for $\Sigma(\pi N)$, compensated both by an increase in δ_π and by the appreciable $\sigma(\pi N)$, could be reconciled with the on-mass-shell result (4.27) only by drastic chenges in $\phi(t)$ over $18m_\pi^2$ or a value of t_0 much closer to the Cheng-Dashen point term $t = 2m_\pi^2$, or both, against, for instance, the predictions of chiral perturbation theory, from which we expect an on-mass-shell result about 14 MeV above the off-mass-shell one (75) (to the lowest order in m_π^2 - but note that our estimate shows that higher orders can be as much important as the lowest one !).

Note that this is largely a consequence of the extreme sensitivity of $\Sigma(\pi N)$ on the value for c: $\Sigma(KN)$ terms are, within larger errors, much stabler with respect to the nuclear corrections.

5. An outlook

Exotic atoms are not for nuclear physicists only. We hope to have shown that they do yield information on meson-nucleon interaction at an energy just below the elastic threshold, where current algebra sum rules can be fruitfully put at work. Also, our information on the other facets of the problem has come to the point where the effects of broken chiral symmetry start to emerge from the remaining errors.

Then, better exotic atom data and better hypernuclear data from kaon and pion factories as well as from the "good old", intermediate-energy machines, like the CERN PS that has just brought us Σ-hypernuclei, will be as useful as better low-energy data or more extended and powerful partial-wave analyses.

On the side of theory, however, the "PCAC errors" in the FF sum rules are "here to stay", and no one, short of summing the whole shebang of QCD perturbation series, could tell what the ∂_{iA} are like, at least in any foreseeable future. On the other hand the systematic dependences on the target quantum numbers, if we can describe them properly, can give us a way to reduce the devastating effects of our ignorance, which we boldly choose to label as our "theoretical uncertainties".

But even at these extremely low energies, so far from the energy regions where PEP, LEP and the collider will soon forge ahead (minus a few hundred keVs, indeed !), the coming years may have in store for us relevant informations on the fundamental constituents of hadronic matter.

References and notes

*. Research supported in part by I.N.F.N., Italy, through the Sezione di Bari.

1. Gell-Mann, M.: 1964, Physics 1, 63.
2. Adler, S.L., and Dashen, R.F. : 1968, "Current Algebras and Applications to Particle Physics", New York.
3. de Alfaro, V., Fubini, S., Furlan, G., and Rossetti, C. : 1973, "Currents in Hadron Physics", Amsterdam.
4. Adler, S.L.: 1968, in "Selected Topics in Particle Physics", Ed. J. Steinberger, New York, p.1.
5. Adler, S.L. : 1965, Phys. Rev. 137, B1022; 1965: Phys. Rev. 139, B1638.
6. Balachandran, A.P., Gundzik, M.G., and Nicodemi, F.: 1966, Nuovo Cimento 44A, 1257.
7. Tomozawa, Y.: 1966, Nuovo Cimento 46A, 707.
8. Weinberg, S.: 1966, Phys. Rev. Lett. 17, 616.
9. Balachandran, A.P., Gundzik, M.G., and Nicodemi, F.: 1968, Nucl.Phys. B6, 557.
10. de Alfaro, V., Fubini,S., Furlan, G., and Rossetti, C.: 1969, Nuovo Cimento 62A, 497.
11. Fubini, S., and Furlan, G. : 1968, Ann. Phys. (N.Y.) 48, 322.
12. Fubini, S.: 1969, in: "Theory and Phenomenology in Particle Physics", ed. A. Zichichi, New York, p. 16.
13. Furlan, G. : 1969, in: "Particle Physics", ed. P. Urban, Vienna, p.52.
14. Weinstein, M. : 1971, in: "Springer Tracts in Modern Physics, No. 60", ed. G. Höhler, Berlin, p. 32.
15. Pagels, H.: 1975, Phys. Rep. 16, 219.
16. Reya, E.: 1974, Rev. Mod. Phys. 46, 545.
17. Altarelli, G., Cabibbo, N., and Maiani, L. : 1971, Nucl. Phys. B34, 621.
18. Osypowski, E.T.: 1970, Nucl. Phys. B21, 615.
19. Cheng, T.P., and Dashen, R. : 1971, Phys. Rev. Lett. 26, 594.
20. von Hippel, F., and Kim, J.K. : 1970, Phys. Rev. D1, 151.
21. Gensini, P. : 1971, Lett. Nuovo Cimento 1, 273; 1971, Nuovo Cimento 6A, 259.

22. Hakim, S.J. : 1972, Nucl. Phys. B48, 265.

23. Höhler, G., Kaiser, F., Koch, R., and Pietarinen, E. : 1979, "Pion-Nucleon Scattering Handbook", Physics Data No. 12-1, Karlsruhe.

24. Di Claudio, B., Rodriguez-Vargas, A.M., and Violini, G. : 1979, Uni-Andes report No. 14 (to appear in Zeit. Phys. C) (see also the previous contribution by Rodriguez-Vargas).

25. Ericson, M., Figureau, A., and Molinari, A. : 1969, Nucl. Phys. B10, 501.

26. Ericson, M., and Rho, M. : 1971, Phys. Lett. 36B, 93.

27. Ericson, T.E.O., and Bernabéu, J.: 1977, Phys. Lett. 70B, 170.

28. Rho, M., and Wilkinson, D.H., eds. : 1979, "Mesons in Nuclei", Amsterdam.

29. D'Auria, R., Napolitano, E., Soliani, G., and Molinari, A. : 1969, Phys. Lett. 29B, 285.

30. Gensini, P. : 1971, Nuovo Cimento 6A, 98.

31. Gensini, P. : 1973, Nuovo Cimento 17A, 557.

32. Seki, R. : 1976, in: "Meson-Nuclear Physics", P.D. Barnes, R. A. Eisenstein, and L.S. Kisslinger eds., New York, p. 80.

33. Feshbach, H., and Villars, F. : 1958, Rev. Mod. Phys. 30, 24.

34. Seki, R., and Cromer, A.H.: 1967, Phys. Rev. 156, 93.

35. Seki, R. : 1969, Phys. Rev. Lett. 23, 1000.

36. Seki, R. : 1972, Phys. Rev. Lett. 29, 240.

37. Krell, M. : 1971, Phys. Rev. Lett. 26, 584.

38. Burhop, E.H.S. : 1969, in : "High-Energy Physics", Vol. 3, E.H.S. Burhop ed., New York, p. 109.

39. Backenstoss, G.: 1970, Ann. Rev. Nucl. Sci. 20, 467.

40. Seki, R. : 1975, Ann. Rev. Nucl. Sci. 25, 241.

41. Hüfner, J. : 1975, Phys. Rep. 21, 1.

42. Tauscher, L. : 1975, in : "High-Energy Physics and Nuclear Structure", D.E. Nagle, et al., eds., New York, p. 541.

43. Tauscher, L. : 1977, in: "Exotic Atoms", G.Fiorentini, and G. Torelli eds., Frascati, p. 145.

44. Powers, R.J. : 1976, in: "Meson-Nuclear Physics", P.D. Barnes, R.A. Eisenstein, and L.S. Kisslinger eds., New York, p. 552.

45. Backenstoss, G.: 1977, in: "Contacts between High-Energy Physics and Other Fields of Physics", P. Urban ed., Vienna, p. 593.

46. Wycech, S. : 1979, in: "Meson-Nuclear Physics", E.V. Hungerford III ed., New York, p. 390.

47. Poth, H. : 1979, "Compilation of Mesic Atoms Data", Physics Data No. 14-1, Karlsruhe.

48. Blomqvist, J., and Molinari, A. : 1968, Nucl. Phys. A106, 545.

49. Hoshi, N., and Fujita, T. : 1976, in: "Meson-Nuclear Physics", P.D. Barnes, R.A. Eisentstein, and L.S. Kisslinger eds., New York, p.540.

50. Becker, F., and Batusov, Yu.A. : 1971, Riv. Nuovo Cimento 1, 309.

51. Ericson, T.E.O., and Locher, M.P.: 1969, CERN report 69-30.

52. Dumbrajs, O. : 1979, Nuovo Cimento 54A, 155.

53. Gensini, P., and Soliani, G. : 1970, Lett. Nuovo Cimento 4, 329.

54. Barbaro-Galtieri, A.: 1972, in: "Proceedings of the XVIth International Conference of High-Energy Physics", J.D. Jackson, A. Roberts, and R. Donaldson eds., Chicago, Vol. 1, p. 159.

55. Particle Data Group: 1978, Phys. Lett. 75B, 1.

56. Povh, B. : 1978, Ann. Rev. Nucl. Part. Sci. 28, 1.

57. Bertini, R. : 1979, in: "Meson-Nuclear Physics", E.V.Hungerford III ed., New York, p. 699.

58. Nagels, M.M., et al. : 1979, Nucl. Phys. B147, 189.

59. Hüfner, J., Tauscher, L., and Wilkin, C. : 1974, Nucl. Phys. A231, 455.

60. Bailey, J., et al. : 1974, Phys. Lett. 50B, 403.

61. Abela, R., et al. : 1977, Phys. Lett. 68B, 429.

62. Block, M. : 1967, Phys. Lett. 25B, 604.

63. Mazur, P.O., Block, M.M., Keren, J., and Meyer, S.L. : 1970, Phys. Rev. D1, 20.

64. Hakim, S.J. : 1972, Nucl. Phys. B36, 436; 1973, Nucl. Phys. B61,116.

65. Callan, C.G. jr.: 1968, Phys. Rev. 169, 1175.

66. Preparata, G., and Weisberger, W.I. : 1968, Phys. Rev. 175, 1965.

67. Gross, D.J., and Wilczek, F.: 1973, Phys. Rev. D8, 3633; 1974, Phys. Rev. D9, 980.

68. Ellis, J., and Sachrajda, C.T.: 1979, CERN report TH.2782 (to be publ. in the Proceedings of the 1979 Cargèse Summer Institute).

69. Segrè, G.: 1971, Phys. Rev. D3, 1954.

70. Cheng, T.P. : 1976, Phys. Rev. D13, 2161.

71. Bjorken, J.D. : 1966, Phys. Rev. 148, 1467.

72. Bjorken, J.D. : 1968, in: "Selected Topics in Particle Physics", J. Steinberger ed, New York, p. 55.

73. Gensini, P. : 1974, Lett. Nuovo Cimento 10, 585.

74. Brown, I.S., Pardee, W.J., and Peccei, R.D. : 1971, Phys. Rev. D4, 2801.

75. Pagels, H., and Pardee, W.J.: 1971, Phys. Rev. D4, 3235.

76. Schnitzer, J.: 1972, Phys. Rev. D5, 1482; 1972, Phys. Rev. D6, 1801.

77. Geddes, H.B., and Graham, R.H. : 1973, Phys. Rev. D7, 1801.

78. Gensini, P.M.: 1977, SLAC report PUB-1944, unpublished.

79. Scadron, M.D., and Jones, H.F. : 1974, Phys. Rev. D10, 967.

80. Jones, H.F., and Scadron, M.D. : 1975, Phys. Rev. D11, 174.

81. Gunion, J.F., McNamee, P.C., and Scadron, M.D. : 1977, Nucl. Phys. B123, 445.

82. Gell-Mann, M., Oakes, R.J., and Renner, B. : 1968, Phys. Rev. 175, 2195.

83. Oakes, R.J. : 1972, in: "Elementary Particle Physics", P. Urban ed., Vienna, p. 518.

84. Renner, B. : 1972, in: "Springer Tracts in Modern Physics" No. 61, G. Höhler ed., Berlin, p. 120.

85. Cocho, G., Garcia, A., and Giambiagi, J.J. : 1971, Phys. Rev. D8, 1820.

DISCUSSION

- O. Dumbrajs :

Could you summarize the main differences in extracting the pion-nucleus scattering length from a pionic atom and the kaon-nucleus scattering length from a kaonic atom? What errors could be expected in both cases.

- P. Gensini :

The main difference lies in the strongly absorptive nature of the scalar K^--nucleus potential. It is precisely this strong absorption which makes the strong correlation between the (complex) energy shift and the (complex) scattering length a highly non-linear one. The correlation reduces to an almost linear one for weak absorption - as in the nucleus case. As far as errors are concerned, only a detailed numerical investigations could answer your second question (and the problems raised by the neglect of c-odd terms in the strong potentials as well).

NEW TECHNIQUES OF ANALYTIC CONTINUATION AND RIGOROUS CORRELATIONS BETWEEN LOW AND HIGH ENERGY SCATTERING(*)

I.Caprini
Institute of Physics and Nuclear Engineering - Bucharest

Introduction

Various methods of analytic continuation of the scattering amplitudes or formfactors have been proposed during the last years, using in a gradually increasing way the powerful tools of functional analysis. These methods certainly led to an improvement of the theoretical knowledge of some hadronic processes like meson-meson, pion-nucleon and kaon-nucleon scattering.

In the present paper I shall discuss two particular techniques which already proved to be useful for the investigation of several hadronic reactions and might be of interest also for the study of the kaon-nucleon interaction. The paper is divided in two sections: in the first one I present a method of analytic continuation which makes essential use of the positivity of the imaginary part of the amplitude. This is related to the physical problem of deriving asymptotic features of the interaction from the values of the amplitude along some suitable directions in the complex energy plane. In the second section I present a modification of the usual application of the maximum modulus theorem for constraining the amplitude at interior points. In our treatment we are able to include the accurate information about the imaginary part of the amplitude, available along a part of the cut. This problem is related to the determination of the $KN\Sigma$, $KN\Lambda$ coupling constants and the zeros in the complex energy plane.

1.

Recent investigations of meson-meson (1) and proton-proton (2) processes showed that for deriving asymptotic features of the scattering a convenient approach is to compute the amplitudes along some suitable unphysical directions in the complex energy plane. It was expected on intuitive grounds that along such directions the asymptotic features might be more precocious than along the physical region, where the process is more complicated due to resonance, inelastic thresholds etc. This expecta

E. Ferrari and G. Violini (eds.), Low and Intermediate Energy Kaon-Nucleon Physics, 363–374.

tion was actually confirmed by our results, which revealed for instance some typically asymptotic features of the forward pion-pion and pion-kaon amplitudes at unphysical energies of relatively low modulus. Such a property might hold for a larger class of amplitudes and therefore an investigation along these lines of the pion-nucleon and kaon-nucleon processes would be of interest.

In order to apply the method it is necessary to make the analytic continuation of the experimental amplitude from the physical region into the complex energy plane. I shall discuss here the analytic continuation in the case when the imaginary part of the amplitude has a definite sign on the cut. This situation often arises in physical problems and is also of a remarkable mathematical interest, since it guarantees the stability of the extrapolation (3, 4), i.e. it ensures a finite extrapolation error which vanishes in the limit of an infinitely accurate input. The problem of extrapolation was solved in (3) in the assumption that the real and the imaginary part of the amplitude are known with the same accuracy. However, in many practical cases, such as the forward KN scattering, the imaginary part is much better known, from total cross section measurements, than the real one. In what follows I shall briefly treat the method of analytic extrapolation with positivity, suitable for the case of unequal errors on the real and imaginary parts of the amplitude.

I consider for simplicity a fixed - t symmetric amplitude $f(\nu^2)$ as a function of the energy variable ν^2. In the ν^2 complex plane $f(\nu^2)$ is real analytic, having a cut for $\nu^2 \geq \nu_o^2$. The above mentioned analyses (1, 2) showed that in this variable the most convenient direction of extrapolation is the unphysical axis $\nu^2 < 0$. As usually the ν^2-plane is mapped onto the unit disk $|z| \leq 1$ by a simple conformal transformation, and in this new variable the direction of interest becomes the real diameter. We assume that $h_{R,i}$ are the experimental values of the real and imaginary parts of the amplitude, known with errors $\varepsilon_{R,i}$ along the region Γ of $|z| = 1$. The analytic continuation of this information is given by the solution $f_o(z)$ of the following convex minimization problem (3, 4)

$$\chi_o^2 = \chi^2(f_o) = \inf_{f\varepsilon C} \chi^2(f)$$
$$\chi^2(f) = \frac{1}{2\pi} \int_\Gamma (f_R-h_R)^2 \, w_R \, d\theta + \frac{1}{2\pi} \int_\Gamma (f_i-h_i)^2 \, w_i \, d\theta \qquad (1.1)$$

In this relation $w_{R,i} > 0$ are suitable weights related to the errors $\varepsilon_{R,i}$ and C denotes the convex cone of the real analytic functions of class H^2 (5), satisfying the property

$$\mathrm{Im}\, f(\theta) \geq 0, \qquad \theta \varepsilon [0, \pi] \qquad (1.2)$$

I mention that the extrapolation in L^2- norm with unequal errors as in (1.1) was treated previously (6) in the simple case without positivity constraints, Γ being the whole boundary, since otherwise the extrapolation is meaningless. We consider in what follows both cases when Γ is the whole boundary or a part of it, which we take for convenience to be the right semicircle $-\pi/2 \leq \theta \leq \pi/2$. It can be shown (3) that the solution $f_o(z)$ of the infimum problem (1.1) exists and is unique. It yields at every interior point z the best analytic continuation of the data, with a finite extrapolation error which can be evaluated following the prescriptions given in (3). For solving (1.1), i.e. for computing $f_o(z)$ we apply the Lagrange duality theorem for convex optimization (7). The Lagrange functional writes as

$$L(f, \delta_o) = \chi^2(f) - \frac{1}{\pi}\int_{-\pi}^{\pi} \mathrm{Im} f \; w_+ \; \delta_o \; d\theta, \qquad (1.3)$$

in terms of the nonnegative function $\delta_o \in L^2 \left[-\pi, \pi\right]$, the generalized Lagrange multiplier. We have also defined

$$w_\pm = \frac{1}{2}(w_R \pm w_i), \quad w_+(\theta) > 0,$$

such that $\chi^2(f)$ becomes

$$\chi^2(f) = \frac{1}{2\pi}\int_\Gamma |f-h|^2 \, w_+ \, d\theta + \frac{1}{2\pi}\int_\Gamma (f-h)^2 \, w_- \, d\theta. \qquad (1.4)$$

According to the Lagrange theorem, the funtion $f_o(z)$ realizes the unconstrained minimum of $L(f, \delta_o)$ and also obeys the alignment condition

$$\mathrm{Im}\, f_o(\theta) = 0, \quad \theta \in S, \quad \mathrm{Im}\, f_o(\theta) > 0, \quad \theta \notin S, \qquad (1.5)$$

where $S = \{ \theta, \ \delta_o(\theta) > 0 \}$ is the support of δ_o. Moreover, δ_o is the solution of the supremum problem (7)

$$\chi_o^2 = \sup_{\delta \geq 0} \ \inf_{f \in H^2} L(f, \delta). \qquad (1.6)$$

The calculations involved by the above theorem are straightforward: one has first to perform the unconstrained minimization of $L(f, \delta_o)$ with respect to f. This gives the expression of $f_o(\theta)$ in terms of the Lagrange multiplier $\delta_o(\theta)$, which fortunately in our case can be written explicitly . By introducing this expression in (1.5), an integral equation for δ_o is finally obtained.

I shall give below only the main steps of these calculations, leaving aside the details (8). Actually, the formulae look completely different in the two cases, with Γ being the whole boundary or a part of it.

When Γ is the whole boundary the solution is obtained rather easily by resorting to the technique of reproducible kernels. Namely, let us introduce the function (6)

$$L(z, y) = G^*(z) \; L_o(z, y) \; G(y) \tag{1.7}$$

where $G(z)$ is the outer function (5)

$$G(z) = \exp\left\{ -\frac{1}{4\pi} \int_{-\pi}^{\pi} \frac{e^{i\theta} + z}{e^{i\theta} - z} \ln w_+(\theta) \, d\theta \right\} \tag{1.8}$$

real-analytic without zeros in $|z| < 1$ and with modulus equal to $1/\sqrt{w_+(\theta)}$ on Γ, while $L_o(z, y) = L_o^*(z, y)$ satisfies the integral equation

$$K_o(z,y) = L_o(z,y) + \frac{1}{2\pi} \int_{-\pi}^{\pi} L_o(z,e^{i\phi}) \; G^2(\phi) \; w_-(\phi) \; K_o(e^{-i\phi},y) \, d\phi \quad , \tag{1.9}$$

where $K_o(z, y) = 1/(1 - z^* y)$.

Then the minimization of $L(f, \delta_o)$ yields the optimal function $f_o(z)$ in the form

$$f_o(z) = \frac{1}{2\pi} \int_{-\pi}^{\pi} \left\{ \left[h(\phi) + i\delta_o(\phi) \right] L(e^{i\phi}, z) \, w_+ + h(\phi) \, L^*(z, e^{-i\phi}) \right\} d\phi . \tag{1.10}$$

The alignment conditions (1.5) becomes (8)

$$\lim_{r \to 1} \left\{ \mathrm{Im} \int_{-\pi}^{\pi} h(\phi) \left[L(e^{i\phi}, re^{i\theta}) \, w_+ + L^*(re^{i\theta}, e^{-i\phi}) \right] w_- \, d\phi \right.$$
$$\left. + \mathrm{Re} \int_S \delta_o \, L(e^{i\phi}, re^{i\theta}) \, w_+ \, d\phi \right\} \geq 0 \tag{1.11}$$

where the equality holds for $\theta \in S$. This is an integral equation for finding the nonnegative function $\delta_o(\theta)$, its properties depending on the functions $w_{\pm}$. Various exactly solvable cases, of interest in applications are discussed in (8). In the particular case $w_R = w_i$, i.e. $w_- = 0$, from (1.9) it follows that $L_o(z, y) = K_o(z, y)$ and (1.11) reduces to the integral equation written in (3). Having found δ_o from this integral equation, we have from (1.10) the expression of the best extrapolation $f_o(z)$, for every point z.

The extrapolation from a part of the boundary is much more complicated than the above case. We develop

$$f(z) = G(z) \sum_{n=0}^{\infty} f_n \, z^n , \tag{1.12}$$

with G defined in (1.8) (in this case we must assign a value to w_+ outside Γ, the results being actually independent of this value). The minimization of L gives the equation

$$(1-T)\left[\frac{f_o-h-i\delta_o}{G}+\frac{f^*-h^*}{G}\frac{w_-}{w_+}\right]=\frac{i}{\pi}\int_{\pi/2}^{3\pi/2}\frac{\delta_o\,d\phi}{G(\phi)\,(1-e^{i(\theta-\phi)})}, \tag{1.13}$$

where T is the singular operator (9)

$$(T\,f)\,(\theta)=-\frac{1}{\pi}\int_{-\pi/2}^{\pi/2}\frac{f(\phi)\;d\phi}{1-e^{i(\theta-\phi)}},\quad \theta\varepsilon\left[-\frac{\pi}{2},\frac{\pi}{2}\right] \tag{1.14}$$

the integral being taken as principal part. The equation (1.13) can be solved by using the spectral properties of the operator T (9). Namely $f_o(\theta)$ for $\theta\ \varepsilon\ [-\pi/2,\pi/2]$ can be written as

$$f_o(\theta)=h(\theta)+\frac{i\delta_o(\theta)}{1-\frac{w_-}{w_+}}+\frac{1}{1-\frac{w_-^2}{w_+^2}}\int_{-1}^{1}\frac{d\tau}{1-\tau}\left[G\psi_\tau-\frac{w_-}{w_+}G^*\psi_\tau^*\right]\left(\psi_\tau,\frac{1}{\pi}\int_{\pi/2}^{3\pi/2}\frac{\delta_o\,d\phi}{G(1-e^{i(\theta-\phi)})}\right) \tag{1.15}$$

where

$$\psi_\tau(\theta)=\frac{1}{\pi\sqrt{(1-\tau^2)\,(1+e^{2i\theta})}}\exp\left\{\frac{i}{2\pi}\ln\frac{1+\tau}{1-\tau}\,\ln\frac{ie^{i\theta}+1}{e^{i\theta}+i}\right\} \tag{1.16}$$

are the generalized eigenfunctions of the operator T which form a complete and orthonormal system (9), and we have used the notation

$$(\psi_\tau,\,g)=\frac{1}{\pi}\int_{-\pi/2}^{\pi/2}\psi_\tau^*(\theta)\;g(\theta)\;d\theta. \tag{1.17}$$

We have now to use the expression (1.15) in the alignment condition (1.5) when $S\subset[-\pi/2,\ \pi/2]$ in order to obtain the equation for δ_o. We also apply the remark (9) that, when $f_o(\theta)$, given on $[-\pi/2,\ \pi/2]$ is the boundary value of an analytic function of H^2-class, then the expression

$$\int_{-1}^{1}\psi_\tau\,(z)\,(\psi_\tau\ ,\ f_o)\;d\tau \tag{1.18}$$

gives the unique extrapolation of f_o inside the disk $|z| < 1$, which allows us to write down the equation for δ_o when S is outside T.

A useful approximation of the above procedure, already applied in (2) is obtained by truncating the series (1.12) at a finite N. Then the function L defined in (1.3) becomes

$$L = \sum_{n,m=0}^{N} U_{nm} f_n f_m - 2 \sum_{n=0}^{N} f_n (h_n + \mu_n),$$

where

$$U_{nm} = \frac{1}{2\pi} \int_{-\pi/2}^{\pi/2} \left[e^{i(n-m)\theta} |G|^2 w_+ + e^{i(n+m)\theta} G^2 w_- \right] d\theta$$

$$h_n = \frac{1}{2\pi} \int_{-\pi/2}^{\pi/2} h(\theta) \left[G^* w_+ e^{-in\theta} + G w_- e^{in\theta} \right] d\theta$$

$$\mu_n = \frac{1}{2\pi} \int_{-\pi}^{\pi} \delta_o(\theta)\, w_+ \, \mathrm{Im}\, (G e^{in\theta})\, d\theta$$

Using these expressions in (1.6) one can prove as in (3) that the solution δ_o of the maximization problem appearing in (1.6) is of the form

$$\delta_o(\theta) = \sum_{k=1}^{p} a_k \left[\delta(\theta-\theta_k) - \delta(\theta+\theta_k) \right], \quad p \le N/2$$

where $\theta_k \in [0,\pi]$ and $a_k \geq 0$ are free parameters. The problem which remains is then a finite optimization with respect to these parameters, which was shown in practice to yield rather easily the solution.

The method described above can be applied for a model independent study of the KN asymptotic total cross sections. Indeed our results (1, 2) showed that the extrapolated values became rather quickly insensitive to the values of the threshold parameters, which are in general poorly known, so we expect the same situation to occur for KN scattering concerning the unphysical region. As in (2) we expect to find a large enough interval in the $\nu^2 < 0$ direction, where the extrapolated values are model independent, exhibit already asymptotic features and have extrapolation errors still relatively low, allowing thus a realiable determination of the high energy parameters.

2.

In this section we investigate the problem of constraining the amplitude at points interior to be analyticity domain. In the case of KN scattering a problem of this type arises in connection with the determination of the KNΣ, KNΛ coupling constants and for locating the zeros of the amplitude in the complex plane. A rigorous approach to these problems is offered by the maximum modulus theorem, which in the general case of an arbitrary number of interior points and the inclusion of derivatives takes the form of the interpolation theory of analytic functions (5,10), in the Carathéodory-Schur and Pick-Nevanlinna formulations. In connection with KN scattering such an analysis was performed by Atkin, Bowcock and Queen (11), who used the data on the modulus of the $K^{\pm}p$ forward amplitudes on the cut, in order to find constraints upon the KNY coupling constants and the position of the zeros of the amplitudes. As it was already remarked from the study of various processes (11, 12, 13) in the modulus representation, where one constructs an analytic function starting from its modulus on the cut rather than its imaginary part, much weaker assumptions at infinity are needed, the main requirements now referring to the knowledge of interior zeros. By majorizing the unknown factors containing them, one obtains essentially model-independent results, as absolute bounds on the values of interest. However, in such an approach a considerable piece of experimental information, namely the imaginary part, known accurately for low and moderate energy forward scattering, is completely ignored. A modification of the method such as to include this information is therefore of interest and may contribute to an improvement of our theoretical understanding of KN scattering. In what follows I shall briefly show how this can be done in a rigorous way.

For simplicity I consider as in the previous section a fixed-t amplitude $f(\nu^2)$ real-analytic in the ν^2 -plane cut along the real axis for $\nu^2 \geq \nu_o^2$. We assume that the imaginary part of f is given along the region Γ_1

$$\operatorname{Im} f(\nu^2) = \rho(\nu^2), \quad \nu_o^2 \leq \nu^2 \leq \nu_1^2 , \tag{2.1}$$

while at higher energies the modulus of $f(\nu^2)$ (or an upper bound on it) is available, from differential cross section measurements or in the frame of a theoretical model

$$|f(\nu^2)| \leq \sigma(\nu^2), \qquad \nu^2 \geq \nu_1^2. \tag{2.2}$$

The problem can be formulated in the general form as : find the largest possible domain $\mathcal{D}$ of the values $f_k = f(z_k)$, k = 1, ... n, taken by the amplitude at n interior points z_k, $|z_k| < 1$, consistent with the input conditions (2.1) and (2.2).

Let us first perform the conformal mapping

$$z = \frac{a - \sqrt{\nu_1^2 - \nu^2}}{a + \sqrt{\nu_1^2 - \nu^2}} , \tag{2.3}$$

such that the ν^2 -plane becomes the interior of the disk $|z| < 1$, cut along the segment $[x_0, 1]$, where

$$x_0 = (a - \sqrt{\nu_1^2 - \nu_0^2})/(a + \sqrt{\nu_1^2 - \nu_1^2}),\ x_0 < 1.$$

Let us write now

$$f(z) = g(z) + \frac{1}{\pi} \int_{x_0}^{1} \frac{\rho(x)\, dx}{x - z} \tag{2.4}$$

By construction the last term is analytic in the z-plane cut between x_0 and 1, the discontinuity over the cut being $\rho(x)$. Then $g(z)$ will be real analytic in $|z| < 1$, and particularly will assume real values on $[x_0, 1]$. It is worthwhile mentioning that all the results exposed be low can be shown to be independent of the particular manner of splitting f in two terms, as in (2.4), with the condition that the discontinuity over the cut $[x_0, 1]$ of one of them is $\rho(x)$. If we define now as usual the outer function (5)

$$G(z) = \exp\left\{ \frac{1}{2\pi} \int_{-\pi}^{\pi} \frac{e^{i\theta} + z}{e^{i\theta} - z} \ln \sigma \, d\theta \right\}, \tag{2.5}$$

which is analytic without zeros in $|z| < 1$ and has modulus equal to σ on $|z| = 1$, the conditions (2.1) and (2.2) can be written in the compact form

$$\left| \tilde{g}(z) - \frac{1}{\pi G(z)} \int_{x_0}^{1} \frac{\rho(x)\, dx}{z - x} \right| \le 1, \quad |z| = 1 \tag{2.6}$$

where $\tilde{g} = g/G$ is a function analytic in $|z| < 1$.

The values of the amplitude $f(z)$ at the points $\{z_k\}$, $k = 1, \dots n$ imply for $\tilde{g}(z)$ the conditions

$$\tilde{g}(z_k) = \frac{1}{G(z_k)} \left[f(z_k) - \frac{1}{\pi} \int_{x_0}^{1} \frac{\rho(x)\, dx}{x - z_k} \right], \quad k = 1, \dots n. \tag{2.7}$$

The problem is therefore to describe the maximum domain of values $\tilde{g}(z_k)$, when the function $\tilde{g}$ satisfies the condition (2.6). It can be shown using simple arguments (12) that this convex domain is explicitely described by the following inequality

$$\inf_{\substack{\tilde{g} \in H^\infty \\ \tilde{g}(z_k) = \text{given}}} \left\| \tilde{g}(\theta) - \frac{1}{G(\theta)\pi} \int_{x_0}^{1} \frac{\rho(x)\, dx}{e^{i\theta} - x} \right\|_\infty \le 1. \tag{2.8}$$

In this relation we used the notation $\|f\|_\infty$ for the essential supremum, i.e. the L^∞-norm (5) of f on $[-\pi, \pi]$. We were thus led to solve a minimum L^∞-norm problem with constraints at interior points, which is a standard problem of functional analysis (5, 10), already applied in particular formulations to a number of physical problems (4, 12, 13, 14). In what follows we briefly summarize the main steps of solving (2.8). Assuming for simplicity that the points z_k are distinct we write $\tilde{g}(z)$ as (12)

$$\tilde{g}(z) = \sum_{k=1}^{n} A_k B_k(z) + h(z) B_{n+1}(z) \tag{2.9}$$

where

$$B_1(z)=1, \quad B_k(z) = B_{k-1}(z) \frac{z - z_{k-1}}{1 - z z_{k-1}}, \quad k = 2, \ldots n+1,$$

are products of Blaschke factors and the coefficient A_k are determined by the system of equations

$$\tilde{g}(z_j) = \sum_{k=1}^{j} A_k B_k(z_j). \tag{2.10}$$

In (2.9) h is an analytic function free of constraints, so that the inequality (2.8) written in terms of h becomes

$$\inf_{h \in H^\infty} \left\| h(\theta) + \sum_{k=1}^{n} A_k \frac{B_k(\theta)}{B_{n+1}(\theta)} - \frac{1}{\pi B_{n+1}(\theta) G(\theta)} \int_{x_0}^{1} \frac{\rho\, dx}{e^{i\theta} - x} \right\|_\infty \leq 1. \tag{2.11}$$

By applying now a duality theorem (5, 7), the minimum norm problem (2.11) is equivalent to the following maximization problem in the Hardy space H^1 which is the dual of H^∞ (5) :

$$\sup_{\substack{F \in H^1 \\ \|F\|_1 \leq 1}} \left| \frac{1}{2\pi i} \int_{|z|=1} F(z) \left[\sum_{k=1}^{n} A_k \frac{B_k(z)}{B_{n+1}(z)} - \frac{1}{\pi G B_{n+1}} \int_{x_0}^{1} \frac{\rho dx}{z-x} \right] dz \right| \leq 1. \tag{2.12}$$

The integral appearing in this relation can be computed easily by applying the residue theorem and by properly deforming the contour of integration around the cut of the last term in the integrand. After some straight-forward calculations we can finally write (2.12) in the form

$$\sup_{\substack{F \in H^1 \\ \|F\|_1 \leq 1}} \sum_{j=0}^{\infty} F_j \, c_{-(j+1)} \leq 1, \tag{2.13}$$

where F_j are the Fourier coefficients of F(z), i.e. $F(z) = \sum_{j=0}^{\infty} F_j z^j$, and the real numbers $c_{-(j+i)}$ have the form

$$c_{-(j+1)} = \sum_{k=1}^{n} z_k^j \left[\frac{z - z_k}{B_{n+1}}\right]_{z=z_k} \frac{f(z_k)}{G(z_k)} - \frac{1}{\pi} \int_{x_0}^{1} \frac{\rho(x)\, x^j}{G(x)\, B_{n+1}(x)}\, dx. \quad (2.14)$$

The evaluation of the supremum (2.13) is again a standard problem (5, 10) : by applying a factorization theorem we can show that it is equal to the norm of the Hankel matrix H, defined as

$$H_{ij} = c_{-(i+j-1)},\; i, j = 1, 2, \ldots \quad (2.15)$$

in terms of the coefficients (2.14). Therefore the domain $\mathcal{D}$ of admissible values $f(z_k)$ is obtained majorizing by 1 this norm, which depends implicitely on $f(z_k)$. We recall that explicit programs for computing norms of Hankel matrices and other quantities of interest in such approaches are described in (15).

We notice finally that a considerable simplification of the problem arises if we replace the condition (2.8) by the weaker one

$$\inf_{\substack{\tilde{g} \in H^\infty \\ \tilde{g}(z_k) = \text{given}}} \left\| w \left[\tilde{g}(\theta) - \frac{1}{\pi G(\theta)} \int_{x_0}^{1} \frac{\rho(x)}{e^{i\theta} - x}\, dx \right] \right\|_2 \leq 1 \quad (2.16)$$

i.e. if we replace the L^∞-norm by an L^2 -norm weighted by an arbitrary analytic function w satisfying the condition $\|w\|_2 \leq 1$ (12). Then the convex domain $\mathcal{D}_w$ of admissible values $f(z_k)$ consistent with (2.16) for a fixed w can be described explicitely as :

$$\sum_{i,j=1}^{n} f(z_i)\, f(z_j) \frac{w(z_i)\, w(z_j)}{G(z_i)\, G(z_j)} \left[\frac{z - z_i}{B_{n+1}}\right]_{z=z_i} \left[\frac{z - z_j}{B_{n+1}}\right]_{z=z_j}$$

$$- 2 \sum_{i=1}^{n} f(z_i) \frac{w(z_i)}{G(z_i)} \left[\frac{z - z_i}{B_{n+1}}\right]_{z=z_i} \frac{1}{\pi} \int_{x_0}^{1} \frac{\rho(x)\, w(x)\, dx}{(1 - x z_i)\, B_{n+1}(x)\, G(x)}$$

$$+ \frac{1}{\pi^2} \int_{x_0}^{1} \int_{x_0}^{1} \frac{\rho(x)\, \rho(y)\, w(x)\, w(y)}{G(x)\, G(y)\, B_{n+1}(x)\, B_{n+1}(y)} \frac{dx\, dy}{1 - xy} \leq 1. \quad (2.17)$$

It can be shown (12) that every domain $\mathcal{D}_w$ yields an approximation of the exact domain $\mathcal{D}$ from outside and moreover the intersection of all these domains coincides with $\mathcal{D}$. This remark is very useful in practice since the description of $\mathcal{D}_W$ given by (2.17) is explicit and very convenient for numerical calculations. We note that the expression (2.17) remains valid also when some points z_k are above the threshold (i.e. $z_k > x_0$), with the modification that instead of $f(z_k)$ we have to put $\mathrm{Re}\, f(z_k)$ and the integrals have to be calculated as principal parts. This situation arises when information about the real part of $f(\nu^2)$ at some points on the cut is included.

The above formalism realizes a model independent correlation of the values of the amplitude at unphysical energies with the imaginary part and the modulus on distinct parts of the cut. This technique can be useful in a variety of situations. For KN scattering we can obtain in this way optimal constraints on the KNΣ, KNΛ coupling constants in a frame more flexible than in (11) and also study the zeros in the complex plane. We mention finally that the technique can be generalized to more complicated kinematical configurations, when more than one analytic function are involved (14) which for KN scattering gives the possibility of including also information about the polarization.

References and notes

*. This talk was presented by Dr. C. Verzegnassi.

1. Ciulli, S., Verzegnassi, C.: 1977, Nuovo Cimento 40A, 24. Caprini, I., Sararu, A., Sararu, M., and Verzegnassi, C.: 1979, Nuovo Cimento 53A, 269.
2. Caprini, I., Sararu, A., Sararu, M., and Verzegnassi, C., On the High-Energy Behaviour of Proton-Proton Scattering, to be published.
3. Caprini, I. : 1974, Nuovo Cimento 21A, 236; 1979, Nuovo Cimento 49A, 307.
4. Ciulli, S., Pomponiu, C., and Sabba-Stefanescu, I.: 1975, Physics Reports 17, 133.
5. Duren, P. : 1970, Theory of H^p- Spaces, Acad. Press.
6. Shepard, H.K., and Shih, C.C. : 1972, Nucl. Phys. B42, 397.
7. Luenberger, D. : 1968, Optimization by Vector Space Methods, John Wiley & Sons.
8. Caprini, I., to be published.
9. Krein, M.G., and Nudelman, P.I. : 1973, Dokl. Acad. Nauk SSSR 209, 537 (in Russian). Koppelman, W., and Pincus, J.D. : 1959, Math. Zeit. 71, 4 , 399.
10. Krein, M.G., and Nudelman, A.A. : 1973, The Problem of Markov Moments and Extremal Problem, Nauka, Moscow (in Russian).
11. Atkin, G.K., Bowcock, J.E., and Queen, N.M.: 1979, Constraints on the KNY coupling constants, Birmingham preprint.

12. Caprini, I., and Dita, P. : 1979, A new method for deriving rigorous results on $\pi\pi$ scattering, preprint ICEFIZ-FT-178-1979, Bucharest, to appear in Journal of Physics A.

13. Raszillier, I.: 1979, Revue Roum. Phys. 24, 873.

14. Caprini, I., Guiasu, I., and Radescu, E., to be published.

15. Caprini, I., et al. : 1977, "A guide to analytic extrapolation II, Ref. TH.-2268; 1979, Computer Physics Comm. 18/3, 305.

POSITIVITY AS A SOURCE OF ASYMPTOTIC BOUNDS IN ELECTRON - POSITRON ANNIHILATION

C. Verzegnassi
Istituto di Fisica Teorica - Università di Trieste

The role played by positivity in allowing the development of an extrapolation strategy in elementary particle physics has been exhaustively illustrated by I. Caprini in her previous talk (1). From this, one has learned that the property of positivity of the imaginary part of certain analytic functions is a very useful one, particularly when such functions are not asymptotically finite, which would be a source of problems in a conventional approach. As an example, the asymptotic behaviour of the proton-proton (and implicitly, of the proton-antiproton) total cross section has been investigated, and upper bounds for the power which fixes its possible logarithmic behaviour have been proposed (2).

In this talk, I will show that something very similar happens in the case of electron-positron annihilation. Here, positivity is at work too, and from this property rather unexpected asymptotic bounds can be derived.

The relevant analytic amplitude to be considered is, in the case of electron-positron annihilation, the so-called hadronic part of the vacuum polarization function $\Pi(s)$. Here

$$s \equiv \frac{K^2}{4m_\pi^2} = x + iy \qquad (1)$$

where K^2 is the virtual photon squared mass. in the complex s plane, $\Pi(s)$ has a cut running from + 1 to ∞, and no other singularities are present. I choose a normalization such that

$$\text{Im } \Pi(x) = \frac{\alpha}{3} R(x) \qquad (2)$$

where

$$R(x) = \frac{\sigma_{e^+e^- \to \text{hadrons}}}{\sigma_{e^+e^- \to \mu^+\mu^-}} \qquad (3)$$

and, in the current philosophy (3)

E. Ferrari and G. Violini (eds.), Low and Intermediate Energy Kaon-Nucleon Physics, 375–379.

$$R(\infty) = \sum_{i=1}^{N} Q_i^2 \equiv R \tag{4}$$

where N is the number of constituent quarks and Q_i their charges.

The first assumption I will need concerns the "generalized first momentum" of $\Pi(s)$, which is defined as :

$$R_1(s) = -s\pi\frac{3}{\alpha}\frac{d}{ds}\;\Pi(s) = -s\int_1^{\infty}\frac{dx'\;\mathrm{Im}\;(x')}{(x'-s)^2} \tag{5}$$

and enjoys the following property :

$$R_1(-\infty) = R \tag{6}$$

If one considers the quantity

$$\frac{d}{dx}\;R_1(-x) \tag{7}$$

in the limit when $x \to \infty$, it can be seen in the framework of a popular theory like quantum chromodynamics that (4)

$$\frac{d}{dx}\;R_1(-x) < 0, \qquad x > x_o \tag{8}$$

i.e. for sufficiently large x values, where the theory is supposed to be valid. I will assume the property eq. (8) to be true in general.

To give an idea of what unexpected consequences a mild assumption like eq. (8) might have, I will consider an oversimplified example of a function $\tilde{\Pi}(s)$ the imaginary part of which has the following form :

$$\mathrm{Im}\;\tilde{\Pi}(x) = \begin{cases} C & 1 < x < \Lambda_1 \\ R_1 & \Lambda_1 < x < \Lambda \\ R_2 & \Lambda < x < \Lambda_A \\ R & \Lambda_A < x \leq \infty \end{cases} \tag{9}$$

The requirement that eq. (8) is asymptotically true, i.e. that the first generalized momentum $\tilde{\Pi}(s)$ is Monotically Asymptotically Decreasing (M.A.D.) in the spacelike region combined with the positivity of C, R_1, R_2, R, leads in a straightforward way to the condition :

$$R < \frac{\Lambda_1 - 1}{\Lambda_A}\;C + \frac{\Lambda - \Lambda_1}{\Lambda_A}\;R_1 + \frac{\Lambda_A - \Lambda}{\Lambda_A}\;R_2 \tag{10}$$

which is, in fact, an upper bound for R. Alternatively, one can suppose for instance that R_1 is known to be larger than C, and one wants to investigate whether R can be larger than R_1. Again, due to the positivity of Im $\tilde{\Pi}(x)$, this will compatible with eq. (8) only if :

$$R_2 > R; \quad (\Lambda_A - \Lambda) > \frac{R - R_1}{R_2 - R}\,(\Lambda - \Lambda_1) \tag{11}$$

In other words, in order that R can be larger than R_1, definite lower bounds on both the "height" and the "width" of Im $\tilde{\Pi}$ in the unknown pre-asymptotic region must exist.

Although the example of eq. (9) might seem unrealistic, it is not, in fact, so far from the actual case of electron-positron annihilation. Here we have, in fact, two regions where the average value of R(x) is approximately constant and a few narrow peaks which, as one immediately verifies, do not contribute to eq. (5) in appreciable way. Consequently, one would expect the general conclusion of eq. (11) to remain true.

To see whether this feeling is correct, we have evaluated (5) $R_1(s)$ at real negative x values using the available data (6) of R(x) for $W \lesssim 32$ GeV, assuming that R(x) jumps to the value R = 5 (corresponding to six constituent quarks) somewhere for $32 \lesssim W \lesssim 100$ GeV, and remains "essentially" constant afterwards. As figure 1 (dotted lines) shows,

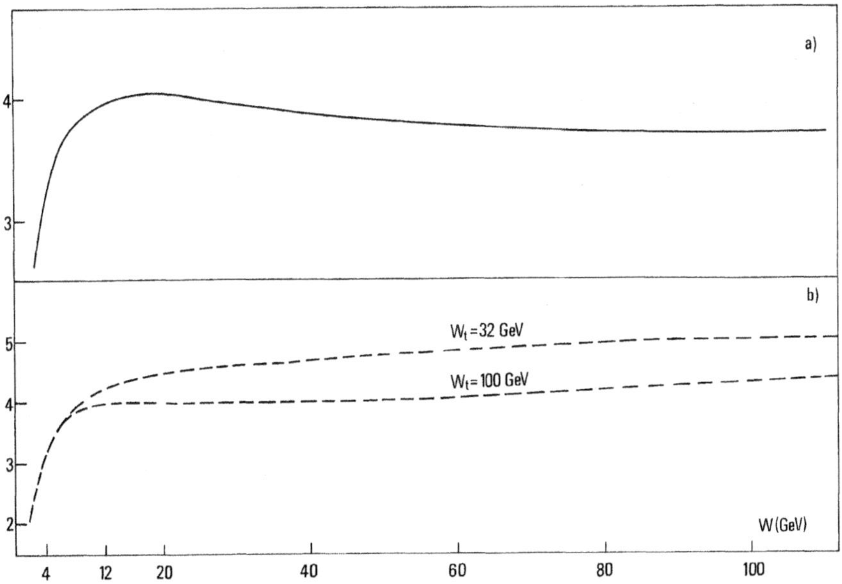

Figure 1 - The extrapolated quantity $R_1(s)$ at real negative s values in the assumptions that: a) R(x) remains essentially equal to the last experimental value (full line); b) R(x) jumps to the value R = 5 at an energy $W \equiv W_t$ between 32 and 100 GeV remaining then constant (dotted lines). Figures are plotted versus the modulus of the corresponding energy W

in no case the related extrapolation has the required spacelike behaviour and points to its asymptotic value from below.

In figure 2 I have plotted the result of our extrapolation assuming that R (x) jumps to a "preliminary" value $\bar{C}$ larger than five, in the two particular cases $\bar{C} = 6$ and $\bar{C} = 8$ (larger $\bar{C}$ values seemed to me unlikely, although they might be easily studied). As one can see, the main feature which comes out is that, for reasonable values of the preliminary bump $\bar{C}$ (e.g. of the order of those of the last resonance), there must be a quite broad region (not smaller than $\sim$ 10 GeV if $\bar{C} \lesssim 8$) where this bump should be seen.

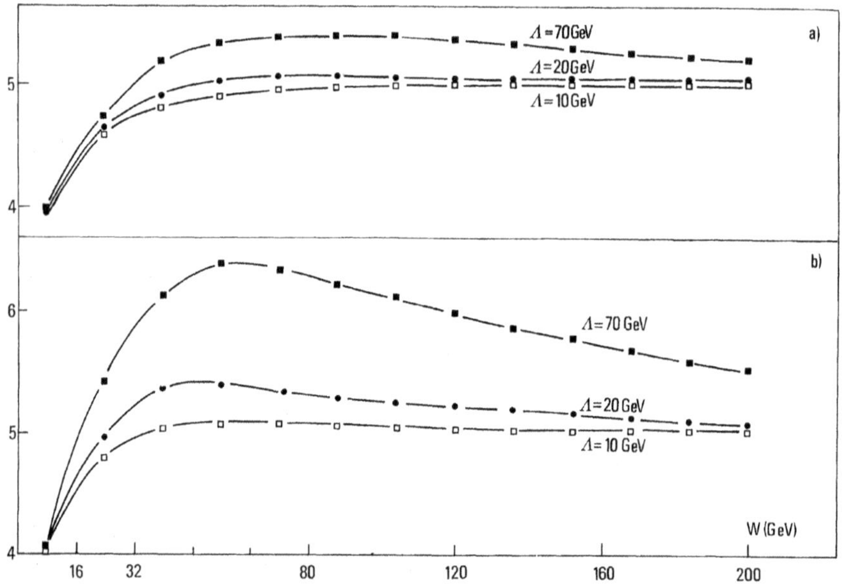

Figure 2 - The extrapolated quantity $R_1(s)$ at real negative s values in the assumptions that : a) R (x) jumps to a preliminary value $\bar{C} = 6$ in a region between 32 and 100 GeV of width Λ (upper figure); b) the same as in the upper figure, with $\bar{C} = 8$ (lower figure), and for W > 100 GeV, R (x) remains equal to 5

In conclusion, I believe to have shown that the requirement of Monotonic Asymptotical Decrease, combined with the positivity property of Im Π (x), generates unexpected links between the preasymptotic and the asymptotic values of R (x). A much more general analysis based on this philosophy is already being performed.

References

1. See, for a complete discussion, Caprini, I. : 1979, Nuovo Cimento 49A, 307.
2. Caprini, I., Sararu, A., Sararu, M., and Verzegnassi, C., "High energy behaviour of proton-proton scattering", submitted to Nuovo Cimento.
3. See, for instance, Biorken, J.D. : 1966, Phys. Rev. 148, 1467.
4. Georgi, H., and Politzer, H.D. : 1973, Phys. Rev. D8, 4000. Gross, D., and Wilczek, F. : 1973, Phys. Rev. D8, 4038.
5. Roberto, V., and Verzegnassi, C., "Spacelike features and asymptotic behaviour of e^+e^- annihilation", to be published on Lett. Nuovo Cimento.
6. Augustin, J.E. : 1979, "e^+e^- Physics below J/ψ resonance ", Invited talk at the EPS Intern. Conf. on High Energy Physics, Geneva; LAL 79/15 preprint.
   Baldini-Celio, R., et al. : 1979, Lett. Nuovo Cimento 24 (10), 324.
   Flügge, G., "Recent experiments at DESY", Lectures presented at the XVIII Int. Universitätswochen für Kernphysik, Schladming, Austria, DESY 79/26 preprint.
   MARK J Collaboration, Barber, D., et al. : 1979, Phys. Rev. Lett. 42, 1113.
   PLUTO Collaboration, Berger, Ch., et al., DESY 79/56 preprint.
   TASSO Collaboration, Brandelix, R., et al.: 1979, Phys. Lett. 83B, 261.
   JADE Collaboration, Bartel, W., et al., DESY 79/64 preprint.
   Wolf, G.: 1979, "High Energy trends in e^+e^- Physics", Rapporteur talk given at the EPS Conference-Geneva.

REMARKS BEARING ON THE INTERPRETATION OF THE $\Lambda(1405)$ RESONANCE

R.H. Dalitz and J.G. McGinley*
Department of Theoretical Physics - Oxford University

1. Introduction

We focus attention here on the $\Lambda(1405)$ resonance not because it has any special importance in itself but because it illustrates particularly well a rather general problem in elementary particle theory. The essential facts about this resonance are relatively well-established (1) and the empirical data concerning the meson-baryon channels linked with it are known in unusual detail, owing to the extensive hydrogen bubble-chamber studies which have been made of the low-energy K^-p scattering and reaction processes over the last two decades (2-5). It was first discovered empirically in 1961, as a $\pi\Sigma$ resonance produced in high-energy reactions of the following types (6, 7)

$$K^- + p \to \Lambda(1405) + \pi^+ + \pi^-, \tag{1.1}$$

$$\pi^- + p \to \Lambda(1405) + K^o. \tag{1.2}$$

It lies 30±40 MeV below the K^-p threshold energy, with width $\Gamma = 38\pm4$ MeV. The only $I=0$ thresholds lying below the $\Lambda(1405)$ mass are the $\pi\Sigma$ thresholds at about 1331 MeV and the $\Lambda\pi\pi$ thresholds at about 1390 MeV. The only decay modes known empirically are $\Lambda(1405) \to \pi\Sigma$(s-wave).

Although many models have been proposed for $\Lambda(1405)$, emphasis has come to be placed on two particular models which represent two extreme points of view. The truth may well lie somewhere between them, or may even lie in some quite other direction. These two extremes may be illustrated as follows :

(1) dynamical models for the interactions between pseudoscalar mesons (P) and baryons (B). The simplest model is based on the attractive forces between P and B resulting from the exchange of vector mesons (V) between them (8-12), as given by SU(3) flavour symmetry. For $\Lambda(1405)$, the relevant channels are $\bar{K}N$, $\pi\Sigma$, $\eta\Lambda$ and $K\Xi$, but the last two have such high thresholds that they play little role in the energy dependence near this resonance. Let us therefore speak here only in terms of the two lowest

E. Ferrari and G. Violini (eds.), Low and Intermediate Energy Kaon-Nucleon Physics, 381–409.

channels ($\pi\Sigma$, $\bar{K}N$), for which we denote the c.m. momenta by (k_1, k_2), respectively. Since the scattering matrix T(E) involves the momentum k_2, which has a square-root branch cut at the $\bar{K}N$ threshold, it is most convenient to make use of the K-matrix, since its elements vary smoothly with energy across any two-particle threshold. We denote this by

$$K_s = \begin{pmatrix} K_{\pi\pi} & K_{\pi K} \\ K_{\pi K} & K_{KK} \end{pmatrix} \qquad (1.3)$$

where the suffix just denotes that we are concerned with the s-wave systems. This matrix may have a pole at a resonance energy E^*. Indeed this is certaintly the case for the corresponding K_d for the $d_{3/2}$-wave at the energy of the resonance $\Lambda(1520)$ in that wave. This $\Lambda(1520)$ state is well established to be a completely normal Breit-Wigner resonance, through the detailed studies of it which have been made most recently by Alston-Garnjost et al. (13). Returning to the $s_{1/2}$-wave, the matrix K_s still holds valid below the $\bar{K}N$ threshold, but the momentum k_2 is now imaginary, given by $+i|k_2|$. The description of the $\pi\Sigma$ scattering in this energy region needs only a one-channel K-matrix, which we term the 'reduced K-matrix' and denote by the symbol K^r. The scattering phase $\delta(\pi\Sigma)$ is then given by

$$k_1 \cot\delta(\pi\Sigma) = -(K_s^r)^{-1} \qquad (1.4)$$

Clearly, this reduced matrix must be expressible in terms of the elements of the full K_s matrix, and the relation is explicitly (14)

$$K_s^r = K_{\pi\pi} + K_{\pi K}(1 - |k_2|\, K_{KK})^{-1}\, |k_2|\, K_{K\pi} \qquad (1.5)$$

There are two ways in which a narrow $\pi\Sigma$ resonance can occur, in this formalism :

(a) The first possibility is that all the elements of K_s have a pole at $E = E^*$. In this case, K_s^r will also have a pole at an energy close to E^* and the phase shift $\delta(\pi\Sigma)$ will pass upwards through $\pi/2$ at this energy. This is the normal situation, as described by Wigner (15) in his reaction-matrix theory of nuclear scattering.

(b) The second possibility is that none of the elements of K_s has a pole, but that the denominator in expression (1.5) has a zero at $E = E^*$, which gives rise to a pole at E^* for the reduced K-matrix K_s^r. This situation would hold if the interactions which couple the channels are relatively weak. In this case K_{KK} would be determined primarily by the interactions in the $\bar{K}N$ channel; K_{KK} is then essentially the $\bar{K}N$ scattering length, and the condition

$$1 - |k_2|\, K_{KK} = 0 \qquad (1.6)$$

may then be recognised as the equation determining the location for a bound state in the $\bar{K}N$ channel. However, $K_{\pi K} \neq 0$, so this bound state can decay to $\pi\Sigma$ and will appear as a resonance in $\pi\Sigma$ scattering. For this reason, we term this $\pi\Sigma$ resonance as an "unstable $\bar{K}N$ bound state".

We emphasize that the above is a much over-simplified view of this resonance mechanism (b). In general, the off-diagonal potentials which couple the channels are just as strong as the diagonal potentials, and the K-matrix element K_{KK} can receive major contributions from these cross-channel couplings. The diagonal potential V_{KK} may well be insufficient to sustain a bound state in the $\bar{K}N$ channel alone, and the other elements of the multichannel potential may be essential for the occurence of a resonance determined by the denominator of (1.5). The distinction between these two possible origins for the resonance pole in K_s^r is clear in any theoretical calculation. Empirically, this distinction would depend on an extrapolation of the elements of K, determined above the $\bar{K}N$ threshold, to energies below this threshold.

(2) An imposed resonance R, in the sense that the existence of R is not due to the interactions within and between the meson-baryon systems, but to other interactions, for example interactions between other, very heavy particles, with high threshold energies, or between confined particles. The most relevant example at present is the three-quark model for baryons, with confined quarks and quark-quark forces with the general characteristics suggested by quantum chromodynamics. Each baryonic state of this type is a member of a large supermultiplet, and has mass and decay properties determined by the general properties which go with this supermultiplet. We know this situation very well for the 56 ground states of three quarks in the quark space (u, d, s), which have total orbital angular momentum $L = 0$ and spin $S = 3/2$ or $1/2$, and whose spin x flavour wavefunctions are permutation symmetric. These states are those comprising the $({}^2 8 + {}^4 10)$ supermultiplet of octet and decuplet states, the notation being ${}^{2S+1}\alpha$ where α denotes the dimensionality of the $SU(3)_f$ representation and f denotes flavour, and they are all known. The first excited configuration of the three-quark system will naturally have $L = 1$, with negative parity. This configuration has mixed permutation symmetry, and this goes with a spin x flavour wavefunction which also has mixed permutation symmetry. This supermultiplet consists of the 70 states belonging to the unitary singlet states ${}^2 1(J=1/2$ and $3/2)$, the octet states ${}^2 8(J=1/2$ and $3/2)$ and ${}^4 8(J = 1/2, 3/2$ and $5/2)$, and the decuplet states ${}^2 10(J = 1/2$ and $3/2)$. The N^* and Λ^* members of all these spin x $SU(3)_f$ multiplets are well-established (1), and we have no doubt about the existence of their strange counterparts, since the $SU(3)_f$ symmetry is well established. The ${}^2 1$ multiplets have no nucleonic members, but well-established candidates exist for them, namely the Λ(1520) resonance for $J = 3/2^-$ and the Λ(1405) resonance for $J = 1/2^-$. As remarked earlier, the Λ(1520) state is of standard Breit-Wigner type, and therefore appears as an imposed resonance; it is therefore natural to expect that the Λ(1405) sould have the same character.

Current theories of baryonic resonance masses according to the three-quark model (16) find difficulty to account for the large mass-difference between these two closely-related Λ states, but this may be due in part to their insistence on the absence of any spin-orbit couplings which would naturally separate the $J = 1/2^-$ Λ(1405) from the $J = 3/2^-$ Λ(1520). There has occasionally been speculation whether there might not exist a second $J = 1/2^-$ Λ-state lying a little below the Λ(1520) state, but the coverage of this mass range in recent experiments (5) has really ruled out such a possibility.

The question has been before us for some time now : how to distinguish between these two interpretations of the origin of the Λ(1405) baryonic state ? Further : whether it is even possible to decide this question, for it might be conceivable that the two mechanisms mentioned above could act together, leading to a single Λ^* resonance state, as observed, but with mass shifted from the three-quark supermultiplet prediction. The special feature about this resonance is that it couples with s-wave meson-baryon channels, where there are no centrifugal barriers; it has long been pointed out (17) that the absence of centrifugal shielding for the $1/2^-$ baryonic states could allow a greater sensitivity of these states to external violations of $SU(3)_f$ symmetry, which can affect their mass value. However, no corresponding sensitivity is apparent in the data for the corresponding cases of $1/2^-$ N^* and Λ^* states.

2. The Data

The empirical data which bear on the analysis of Λ(1405) are of four kinds at present :

(i) K^-p elastic and charge-exchange scattering cross sections for laboratory momenta p_{Lab} in the range 100-250 MeV/c, and the cross sections for the reactions leading to the πΛ and the three πΣ final states, in the same momentum range. To this, we must add the relative rates for these four reactions, determined for K^- mesons coming to rest in liquid hydrogen, since we believe that these reactions are due to the s-wave K^-p absorptive interactions at the K^-p threshold energy.

(ii) The energy dependence of the total cross section for πΣ elastic scattering, as deduced from the πΣ spectra observed as final states following Λ(1405) production processes. The reaction (1.2) appears especially suitable for this purpose since the hadronic interactions of the K^0 meson are much less pronounced, and are therefore expected to give less distorsion to the final state spectra, than those resulting from final pions.

(iii) The energy-level shift and width for the 1p→1s X-ray emission, for the K^-p atoms formed when K^- mesons are stopped in liquid hydrogen. These are given by the formula

$$\varepsilon - i\Gamma/2 = -412.11\, a_K^C \text{ eV fm}^{-1} \qquad (2.1)$$

where a_K^C denotes the zero-energy K^-p scattering length, the superfix C referring to inclusion of the simplest Coulomb correction (10), and is given by

$$a_K^C = \left\{\frac{1+|k_0|\,(a_1+a_0)/2}{(|k_0|a_0a_1 + (a_1+a_0)/2)} + \frac{2}{B}\,(2\gamma + \ln(\frac{2d}{B}))\right\}^{-1} \tag{2.2}$$

where B denotes the Bohr radius for kaonic hydrogen, γ is Euler's constant, d is a cut-off radius of about 0.4 fm, k_0 is the (purely imaginary) momentum in the $\bar{K}^0n$ channel at the K^-p threshold, and a_I denotes the $\bar{K}N$ scattering length for isospin I in the limit of charge-independence. The only experimental value available to date is that of Davies et al. (18),

$$a_K^C = +0.10\pm0.15 + i\left(0.0^{+0.28}_{-0.0}\right)\ \text{fm}. \tag{2.3}$$

This value is incompatible with the data under (i) above, which leads to an expectation of the order of (-1.0 + i0.7) fm, and it is not known how this discrepancy is to be interpreted. An extension of this experiment has been completed by the same group and will soon reach a conclusion comparable with (2.3). It is very desirable to have this conclusion confirmed by an independent experiment.

(iv) The real parts of $K^\pm p$ elastic scattering amplitudes at 0°, denoted by $D_p^\pm$, have been determined from very accurate differential cross section measurements in the Coulomb interference region for a number of incident $K^\pm$ laboratory momenta p_K. Baillon et al. (19) have reported and analysed data for p_{K^-} in the range 0.9 to 2.6 GeV/c and p_{K^+} in the range 1.2 to 2.6 GeV/c, and also (20) for three higher $K^\pm$ momenta running up to 10 GeV/c. Jenni et al. (21) have obtained D_n^- values for K^-n elastic scattering, for p_{K^-} in the range 1.2 to 2.6 GeV/c, by the use of a deuterium target and the measurement of K^-d differential cross sections for both elastic and deuteron-breakup inelastic scattering. All these data have relevance for our present topic through their use in forward dispersion relations, which provide additional constraints on the $\bar{K}N$ scattering amplitudes in the neighbourhood of the $\bar{K}N$ threshold and the Λ(1405) resonance, as has been emphasized by Martin (22).

3. Elementarity

The imposition of a resonance R on a system is closely connected with the question of 'elementarity', which has been discussed in some detail by Weinberg (23) for the deuteron and extended to the Λ(1405) by Rajasekaran (24). Weinberg introduces a deuteron field $\psi_d(x)$ coupled with the s-wave np system (we omit mention of spin, which is an irrelevant complication here). Owing to this coupling, the deuteron field becomes renormalized by a factor Z. If the deuteron is a composite of n and p, the deuteron field $\psi_d(x)$ will become distributed over all np states, then giving an infinitesimal contribution to the physical deuteron state; in other words we then have $Z = 0$. However, if $Z \neq 0$, this means that the deuteron field $\psi_d(x)$ contributes a finite part, of probability Z, to the physical deuteron state, which is then not completely composite.

Weinberg pointed out that if the binding energy b is small and the composite contributions are s-wave, that it is possible to write down model-independent expressions for the np effective range parameters,

$$a = (2(1-Z)/(2-Z))B + O(\rho), \qquad (3.1.1)$$

$$r_{eff} = (-Z(/(1-Z))B + O(\rho), \qquad (3.1.2)$$

where $B = (2\mu b)^{-1/2}$, μ is the np reduced mass, and ρ is the range parameter for the np potential. In particular, when Z is appreciable, (3.1.2) indicates that r_{eff} will be negative and large, since $B = 4.3$ fm. In fact, r_{eff} is known to have the positive value +1.75 fm, comparable in magnitude with $\rho = 1/m_\pi$, the range parameter appropriate to nuclear forces. Weinberg concludes that the data are consistent with $Z = 0$, i.e. with compositeness for the deuteron, but put a strong limit on Z, in any case.

A more general procedure for imposing an elementary field is by the introduction of CDD poles (25) in the S-matrix elements. This has been illustrated explicitly by various authors (26, 27), by extensions of the Lee model. Weinberg explicitly rejects this possibility, since it leads to an expression for $k\cot\delta$ which is of the following form, for the simplest case,

$$k\cot\delta = -K^{-1} = -1/\alpha + \tfrac{1}{2}\beta k^2 + \gamma/(1-\eta k^2) \qquad (3.2)$$

Since the effective range expansion with $\gamma = 0$ fits the $I = 0$ np s-wave scattering data up to $k^2/2\mu \simeq 10$ MeV, an energy large relative to the deuteron binding energy, $b = 2.23$ MeV, Weinberg concludes that an additional pole term in (3.2) would violate its observed linear dependence on k^2, unless the pole is so far away that it could not reasonably be regarded as the origin of the physical deuteron state.

Rajasekaran (24) developed a formalism for the multichannel case appropriate to Λ(1405), requiring that the K-matrix should have a pole at E_K, close to the resonance energy E^*, and that $\det(K^{-1})$ should increase through zero, with increasing energy E, at the pole E_K. Although the existence of such a pole can arise in many ways, and is not necessarily an indicator of 'elementarity', the absence of a K pole near the resonance energy E^* was considered to be an indicator of compositeness (however, see Sec. 5 below).

For Λ(1405), the binding energy relative to the $\bar{K}N$ threshold is $b \simeq 30$ MeV, which is not so small; on the other hand, the range parameter of the meson-baryon s-wave forces is given by $\rho \simeq 1/m_V$ and so the ratio ρ/B for Λ(1405) is smaller by about a factor 2 than that for the deuteron.

Empirically, and not surprisingly, the energy-dependence of the two-channel K-matrix for the ($\pi\Sigma$, $\bar{K}N$) channels is much less well known than that for the one-channel I = 0 np state, as we shall discuss in the next section.

4. Phenomenological Fits to the Data

The early K^-p interaction data (2) was for the K^- laboratory momentum range 0 - 250 MeV/c and was fitted quite adequately with a constant K-matrix. The new body of low-energy K^-p data reported by Kim (2) in 1966 was analysed together with other data up to momentum 550 MeV/c using for the s-wave an energy-dependent K-matrix given for I = 0 by the linear form,

$$K^{-1} = K_t^{-1} + \sigma(E-E_t), \qquad (4.1)$$

advocated by Shaw and Ross (28), K_t and σ both being symmetric 2 x 2 matrices and E_t being the K^-p threshold energy. When this fitted K-matrix was continued below this threshold, it was found that it led to the prediction of a Λ^* resonance at 1403 MeV, with width 50 MeV, in striking accord with observation. However, it was unclear whether the description of this K^-p interaction data really required the introduction of as many additional parameters as are contained in σ. Acceptable fits to the data for the momentum range 0 - 300 MeV/c with a constant K-matrix were reported by Martin and Sakitt (29) and by Martin and Ross (30), and these fits also predicted a Λ^* resonance below the $\bar{K}N$ threshold, although not with such a good fit to the resonance parameters.

In 1971, Martin et al. (31) and Barbaro-Galtieri (32) made fits to the data for the momentum range 0 - 300 MeV/c, adding to the constant K-matrix a single pole term, thus

$$K = K_t + c\tilde{c}/(E-E^*), \qquad (4.2)$$

where c is a real 1 x 2 column matrix and $\tilde{c}$ denotes its transpose, and the energy E^* is adjustable. Despite the three additional parameters, the fit was very little improved; the energy E^* moved out of the momentum range covered by the data, so that the only role of the pole term was to give some energy-dependence to K in the region where it was being used. The logical step then was to include the $\pi\Sigma$ production spectra below the $\bar{K}N$ threshold, in the K-matrix analysis. This was done by Chao et al. (33), making use of $\pi^-p \rightarrow \pi^\pm\Sigma^\mp K^0$ data for $p_\pi \approx 2$ GeV/c, obtained by Thomas et al. (34) and shown on figure 1. Their analysis also used a constant K-matrix K_t and they found that this constraint gave quite a strong discrimination between different solutions for K_t, which gave an equally good fit to all the above threshold data.

More recently, the accurate data on $D_p^\pm$ and D_p^- referred to in Sec.3 have added further constraints on the low-energy $\bar{K}N$ scattering amplitudes and on the $\Lambda(1405)$ region. Also, other kinds of data have become available,

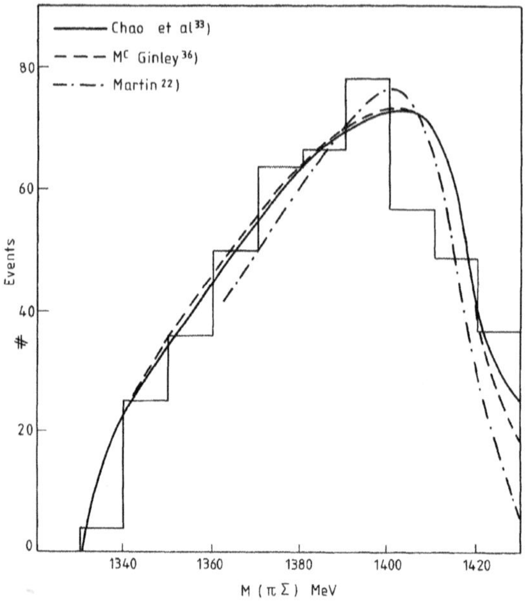

Figure 1 - A plot of the final-state $\pi\Sigma$ mass distribution observed by Thomas et al. (34) for the reaction (1.2) at $p_\pi \simeq 2$ GeV/c, is compared with the spectra as fitted by Chao et al. (33), as predicted by Martin (22) from his fit to much of the other data listed in Sec. 2, and as predicted from the fit by McGinley (36) to all of the other data listed in Sec. 2

e.g. on the forward cross section for $K^-p \to \bar{K}^0 n$ and on scattering and reaction cross sections for $K_L^0 p$ collisions. All these data (excluding $\pi\Sigma$ spectra below $\bar{K}N$ threshold) were taken together in an extensive analysis by Martin (22) in 1976, using the linear parametrisation (4.1) for the I = 0, the I = 1 K-matrix being constant. The energy dependence of the I = 0 matrix is quite crucial for obtaining a good fit to the data; explicitly, it is given by

$$K^{-1} = \begin{pmatrix} -0.10 & -0.92 \\ -0.92 & 1.60 \end{pmatrix} + \begin{pmatrix} 1.34 & -0.98 \\ -0.98 & 0.22 \end{pmatrix} (E-E_t) \quad \begin{matrix} :\bar{K}N \\ :\pi\Sigma \end{matrix} \qquad (4.3)$$

where the first matrix has unit fm^{-1} and the second is dimensionless, energy being measured in MeV. The $\pi\Sigma$ spectrum predicted by Martin's fit was a good fit to the data of Thomas et al. (34), which had been used as input in the earlier fit by Chao et al. (33). The charge-independent $\bar{K}N$ elastic scattering amplitudes obtained by Martin for the $\bar{K}N$ energy were then

$$A_0 = (-1.66 + 0.75i)\mathrm{fm}, \quad A_1 = (0.35 + 0.66i)\mathrm{fm} \qquad (4.4)$$

Another analysis in the same spirit as that of Martin, but using the threshold data of Nowak et al. (4) and updating some of the above-threshold data (cf. Goossens, (35)), has recently been given by McGinley (36), who also concluded that, for $I = 0$, an energy-dependent K-matrix was essential for a good fit, if the dispersion relation constraints were imposed. McGinley then sought the best fit for an $I = 0$ K-matrix of the form (4.2), but finally came to the conclusion that the best fit of this general class was of the form

$$K = -c\tilde{c}/(E - E^{*}) \qquad (4.5)$$

with $\tilde{c} = (0.94, -0.47)$ and $E^{*} = 1530$ MeV. We note that this expression has a pole whose residue has the wrong sign (causality requires that the diagonal parts of the residue should be positive); however, this pole lies at an energy well above the energy range for the data being fitted, so that we need not be concerned about the unphysical nature of (4.5) in the vicinity of E^{*}. The data has chosen the wrong sign for (4.5) because it requires a slope for K vs. E which cannot otherwise be achieved with the form (4.2). This K-matrix (4.5) has no inverse, since one of its eigenvalues is identically zero, so that it cannot be written in the form (4.1). McGinley's fit gives a rather broad Λ(1405), but the $I = 0$ $\pi\Sigma$ spectrum calculated from it, given by the form $(\sin\delta(\pi\Sigma))^2/k_1$ and plotted on figure 1, compares quite well with the data, as well as with the fits obtained by Chao et al. (33) and the prediction by Martin (22). The threshold scattering lengths obtained by McGinley are

$$A_0 = (-1.59 + 0.65i)\mathrm{fm}, \quad A_1 = (-0.02 + 0.56i)\mathrm{fm} \qquad (4.6)$$

the standard error on each number quoted being ±0.05. The value found for $\mathrm{Re}A_1$ by McGinley differs by many standard deviations from that of Martin, given in eq. (4.4), but the other three parameters are in tolerable agreement.

McGinley found one type of fit of the form (4.2) which is worthy of note. In this fit, the elements of c were small, of order 0.03, and the energy E^{*} coincided with the $\bar{K}N$ threshold energy. The pole term contributed significantly only within ±1 MeV of this threshold, i.e. it affects only the comparison with the at-rest data. The function of this pole, with its three parameters, is just to bridge the gap between the threshold values suggested by the analysis of all the data above threshold and the dispersion relation expressions and the data. The same mechanism acts if one endeavours to impose on the K-matrix elements the threshold constraint (2.3) obtained from the K^-p X-ray measurements, as Kumar and Nogami (37) have also found recently. K-matrix solutions which have this characteristic are to be regarded as pathological, for no reasonable physical model could ever give rise to such a special feature, centred only on such a limited energy range.

5. Potentials and Imposed Particles

We start with the one-vector-meson exchange potential used by Dalitz, Rajasekaran and Wong (DRW) (10), already mentioned above. This involves the four meson-baryon channels $\pi\Sigma$, $\bar{K}N$, $\eta\Lambda$ and $K\Xi$, labelled $i=1$ to 4. The Schrodinger equation takes the form

$$[E - T_i(k)]\ \psi_i(\underline{k}) = \Sigma_j \int V_{ij}(\underline{k},\underline{k}')\ \psi_j(\underline{k}')\ d^3\underline{k}' \tag{5.1}$$

where $T_i(k)$ denotes the meson-baryon energy operator,

$$T_i(k) = (m_i^2 + k^2)^{1/2} + (M_i^2 + k^2)^{1/2}, \tag{5.2}$$

and (m_i, M_i) are the (meson, baryon) mass values for channel i. The potential adopted has the form

$$V_{ij}(\underline{k},\underline{k}') = \frac{-f^2}{4\pi}\cdot C_{ij}\left\{m_V^2 + (\underline{k}-\underline{k}')^2\right\}^{-1} \tag{5.3}$$

where the matrix C for the vector-meson-baryon couplings according to $SU(3)_f$ symmetry, has the following symmetric form

$$C = \begin{pmatrix} -2 & -\sqrt{6}/4 & 0 & -\sqrt{6}/4 \\ -\sqrt{6}/4 & -3/2 & -3\sqrt{2}/4 & 0 \\ 0 & -3\sqrt{2}/4 & 0 & 3\sqrt{2}/4 \\ -\sqrt{6}/4 & 0 & 3\sqrt{2}/4 & -3/2 \end{pmatrix} \tag{5.4}$$

In the DRW calculations, eq. (5.1) was changed into a standard Schrodinger form by rewriting its left-hand side in the form (5.5) and then approximating $\mu_i(k)$ by $\mu_i(k_i(E))$, where $k_i(E)$ denotes the physical

$$(E - m_i - M_i - k^2/2\mu_i(k)) \rightarrow E - m_i - M_i - k^2/2\mu_i(k_i(E)) \tag{5.5}$$

momentum in channel i for energy E. In other words, the operator $\mu_i(k)$ was replaced by an E-dependent c-number, leaving the k^2 in the numerator as an operator, so that the equations can be solved in co-ordinate space, by standard Schrodinger methods. Using a mean vector meson mass, $m_V \simeq 890$ MeV/c, these equations have only one free parameter, the coupling constant $f^2/4\pi$, which is to be adjusted to place the $\Lambda(1405)$ resonance, defined by $\delta(\pi\Sigma) = 90°$, at the observed mass value. The result obtained was $f^2/4\pi = 1.25$, quite a reasonable value for this coupling constant, in view of the typical estimates $f^2/4\pi \simeq 2$, derived from other vector meson phenomena. The width then predicted for $\Lambda(1405)$ was 27 MeV, somewhat less than the observed value. The K-matrix for the $(\pi\Sigma, \bar{K}N)$ channels was well fitted by the linear form (4.1), and it had no pole on the real energy axis, so that this calculation provides a practical example of an "unstable $\bar{K}N$ bound state" resonance.

Although the largest attractive eigenvalue for the matrix C occurs for the unitary singlet state, the eigenvalues for the octet states being a factor of two smaller, this calculated Λ(1405) state has substantial octet admixtures, with total intensity 15%. Indeed, the ratio of the coupling strengths of $\bar{K}N$ and $\pi\Sigma$ to this Λ(1405) state is calculated to be $g(\Lambda^*\pi\Sigma)^2 / g(\Lambda^*\bar{K}N)^2 = 0.23$, very far different from the ratio 3/2 expected for a unitary singlet, but this is in good qualitative accord with the empirical estimate, as derived from the observed $\Lambda(1405) \to \pi\Sigma$ width and the $I = 0$ $\bar{K}N$ zero-energy scattering amplitude.

We now consider the effect on the K-matrix for such a system of meson-baryon channels, of imposing a particle R with mass m_R and amplitudes g_i for its s-wave coupling $R \to m_i B_i$ to the meson (m) and baryon (B) of the i-th channel when the potential interactions V acting within and between these channels are turned off. For K-matrix, we are concerned with a Green's function with standing waves, for which we shall use the notation $G_V(E; \underline{r}', \underline{r})$. With this notation, the free-particle Green's function is $G_0(E; \underline{r}', \underline{r})$ and it is explicitly

$$\left[G_0(E; \underline{r}', \underline{r})\right]_{ij} = \delta_{ij} P \int \frac{\exp(i\underline{k} \ (\underline{r} - \underline{r}'))}{(E - T_i(k))} \frac{d^3\underline{k}}{(2\pi)^3} \tag{5.6}$$

where P denotes a principal value integration. For simplicity, we shall adopt the approximation (5.5) for the rest of Sec. 5, neglecting even the E-dependence of μ_i. The Green's function $G_V(E; \underline{r}', \underline{r})$ is depicted by figure 2(a), the propagators between successive interaction V being the freeparticle propagator just specified and the sum going over all numbers of V interactions, including zero. $G_V(E; \underline{r}', \underline{r})$ is completely defined by this characterisation and we may assume it to be a known function. The inclusion of the particle R is indicated symbolically in figure 2 (b, c); to write it out explicitly, we need two quantities which can be constructed from $G_V(E, \underline{r}', \underline{r})$, namely the two elements depicted separately in figure 3. These are

(i) the vertex element v_V given by

$$v_V(E; \underline{r}) = \int G_0(E; 0, \underline{r}') \, G_V(E; \underline{r}', \underline{r}) \, d^3\underline{r}' \tag{5.7}$$

(ii) the self-energy element given by

$$\Sigma_V(E) = \iint G_0(E; 0, \underline{r}') \, G_V(E; \underline{r}', \underline{r}) \, G_0(E; \underline{r}, 0) \, d^3\underline{r} \ d^3\underline{r}' \tag{5.8}$$

If we now denote by Γ_n the value of the graph in which the particle R occurs n times, so that figure 2 (a, b, c) represent Γ_0, Γ_1 and Γ_2, respectively, then the complete Green's function $G_V^R(E; \underline{r}, \underline{r}')$ is obtained by summing all Γ_n,

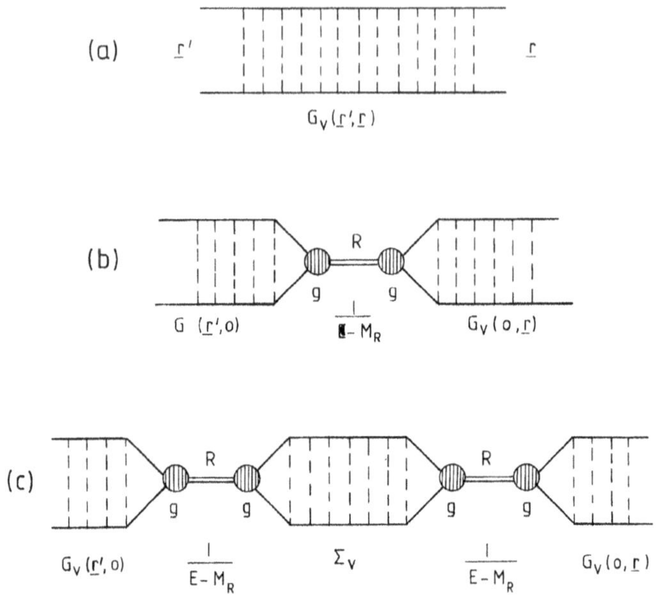

Figure 2 - Graphs illustrating typical contributions to the Green's function $G_V^R(E; \underline{r}', \underline{r})$ for standing waves, for (a) the action of an instantaneous potential V alone, (b) the action of V when an imposed state R with mass M_R and coupling g for $R \to mB$ appears in only one intermediate state, (c) the same when R is present in two intermediate states. These contributions are the first three in an infinite series of such terms. The action of the potential V is represented by the dashed lines. Time proceeds from left to right

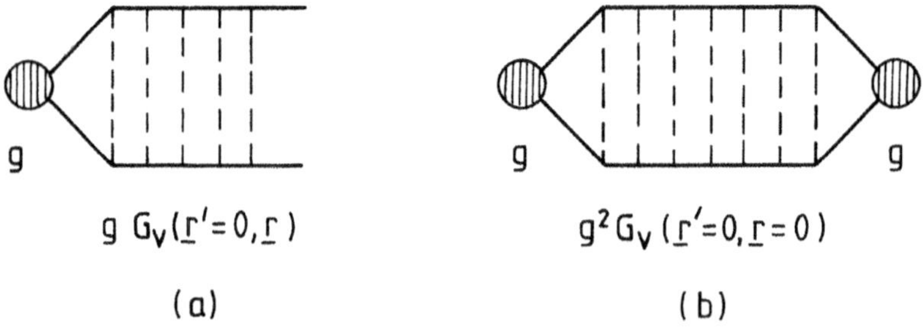

Figure 3 - Depicts two elements which appear repeatedly in the graphs of figure 2: (a) the vertex element $gv_V(E, \underline{r})$, and (b) the R self-energy element $g^2\Sigma_V(E)$

$$G_V^R(E;\ \underline{r}',\underline{r}) = \Sigma_{n=0}^{\infty}\Gamma_n,$$

$$= G_V(E;\underline{r}',\underline{r}) + v_V(E,\underline{r}')\left\{\frac{1}{E-m_R}+\frac{1}{E-m_R}g^2\Sigma_V(E)\frac{1}{E-m_R}+\ldots\right\}v_V(E,\underline{r}),$$

$$= G_V(E;\underline{r}',\underline{r}) + v_V(E,\underline{r}')\frac{1}{(E-m_R-g^2\Sigma_V(E))}v_V(E,\underline{r}) \qquad (5.9)$$

The quantity $\Sigma_V(E)$ is necessarily definite, since this is the case even for $V = 0$, when it is given by the simple graph $\Sigma_0(E)$ shown on figure 4.

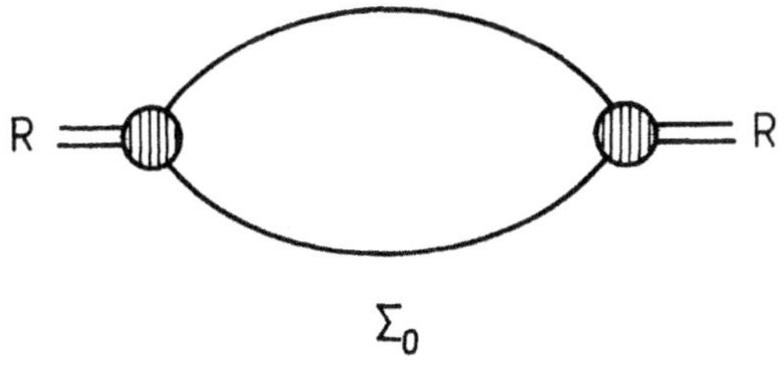

Figure 4 - The simplest R self-energy graph, effective even in the absence of V and denoted by $\Sigma_0(E)$

This corresponds to the quantity $G_0(E;\ 0,\ 0)$, obtained from (5.6) as the principal value integral

$$\Sigma_0(E) = P\int^{\Lambda}\frac{1}{E-k^2/2\mu}\ \frac{d^3\underline{k}}{(2\pi)^3} = \Lambda, \qquad (5.10)$$

where Λ is an appropriate cut-off, in the non-relativistic limit. This infinite term has to be taken care of by a mass renormalization. If this is the only divergent term, we simply replace $(m_R + g^2\Lambda)$ by m_R and proceed, replacing $\Sigma_V(E)$ by $\delta\Lambda_V(E)$, the part of it which is due to the potential V, thus

$$\delta\Sigma_V(E) = (\Sigma_V(E) - \Sigma_0(E)) \qquad (5.11)$$

For non-singular potentials, $\delta\Sigma_V$ is finite. For singular potentials, such as the Hulthen or Yukawa potentials, $\Delta\Sigma_V(E)$ will be logarithmically infinite, and it will then prove necessary to renormalize the coupling constant g as well.

The quantity of interest to us here is the K-matrix, given by

$$K_V^R(\underline{k}',\underline{k}) = \int e^{-i\underline{k}'\cdot\underline{r}'} G_V^R(E;\underline{r}',\underline{r})\, e^{i\underline{k}\cdot\underline{r}}\, d^3\underline{r}'\, d^3\underline{r} \tag{5.12}$$

The equation (5.7) then gives us the following expression,

$$K_V^R(\underline{k}',\underline{k}) = K_V(\underline{k}',\underline{k}) + v_V^R(k')\frac{1}{(E-m_R - g^2\delta\Sigma_V)} v_V^R(k). \tag{5.13}$$

To illustrate the points of interest without substantial calculation, we shall consider in explicit detail the case of a two-channel separable potential, using in place of (5.3) the form

$$V_{ij}(\underline{k},\underline{k}') = \lambda C_{ij}\gamma(k)\gamma(k'), \tag{5.14}$$

then all of the steps in the calculation outlined above can be carried out explicitly, using full relativistic kinematics if appropriate and without divergences if $\gamma(k)$ has a sufficiently soft form. First we construct the K_V matrix, for arbitrary $\underline{k}$ and $\underline{k}'$ and energy E,

$$K_V(E;\underline{k}',\underline{k}) = \gamma(k')\gamma(k)F(E)/D(E), \tag{5.15}$$

where F(E) denotes the symmetric matrix

$$F(E) = \lambda\begin{pmatrix} c_{11} & c_{12} \\ c_{12} & c_{22}\end{pmatrix} + \lambda^2(c_{11}c_{22}-c_{12}^2)\begin{pmatrix} I_2 & 0 \\ 0 & I_1\end{pmatrix}, \tag{5.16}$$

and D(E) is the determinant

$$D(E) = \begin{vmatrix} 1+\lambda I_1 c_{11} & \lambda I_1 c_{12} \\ \lambda I_2 c_{21} & 1+\lambda I_2 c_{22}\end{vmatrix} \tag{5.17}$$

The quantities I_i are given by the principal value integrals

$$I_i(E) = P\int \frac{\gamma(k)^2}{T_i(k) - E} \cdot \frac{d^3\underline{k}}{(2\pi)^3} \cdot \qquad (5.18)$$

Of course, with an s-wave separable potential, K_V has only s-wave elements. Next, we construct the vertex element, given by the following matrix :

$$v_V(E, \underline{k}) = (g_1, g_2)\left\{1 + P\int \frac{1}{T(k')-E} \frac{d^3\underline{k}}{(2\pi)^3} K_V(E; \underline{k}', \underline{k})\right\} \qquad (5.19)$$

From (5.15), it is clear that this involves the integral H_i, given by

$$H_i(E) = P\int \frac{\gamma(k')}{T_i(k') - E} \cdot \frac{d^3\underline{k}'}{(2\pi)^3}, \qquad (5.20)$$

whence we have

$$v_V(E, \underline{k}) = (g_1, g_2) + \gamma(k)\,(g_1 H_1, g_2 H_2)\,F(E)\,/\,D(E) \qquad (5.21)$$

Finally, we have, following (5.8) and (5.11),

$$\delta\Sigma_V^R(E) = D(E)^{-1}(g_1 H_1, g_2 H_2)\,F(E)\begin{pmatrix} H_1 g_1 \\ H_2 g_2 \end{pmatrix} \qquad (5.22)$$

In the limit $\{g_i\} \to 0$, we have only potential interactions. For definiteness, we shall consider the case

$$\gamma(k) = (m_V^2 + k^2)^{-1} \qquad (5.23)$$

If we set the K_V-matrix pole at E = 1405 MeV, the appropriate value of λ can be deduced from eq. (5.15), the condition $D(E) = 0$ giving an eigenvalue equation for λ,

$$(1 + \lambda I_1 c_{11})\;(1 + \lambda I_2 c_{22}) - \lambda^2 I_1 I_2 c_{12}^2 = 0. \qquad (5.24)$$

This equation has two and only two roots (if $c_{12}^2 - c_{11}c_{22} \neq 0$, as is the case here). With the parameters of interest, these roots are both real; with m_V = 800 MeV, they are λ_1 = 0.060 GeV^2 and λ_2 = 0.133 GeV^2. Both of these values give the correct resonance behaviour at E = 1405 MeV, i.e. the phase shift $\delta(\pi\Sigma)$ rises through 90° at this energy. This behaviour has a simple interpretation. For sufficiently small λ, there is no K-matrix pole. As λ increases to a critical value λ^* = 0.055 GeV^2, the first pole appears at the $\bar{K}N$ thereshold, and this moves down to 1405 MeV as λ increases to λ_1. As λ increases further, this pole moves down in energy, ultimately appearing as a bound state below the $\pi\Sigma$ threshold, while a second pole appears at the $\bar{K}N$ threshold at potential strength $\lambda^{**} = 0.12\ GeV^2$ and moves down to 1405 MeV as λ increases to λ_2. As λ increases further, there are ultimately two $\pi\Sigma$ bound states, and never more than this. This behaviour is characteristic of a simple functional form for $\gamma(k)$. The number of resonance poles or bound states is then less than or equal to the number of channels. For a single channel, at most one bound state is possible. If there are n independent channels, there are then at most n bound states possible. When the n channels are coupled weakly, the total number of bound states and resonances remains the same, although they become somewhat displaced and some former bound states may now become resonances. When the couplings between channels are increased, the bound state and resonance energies are changed strongly, and there may be an increase or decrease in the number of bound states and resonances, but never beyond a total of n.

In the limit $\lambda \to 0$, $K_V \equiv 0$ and the K_V^R-matrix reduces to

$$\begin{pmatrix} g_1 \\ g_2 \end{pmatrix} \frac{1}{E - m_R} (g_1, g_2) \tag{5.25}$$

after carrying out the mass renormalization due to the graph Σ_0 of figure 4. This has two eigenstates, one with zero eigenvalue, given by the form ν_0 of (5.26a), and the other orthogonal to it, given by (5.26b), the latter having

$$\text{(a)}\quad \nu_0 = \begin{pmatrix} g_2 \\ -g_1 \end{pmatrix} \qquad \text{(b)}\quad \nu = \begin{pmatrix} g_1 \\ g_2 \end{pmatrix} \tag{5.26}$$

eigenvalue $(g_1^2 + g_2^2)/(E-m_R)$. The T-matrix corresponding to (5.25) then has the Breit-Wigner form usual for s-waves,

$$T = \begin{pmatrix} g_1 \\ g_2 \end{pmatrix} \frac{1}{(E-m_R + i\,(g_1^2 k_1 + g_2^2 k_2))} (g_1, g_2) \tag{5.27}$$

When both the particle R and the potential V are effective, the matrix K_V^R is given by (5.13). For illustration, we consider a case where the K_V-matrix pole is placed at $m_R = 1405$ MeV when $\lambda = 0$. The parameters g_1 and g_2 were chosen to be such that $g_1/g_2 = \sqrt{(3/2)}$, as for a unitary singlet, and that the resonance had width $\Gamma = 2g_1^2 k_1 = 18$ MeV. We note that $k_2 = +i|k_2|$ at 1405 MeV, so that in the denominator of (5.27), the k_2 term does not contribute to the width, but only to the location of the resonance, placing it at mass $(m_R + g_2^2|k_2|) = 1415$ MeV. Next, we turn on the potential V. As λ increases from zero, the state R moves to lower energy, as expected when an additional attractive potential comes into play. The situation is depicted in figure 5.

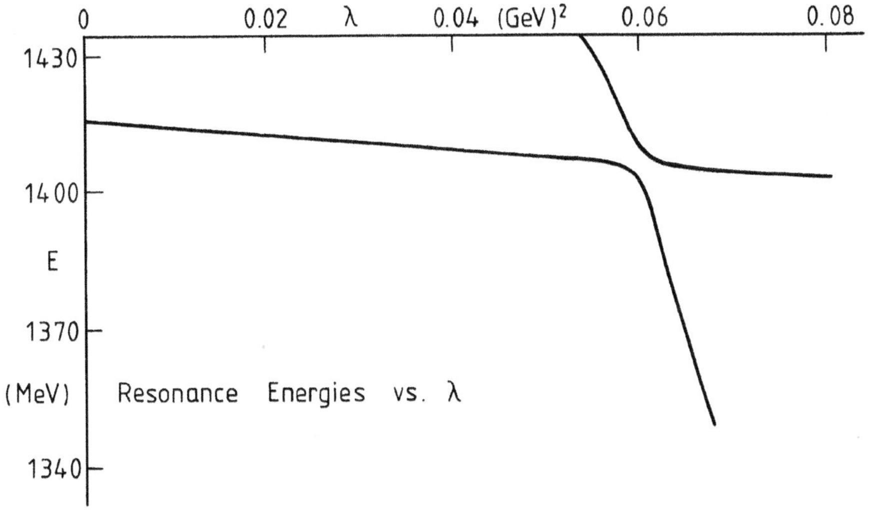

Figure 5 - The locations of the K-matrix poles for the R-related state and the 'unstable $\bar{K}N$ bound state' for the separable potential (5.14) with form factor (5.23) are plotted vs. the potential strength λ. This figure illustrates the Wigner 'no crossing' theorem in action for these two eigenvalue trajectories, despite their totally different origins.

When λ reaches a critical value of 0.05 GeV^2, for the above parameters, a second πΣ resonance appears at the $\bar{K}N$ threshold and moves rapidly to lower energies as λ increases further. As we see in figure 5, the two resonance trajectories do not cross. For the present choice of parameters, both trajectories take on a very rapid λ-dependence over the small range $0.060 \leq \lambda \geq 0.062$, after which the first trajectory falls rapidly with increasing λ, while the second trajectory then falls only gradually. The physical quantities calculated with this model such as forward amplitudes, cross sections, etc., do not reflect the dramatic λ dependence of the eigenenergies shown in figure 5. This is only an artifact of the formalism (38, 39) for there is no long-range parameter in the interactions to match the small mass range over which these dramatic variations occur. In fact, this behaviour is an example of Wigner's "no-crossing" theorem for eigenvalues as a function of any parameter in the Hamiltonian or the boundary conditions. For λ in the range (say) $0.055 \leq \lambda \geq 0.060$, the limit $\{g_i\} \to 0$ for all the $R \to m_i B_i$ couplings leads to the disappearance of the lower resonance, while the energy of the upper resonance is very little affected: for $\lambda > 0.065$ (say), this limit leads to the disappearance of the upper resonance, while the energy of the lower resonance is very little affected. In the region $\lambda > 0.065$, the slope of E_2 vs. λ is approximately equal to the slope of E_1 vs. λ in the region $\lambda < 0.055$, and in both regions the corresponding wavefunctions are dominated by the state R. Only in the "no-crossing" region are the meson-baryon states strongly mixed with the state R. However, their effect on each other is not negligible outside this region; for example, the R-related resonance is shifted downward in mass by 15 MeV as λ increases from 0 to 0.10. The behaviour of these two states, and their interaction, is completely normal, showing no reflection of their completely unrelated mechanisms of origin.

Although the use of the K-matrix is most convenient for the parametrization of scattering and reaction data over limited energy ranges about threshold energies, or across narrow resonances, or both, it must always be remembered that the physical phenomena observed depend most directly on the T-matrix and the location and strength of its poles and branch cuts. The relationship between K-matrix poles on the real axis and T-matrix poles in the complex energy plane is not simple, unless we have to do with narrow resonances. Most of our intuition about this relationship comes from our experiences with these limiting cases, and it can often lead us astray in physical situations where the resonances are not narrow. We shall now illustrate these cautionary remarks by reference to the above calculations with separable potentials:

(i) non-existence of K-matrix poles when a state R is imposed, within the physical range of interest. From eq. (5.11), we see that the K-matrix poles associated with R are given by the eigenvalue equation

$$E - m_R - \delta\Sigma_V^R = 0 \qquad (5.28)$$

The funtion $(E - m_R - \delta\Sigma_V^R(E))$ has been plotted vs. E on figure 6, for the case $\lambda = 0.038\ \mathrm{GeV}^2$, and we see that it has no zeros. It follows that <u>the two-channel K_V^R-matrix</u> for this value of λ has no pole, despite the imposed resonance. The cross section $\sigma(\pi\Sigma)$ for this case is plotted as curve (a) on figure 7, and this shows a clear resonance peak at about 1405 MeV, as we expect from the discussion above and from figure 5. The explanation is that the element $(K_V^R)_{\bar{K}\bar{K}}$ is large and leads to a pole in the one-channel reduced K-matrix $(K_V^R)^r$, given by the form (1.5), at an energy controlled by primarily the mass m_R of the imposed state R. If R is decoupled from the meson-baryon systems by taking the limit $\{g_i\} \to 0$ the $\pi\Sigma$ resonance shown in figure 5 becomes narrower and narrower until finally it vanishes; as we know from the discussion above, the potential V for $\lambda = 0.038\ \mathrm{GeV}^2$ is not sufficiently strong to support a resonance.

(ii) The existence of K_V-matrix poles, generated by attractive s-wave potentials for $n \geq 2$ channels, not leading to any resonance. To provide an explicit example, we take $\lambda = 0.045\ \mathrm{GeV}^2$ and set $\{g_i\} = 0$, so that the state R no longer contributes to the scattering and reaction

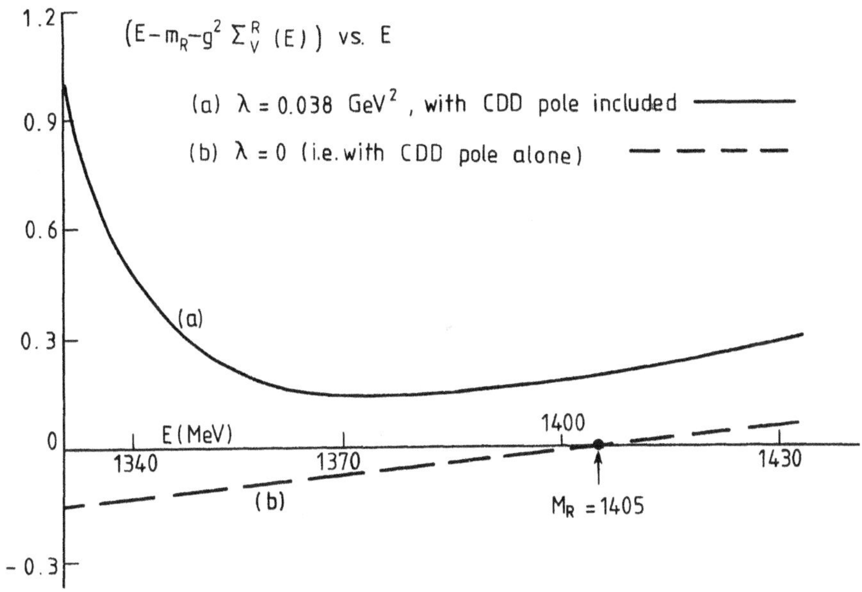

Figure 6 - The denominator function (5.28) from eq. (5.13) which includes the action of the R particle is plotted vs. energy for two strengths of the potential V: (a) for $\lambda = 0.038\ \mathrm{GeV}^2$, where this function has no zero, and (b) for $\lambda = 0$, where the zero is at $m_R = 1405$ MeV

processes. We are now concerned only with K_V, given by eq. (5.15). We know that there is no resonance state, since this value of λ lies below the critical value λ^* at which the first resonance appears at the $\bar{K}N$ threshold. Nevertheless, the K_V-matrix elements have a pole at 1405 MeV, since their denominator D(E), given explicitly by (5.17), has a zero at 1405 MeV for $\lambda = 0.045\ \mathrm{GeV}^2$. The $\pi\Sigma$ cross section for this value of λ is given by curve (b) on figure 7 and varies smoothly with energy, with no resonance behaviour.

These model calculations illustrate the fact that the relationship between K-matrix poles and resonances is simple only if these resonances are sufficiently narrow. Let us consider again our two-channel K-matrix. It is necessarily a real, symmetric matrix when the Hamiltonian is hermitian and invariant under time-reversal (40). It will therefore have two eigenvalues, κ_α and κ_β, with two corresponding eigenstates ϕ_α and ϕ_β, which are orthogonal. Let us then denote ϕ_α and ϕ_β by the following matrices in channel space:

$$\text{(a)}\quad \phi_\alpha = \begin{pmatrix} \alpha_1 \\ \alpha_2 \end{pmatrix} \qquad \text{(b)}\quad \phi_\beta = \begin{pmatrix} \alpha_2 \\ -\alpha_1 \end{pmatrix} \tag{5.29}$$

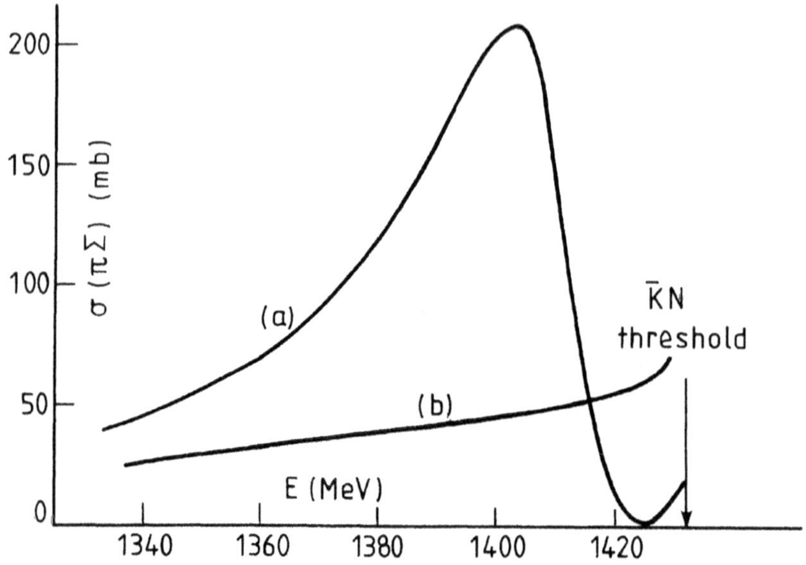

Figure 7 - The $\pi\Sigma$ cross section is given vs. c.m. energy E for two instructive cases: (a) for R imposed with M_R = 1405 MeV, and λ = 0.038 GeV^2. The denominator function (5.28) has no zero, yet the calculated cross section shows a strong resonance near 1405 MeV. (b) for λ = 0.045 GeV^2, and no R particle. The two-channel K_V matrix has a pole at about 1405 MeV, but the corresponding cross section clearly has no resonance

with $\alpha_1^2 + \alpha_2^2 = 1$. Shaw and Ross have shown with a great degree of generality (28, 40) that the matrix K^{-1} is linear in the energy, as given in eq. (4.1), in any sufficiently limited energy range. As the energy E varies, it is possible for the eigenvalue κ_α^{-1} (say) to pass through zero, say at energy E_K, so that in this neighbourhood of E it can be written

$$\kappa_\alpha^{-1} = 2(E_K - E)/\tau. \tag{5.30}$$

Since K may be expressed in terms of its eigenvalues and eigenvectors, we can write it in channel space as

$$K = \kappa_\alpha \phi_\alpha \bar{\phi}_\alpha + \kappa_\beta \phi_\beta \bar{\phi}_\beta = \begin{pmatrix} \alpha_1^2 & \alpha_1\alpha_2 \\ \alpha_1\alpha_2 & \alpha_2^2 \end{pmatrix} \frac{\tau}{E_K - E} + \kappa_\beta \begin{pmatrix} \alpha_2^2 & \alpha_1\alpha_2 \\ -\alpha_1\alpha_2 & \alpha_1^2 \end{pmatrix} \tag{5.31}$$

where α_1, α_2, κ_β and τ are all non zero at E_K, and can be expanded about E_K. This is of the form (4.2) and its derivation has been aimed to demon strate the motivation for the use of that form in phenomenological analyses. When the T-matrix is calculated from the K-matrix (5.31), using the relationship

$$T = K(1 - ikK)^{-1}, \tag{5.32}$$

the T-matrix takes the form

$$T = \begin{pmatrix} \alpha_1^2 & \alpha_1\alpha_2 \\ \alpha_1\alpha_2 & \alpha_2^2 \end{pmatrix} \frac{\tau}{(E_K - E) - i\tau(k_1\alpha_1^2 + k_2\alpha_2^2)} + \text{(terms finite near } E=E_K) \tag{5.33}$$

which shows that T necessarily has a pole in the complex energy plane, at the energy value

$$E_{res.} = E_K - i\tau(k_1\alpha_1^2 + k_2\alpha_2^2), \tag{5.34}$$

This pole does not occur on the physical sheet of the energy plane, but on an adjacent unphysical sheet. This remark holds rigorously only for sufficiently small τ.

The reduced K-matrix K^r for channel 1 may be obtained from (5.31), using the relation (1.5), as follows

$$K^r = \frac{\tau\alpha_1^2}{E_K-E} + \kappa_\beta\alpha_2^2 - |k_2|\alpha_1^2\alpha_2^2\left(\kappa_\beta - \frac{\tau}{E_K-E}\right)^2\left\{1+|k_2|\left(\kappa_\beta\alpha_1^2 + \frac{\tau\alpha_2^2}{(E_K-E)}\right)\right\}^{-1} \tag{5.35}$$

and this is finite at $E = E_K$. However K^r has a pole at energy E^r given by the vanishing of the last denominator, thus for

$$E^r = E_K + \tau\alpha_2^2|k_2| / (1+|k_2|\kappa_\beta\alpha_1^2). \tag{5.36}$$

This is an implicit equation for E^r, since τ, k_2, α_1, α_2 and κ_β are all finite functions of E^r, and it does not necessarily have a solution. However, if τ is sufficiently small, eq. (5.36) can be solved by successive approximations. In other words, a resonance energy (for which $\delta(\pi\Sigma) = 90°$) necessarily exists, if τ is sufficiently small.

It is difficult to give any general criterion as to how small τ should be in order that the expressions (5.34) and (5.36) should lead to reliable predictions about the existence or non-existence of any of these poles E_K, E_{res} and E^r given the existence of any one of them. The model calculations presented above do demonstrate that the relationships valid for sufficiently small τ do not necessarily hold in all situations where the parameters are chosen to have physically reasonable values. Although the case of small τ is pedagogically useful, since it highlights the difference between possible physical situations, it must not be assumed that the actual physical situation necessarily conforms to any of the extreme situations illustrated by that limit.

6. Relativistic Calculations

We return now to discuss the effect of using the relativistic energy expression (5.2) for $T_i(k)$ rather than the approximation (5.5). First we consider the two-channel separable potential (5.14). The formal manipulations proceed as before, the integrals I_i and H_i being re-evaluated with the relativistic $T_i(k)$. Again, the resonance (defined by $\delta(\pi\Sigma) = 90°$) occurs at 1405 MeV for two values of the potential strength λ, for $\lambda = 0.038\ \mathrm{GeV}^2$ and $\lambda = 0.082\ \mathrm{GeV}^2$. These are quite different from the strengths deduced with (5.5), and the other properties of the scattering are also different.

Resonance	T(k)	λ	ReA_t (fm)	ImA_t (fm)	Γ(MeV)
(1)	approx. (5.5)	0.060	-1.77	1.15	50
	rel$^{\underline{c}}$. (5.2)	0.038	-0.28	1.44	100
(2)	approx. (5.5)	0.133	-0.95	0.16	94
	rel$^{\underline{c}}$. (5.2)	0.082	-1.31	0.31	36

Physically, only the first resonance is of interest to us, as a model for $\Lambda(1405)$, but we give the calculated quantities also for the second resonance, to accentuate the difference. The reason for this sensitivity is that the integrals I_i and H_i are principal value integrals, involving a cancellation between large quantities of opposite sign. For large k, above the singular point in these integrals, the approximation (5.5) gives $T(k) = k^2/2\mu$ whereas the correct expression gives $T(k) \to 2k + (m^2 + M^2)/k + ..$ The convergence of these integrals at large momenta, beyond the singular point, is much slower in the relativistic case that with the approximation (5.5), and so the values obtained for them differ widely. For the physical case, it is essential to use the relativistic form for T(k). However, the correspondence with the experimental data then appears rather less satisfactory, as is shown by the table above.

Other, more technical difficulties also arise, if we turn now to the DRW problem, the use of multichannel potentials such as those generated by vector-meson exchange. With relativistic T(k), the singularities become much more complicated. Since the K-matrix involves principal value integrations, analytic continuations become uncertain or impossible and the complications quickly become so great that it is best to abandon the K-matrix and to work directly with the T-matrix. Also, the dynamical equations (5.1) are integral equations and cannot be treated with the well-known Schrodinger methods. In this four-channel problem, where the thresholds range so widely in energy, there are further problems because the left-hand cuts due to the finite range of the potentials in the upper channels strongly overlap the right-hand unitarity cuts for the lower channels. We shall not go into details here about the means followed to cope with these difficulties of analyticity, but they have been surmounted, as have also the problems of numerical mathematics posed by the singular nature of these integrals, due to the factor $\lim_{\varepsilon\to 0}(E - t(k) + i\varepsilon)^{-1}$ in them.

On further difficulty then arises. For large p, the integral equations (5.1) for the vector-meson-exchange potential (5.3) are homogeneous in p. This means that the wavefunction $\psi(p)$ necessarily has the asymptotic form

$$\psi_i(p) \rightarrow a_i p^{\alpha}, \tag{6.1}$$

where i refers to channel space and the index α is determined entirely by the large momentum characteristics of the potential. Its behaviour is reminiscent of that of the wavefunction for a Schrodinger equation with a potential of the form C/r^2 at short distances, where the indicial equation involves the strength of the potential and there does not necessarily exist any acceptable wavefunction for all values of C. This behaviour has also been seen in other relativistic equations, for example in the Tamm-Dancoff equations for pion-nucleon scattering (41) and in the Bethe-Salpeter equation for nucleon-nucleon scattering (42). The indicial equation for eq. (5.1) can be obtained by inserting (6.1) on both sides and requiring consistency. This leads to the following equations to be satisfied by α and the matrix (a),

$$\left\{I + \frac{f^2}{4\pi} \; \frac{\tan(\alpha+1)\pi/2}{2(\alpha+1)} \; C\right\}(a) = 0 \tag{6.2}$$

where C denotes the matrix of coefficients in expression (5.3). Their consistency requires that

$$\det\,(I + \lambda C) = 0, \tag{6.3}$$

where $\lambda = (f^2/8\pi)\,(\tan(\alpha+1)\pi/2)/(\alpha+1)$. in other words, $(-1/\lambda)$ must be one of the eigenvalues of the SU(3) matrix C. For Λ(1405), which has $I = Y = 0$, the relevant SU(3) representations are 1, 8_s, 8_a and 27, for which C has the eigenvalues $C_\rho = -3, -3/2, -3/2$ and $+1$, respectively. The index α_ρ which goes with SU(3) representation ρ is then given by

$$\phi(\alpha_\rho) = \frac{\tan\pi(\alpha_\rho+1)/2}{\pi(\alpha_\rho+1)/2} = -\frac{4C_\rho}{\pi}\left(\frac{f^2}{4\pi}\right)^{-1}. \tag{6.4}$$

The function $\phi(\alpha)$ is plotted on figure 8.

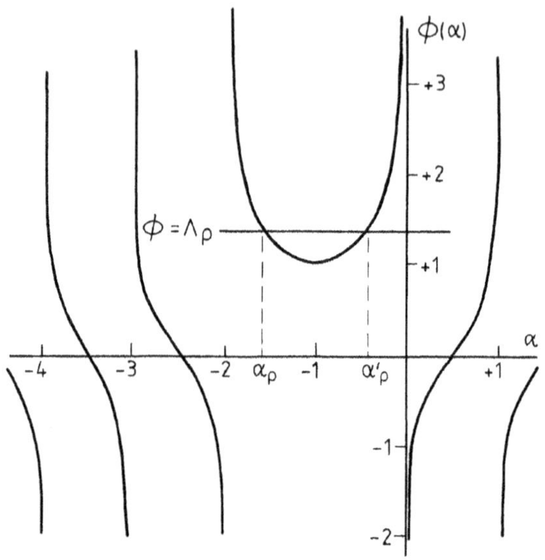

Figure 8 - A plot of the function $\phi(n)$ from which the exponent α of the asymptotic form $\psi(p) \sim p^{\alpha}$ of the wavefunction satisfying the relativistic Schrodinger equation (5.1) may be determined by the equation $\phi(\alpha) = \Lambda_\rho$, where Λ_ρ is related with the strength of the potential (5.3) for the unitary representation ρ.

For given $f^2/4\pi$ and C_ρ, the possible values of α_ρ are determined by the horizontal line $\phi(\alpha) = \Lambda_\rho$ shown there. The physically appropriate value is that reached by starting at small $f^2/4\pi$, α_ρ is given by perturbation theory and so has the value -2, as can be deduced directly from the form of (5.1). Hence, for repulsive potentials, α_ρ lies on the branch between -2 and -3, and there always exists one such value. For attractive potentials, α_ρ lies on the branch between -2 and -1, but there does not exist any such α_ρ if $\Lambda_\rho < 1$, i.e. if $f^2/4\pi$ is too large. The least favourable case is the unitary singlet representation, for which $C_\rho = -3$. There is then an acceptable wavefunction only if

$$f^2/4\pi < 4/3\pi, \tag{6.5}$$

and the $\Lambda(1405)$ state will certainly be dominantly singlet, although it may include some octet and 26 admixtures.

We note that this condition (6.5) does not depend in the energy of the state nor on the range of the potential acting. Solution of the integral equation for potential range $1/m_V$ shows that the limit (6.5) does not allow the potential to be strong enough to generate a Λ(1405) state. Of course, this limit (6.5) lies far below the values $f^2/4\pi \simeq 2$ derived for this vector-meson-baryon coupling from other phenomena. The equations (5.1) simply do not allow any solution corresponding to Λ(1405), for physically reasonable parameters. However, if we increase the potential range by decreasing m sufficiently, it becomes possible to obtain a resonant state at 1405 MeV; we did carry the calculations through for such a case, in order to test our method of solution, but the calculation no longer bears on the physical situation.

This difficulty is not of fundamental origin, as things now stand, since the equation (5.1) adopted has not been derived from any fundamental theory. It is easy to write down a modified equation which would have a solution, but this is essentially equivalent to cutting the momenta off above some momentum k_{max}, either sharply or with some form factor. In phenomenological terms this means the introduction of one further parameter, which would leave this model without significant predictive power.

7. Conclusion

The only thing which is clear from the above discussion is that we do not yet know what is the proper interpretation for the state Λ(1405). We do not know whether its interpretation as three-quark baryonic state is or is not consistent with its appearance as an "unstable $\bar{K}N$ bound state", the conclusion which the experimental data point to, rather clearly. In the above remarks, we have tried to indicate and illustrate in a simplified way some of the difficulties which we have found along our way, rather than to insist on models which are realistic in all respects. Few of these difficulties are fundamental in origin, but they do complicate the treatment a great deal. The large question which is still left open in the above is that of the relationship between "elementarity", as defined by the $Z \neq 0$ condition, and the occurrence of poles in the K-matrix, the reduced K-matrix and the T-matrix.

How shall be able to go furher in the future? Can we hope for further clarification on the experimental side? There is some hope of measuring πΣ scattering from the one-pion-exchange pole in $\Sigma N \rightarrow \pi\Sigma N$ and related processes, where a Σ-particle beam is scattered by a virtual pion in the pion cloud around the target proton. Also, we have not yet used fully the possibilities for the study of these pion-hyperon interactions in final states in other high energy collision processes; there is probably more information hidden there than we have yet appreciated. Above all, it is clear that we need improved data, especially on the $\bar{K}N$ interaction, in all of its aspects. For example, we can look forward hopefully to more accurate determinations of the level shift and width of the 1s level in the K^-p atom, for these would be immensely valuable (43).

* Present address : Burrough's Machines Ltd., Detroit Way, Glenrothes, Fife, Scotland.

References

1. Partcle Data Group : 1980, Rev. Mod. Phys. 52, S1.
2. Humphrey, W., and Ross, R. : 1962, Phys. Rev. 127, 1305.
3. Kim, J.K.: 1967, Phys. Rev. Letters 19, 1074; see also Columbia University: 1966, Rept. Nevis 149.
4. Nowak, R., Armstrong, J., Davis, D.H., Miller, D.J., Tovee, D.N., Bertrand, D., Goossens, M., Vanhomwegen, G., Wilcquet, G., Abdullah, M., Bedford, N.H., Ciborowski, J., Major, J.V., Zakrzewski, J.A., Kielczewaska, D., Tymieniecka, T., and Rondio, E. : 1978, Nucl.Phys. B139, 61.
5. Miller, D.J.: 1980, Nukleonika, in press.
6. Alston, M.H., Alvarez, L.W., Eberhard, P., Good, M.L., Graziano, W., Ticho, H.K., and Wojcicki, S.G.: 1961, Phys. Rev. Letters 6, 698.
7. Alexander, G., Kalbfleisch, G.R., Miller, D.H., and Smith, G.A. : 1962, Phys. Rev. Letters 8, 447.
8. Sakurai, J.J.: 1960, Ann. Phys. (N.Y.) 11, 1.
9. Arnold, R.C., and Sakurai, J.J.: 1962, Phys. Rev. 128, 2808.
10. Dalitz, R.H., Wong, T.C., and Rajasekaran, G. : 1967, Phys. Rev. 153, 1617.
11. Wyld, H.W.: 1967, Phys. Rev. 155, 1649.
12. Logan, R.K., and Wyld, H.W.: 1967, Phys. Rev. 158, 1467.
13. Alston-Garnjost, M., Kenney , R.W., Pollard, D.P., Ross, R.R., Tripp, R.D., Nicholson, H., and Ferro-Luzzi, M.: 1978, Phys. Rev. D18, 182.
14. Dalitz, R.H., and Tuan, S.F.: 1960, Ann. Phys. (N.Y.) 10, 307.
15. See, for example, Sachs, R.G.: 1953, Nuclear Theory, Addison-Wesley Publ. Co., Cambridge (Mass.).
16. Isgur, N., and Karl, G.: 1978, Phys. Rev. 18D, 4187.
17. Dalitz, R.H.: 1965, Proc. Roy. Soc. A288, 183.
18. Davies, J.D., Pyle, G.J., Squier, G.T.A., Batty, C.J., Biagi, S.F., Hoath, S.D., Sharman, P., and Clough, A.S.: 1979, Phys. Lett. 83B, 55.
19. Baillon, P., Bricman, C., Ferro-Luzzi, M., Jenni, P., Perreau, J., Tripp, R., Ypsilantis, T., Declais, Y., and Seguinot, J.: 1974, Phys. Lett. 50B, 377; 1976, Nucl. Phys. B105, 365.
20. Baillon, P., Declais, Y., Ferro-Luzzi, M., Jenni, P., Perreau, J., Seguinot, J., and Ypsilantis, T.: 1976, Nucl. Phys. B107, 189.

21. Jenni, P., Baillon, P., Bricman, C., Ferro-Luzzi, M., Perreau, J., Tripp, R., Ypsilantis, T., Declais, Y., and Seguinot, J.: 1976, Nucl. Phys. B105, 1.

22. Martin, A.D.: 1976, Phys. Lett. 65B, 346; 1977, Nukleonika 22, 857.

23. Weinberg, S.: 1963, Phys. Rev. 137, B672.

24. Rajasekaran, G.: 1972, Phys. Rev. D5, 610.

25. Castillejo, L., Dalitz, R.H., and Dyson, F.J.: 1956, Phys. Rev. 101, 453.

26. Dyson, F.J.: 1957, Phys. Rev. 106, 157.

27. Vaughn, M.T., Aaron, R., and Amado, R.D.: 1961, Phys. Rev. 124, 1258.

28. Shaw, G., and Ross, M.: 1962, Phys. Rev. 126, 806.

29. Martin, B.R., and Sakitt, M.: 1969, Phys. Rev. 183, 1345.

30. Martin, A.D., and Ross, G.C.: 1970, Nucl. Phys. B16, 479.

31. Martin, A.D., Martin, B.R., and Ross, G.C.: 1971, Phys. Lett. 35B, 62.

32. Barbaro-Galtieri, A.: 1972, Lawrence Berkeley Lab. Rept. 555.

33. Chao, Y., Kraemer, R.W., Thomas, D.W., and Martin, B.R. : 1973, Nucl. Phys. B56, 46.

34. Thomas, D.W., Engler, A., Fisk, H., and Kraemer, W.: 1973, Nucl. Phys. B56, 15.

35. Goossens, M.: 1978, Ph.D. dissertation, University College, London.

36. McGinley, J.G.: 1979, "$\bar{K}$-nucleon scattering and models of the Λ(1405) resonance", D.Phil. dissertation, Oxford University.

37. Kumar, K.S., and Nogami, Y.: 1979, McMaster University preprint.

38. Dalitz, R.H., and Moorhouse, R.G.: 1970, Proc.Roy.Soc. A318, 245.

39. McVoy, K.: 1967, in 'Fundamental in nuclear theory' (I.A.E.A., Vienna), 419.

40. Dalitz, R.H.: 1962, 'Strange Particles and Strong Interactions' (O.U.P., Bombay).

41. Dalitz, R.H., Sundaresen, M.G., and Bethe, H.A.: 1956, Proc. Camb. Phil. Soc. 52, part. 2, 251.

42. Mandelstam, S.: 1956, Proc. Roy. Soc. A237, 496.

43. Note added July 1980. At CERN, Izycki et al, (44) have now reported a measurement of the level shift and width for the 1s level of the K^-p atom, with the result

$$\Delta E = + 270 \pm 80 \text{ eV}, \qquad \Gamma = 560 \pm 260 \text{ eV}.$$

This corresponds to $a(K^-p) = + 0.66 + i0.70$ fm. The real part has the same magnitude as expectation, but the wrong sign.

44. Izycki, M., Backenstoss, G., Tauscher, L., Blüm, P., Guigas, R., Hassler, N., Koch, H., Poth, H., Fransson, K., Nilsson, A., Pavlopoulos, P., and Zioutas, K. : 1980, "Results of the search for K-series X-rays from kaonic hydrogen", preprint CERN-EP/80-45.

DISCUSSION

- R. E. Cutkosky :

I would like to point out some similarities to the method we have used to parametrize πN resonances. We also used separable potentials to describe the background potential effects. Our integrals are more convergent because we put in the size of the quark model states. We found, that to describe the energy dependence over a broad range, in all partial waves repulsive effects seemed to dominate at short distances, even if long-range attractions were evident at low energies. This makes it easier to interpret the physical resonances as really coming from the quark-model states put in.

We also found the K-matrix parameters coming from the fits to be very hard to interpret, and we decided not to publish them. We had considered that the "bare"energies and couplings of the quark-model resonances might be used in comparison with models. However, the level shifts, and the mixings included between different quark-model states by the couplings to various meson-baryon channels, turned out to be so large and so dependent on how many resonances and how many channels were included explicitly that we have not known how to interpret the results. Only the T-matrix poles, and to some extent, the conventional mass and width parameters,were relatively stable.

- R. H. Dalitz :

Of course, our situation has been much simpler, in some ways. We have been concerned with a smaller range of energies, a range in which there is only one known resonance. Further, there is only one channel open at this resonance energy, although this is indeed a channel not easily accessible by experiment.

- G. C. Oades :

Can you say anything about the relative sizes of the Λ(1405) couplings to the $\bar{K}N$ and πΣ channels in McGinley's calculations?

- R. H. Dalitz :

This question can only be answered roughly, since this ratio is really given by the residues of the T-matrix resonance pole, whereas McGinley's calculations have been confined to the real axis. If you will be satisfied by the crude estimation procedure given in ref.10, then the ratio is about $g(\Lambda^*\bar{K}N)/g(\Lambda^*\pi\Sigma) = -1.24$. However, McGinley's relativistic calculation is not physically realistic; the existence of a solution requires a potential range more than twice that appropriate to vector meson exchange. The old calculation of ref. 10, which uses a physically realistic range, gave -2.6 for this ratio, but McGinley found that calculation to be unjustified, because the πΣ channel is highly relativistic.

THE ANTISYMMETRIC SPIN-ORBIT INTERACTION BETWEEN QUARKS

F.E. Close
Theoretical Physics Division, Rutherford and Appleton Laboratories - Chilton, Didcot

R.H. Dalitz
Theoretical Physics Department - Oxford University

1. Introduction

Two possible origins are usually considered for the spin-orbit component of the quark-quark interaction in baryons :

(i) the one-gluon-exchange potential V_g, given by QCD (1). In a baryon, any two quarks are necessarily in a unitary $\bar{3}_C$, where C denotes colour, state and their leading central interaction is given by

$$v_g(ij) = -(\tfrac{2}{3}\alpha_s)\frac{1}{r_{ij}} \tag{1.1}$$

With vector coupling for the quark-gluon interaction, and assuming $SU(3)_f$ symmetry, so that the quarks all have the same mass m, the spin-orbit component of V_g, obtained through its non-relativistic reduction, takes the well-known form (2)

$$V_g^{SO}(ij) = \frac{1}{m^2 r_{ij}}\frac{dv_g}{dr_{ij}}\left\{3\underline{L}_{ij}\cdot\underline{S}_{ij}^{+} - \underline{K}_{ij}\cdot\underline{S}_{ij}^{-}\right\} \tag{1.2}$$

where $\underline{L}_{ij} = (\underline{r}_i - \underline{r}_j) \times (\underline{p}_i - \underline{p}_j)$, $\underline{K}_{ij}$ is given by

$$\underline{K}_{ij} = (\underline{r}_i - \underline{r}_j) \times (\underline{p}_i + \underline{p}_j), \tag{1.3}$$

and $\underline{S}_{ij}^{\pm} = (\underline{S}_i \pm \underline{S}_j)$, where $\underline{S}_i = \underline{\sigma}_i/2$ is the intrinsic spin operator for quark i.

(ii) The confinement potential V_C. The form of V_C is unknown at present and it may well turn out to have a complicated three-particle

E. Ferrari and G. Violini (eds.), Low and Intermediate Energy Kaon-Nucleon Physics, 411–418.

structure for a three quark system. For the present it is generally assumed to be a sum of the three qq interactions, thus

$$V_c = \sum_{i>j} V_c(ij). \tag{1.4}$$

We denote the radial part of $V_c(ij)$, before non-relativistic reduction, by $v_c(r_{ij})$, where $v_c(r)$ increases infinitely as $r \to \infty$, linearly, quadratically or even logarithmically, for example. It is generally assumed that $V_c(ij)$ has scalar coupling; in this case, the non-relativistic reduction leads to the two-body spin-orbit coupling (3),

$$V_c^{SO}(ij) = -\frac{1}{4m^2 r_{ij}} \frac{dv_c}{dr_{ij}} \left\{ \underline{L}_{ij} \cdot \underline{S}^+_{ij} + \underline{K}_{ij} \cdot \underline{S}^-_{ij} \right\} \tag{1.5}$$

If there are three-body terms in V_c, then there will be corresponding three body terms in its spin-orbit component V_c^{SO}, after the non-relativistic reduction, but such terms have not yet been invoked in the literature.

The terms of (1.2) and (1.5) involving $\underline{K}_{ij}$ have a form unusual in nuclear physics. They do not occur in the list of permissible forms made for the NN interaction by Eisenbud and Wigner (4) and indeed these terms do violate condition (4) in their list of symmetry constraints which the NN interaction operator must satisfy. This condition expresses invariance under Galilean transformations; under such a transformation for velocity $\underline{v}$, we have $\underline{p}_i \to (\underline{p}_i + m_i\underline{v})$, so that

$$\underline{K}_{ij} \to \underline{K}_{ij} + (m_i + m_j)\,(\underline{r}_i - \underline{r}_j) x \underline{v} \tag{1.6}$$

When particles i and j are within a system of n particles, this operator $\underline{K}_{ij}$ can be made Galilean invariant by the <u>ad hoc</u> replacement of (1.3) by the expression

$$\underline{K}_{ij} = (\underline{r}_i - \underline{r}_j) \times (\underline{p}_i + \underline{p}_j - \frac{(m_i + m_j)}{M}\underline{P}) \tag{1.7}$$

where $\underline{P}$ and M denote the total momentum and total mass of the system, which involves the other (n-2) particles in a symmetrical way. It is natural to calculate the total energy of the n particles in their c.m. frame, and in this frame the expressions (1.3) and (1.7) coincide.

For n = 3, the case of the baryon, expression (1.7) involves the coordinates of all three quarks. For this reason, the spin-orbit interactions involving K_{ij} have sometimes been referred to as "three-body spin-orbit terms". However, with the assumptions made above, it is clear that the physical processes of particle exchange which give rise to V_g and v_c involve only the two particles i and j, so that this name is inappropriate and misleading.

Let us confine attention to the SU(3)-symmetric case of equal mass m for all three quarks, in which case half of the terms of (1.2) and (1.6) drop out. Adopting the usual co-ordinates

$$\underline{\rho} = (\underline{r}_1 - \underline{r}_2)/\sqrt{2}, \qquad \underline{\lambda} = (\underline{r}_1 + \underline{r}_2 - 2\underline{r}_3)/\sqrt{6}, \tag{1.8}$$

with the notation $\underline{p}_\rho$, $\underline{p}_\lambda$ for the corresponding internal momenta, these spin-orbit interactions (1.2) and (1.5) take the forms

$$V_g^{SO} = \frac{3}{4m^2\rho}\frac{dV_g}{d\rho}\left\{\underline{S}_{12}^{+}\cdot\underline{\rho}\times\underline{p}_\rho - \frac{1}{3\sqrt{3}}\underline{S}_{12}^{-}\cdot\underline{\rho}\times\underline{p}_\lambda\right\}, \tag{1.9.1}$$

$$V_c^{SO} = -\frac{1}{4m^2\rho}\frac{dv_c}{d\rho}\left\{\underline{S}_{12}^{+}\cdot\underline{\rho}\times\underline{p} + \frac{1}{\sqrt{3}}\underline{S}_{12}^{-}\cdot\underline{\rho}\times\underline{p}_\lambda\right\}. \tag{1.9.2}$$

The case of most interest to us here is the doublet of Δ states in the N = 1 excited three-quark configuration $^2P[\overline{10}]$. The wave functions of these states are indicated by the form (5)

$$\left\{\rho\,|(10)\,1\,(0\tfrac{1}{2});\ J> + \lambda\,|\,(01)\,1\,(1\tfrac{1}{2})\,\tfrac{1}{2};\ J>\right\} \tag{1.10}$$

where the notation used is $|(L_{12}\,L_3)\,L\,(S_{12}\,S_3)\,S;J>$. With this wavefunction, the expectation value for the $\underline{S}_{12}^{+}\cdot\underline{L}_{12}$ term of the spin-orbit interactions (1.9) is zero; the term ρ^2 vanishes because it has $S_{12} = 0$, the term λ^2 because it has $L_{12} = 0$, and the $\rho\lambda$ cross term vanishes because the operator $\underline{S}_{12}^{+}$ cannot change S_{12}. This has been known for the orbital configuration (1s) (1p) for a long time (6). The inclusion of higher configurations (e.g. (1d) (1p)) would lead to a non-zero expectation value, in general, arising from λ^2 terms for which $L_{12} \neq 0$. In contrast, $\underline{S}_{12}^{-}.\underline{K}_{12}$ has a non-zero expectation value for the wavefunction (1.10), although the ρ^2 and λ^2 terms both vanish again, this time because the operator $\underline{S}_{12}^{-}$ is non-zero only for the transitions $S_{12} = 0 \rightleftarrows S_{12} = 1$. The $\rho\lambda$ cross term satisfies this last condition, but its non-vanishing requires further that there be operators giving non-zero matrix-elements for both $L_{12} = 0 \rightleftarrows L_{12} = 1$ and $L_3 = 0 \rightleftarrows L_3 = 1$ and this is the case for the space factor $\underline{\rho}\times\underline{p}_\lambda$. Gromes and Stamatescu (7) first evaluated the expectation value for $\underline{S}_{12}^{-}\cdot\underline{K}_{12}$ for the non-strange states of the N = 1, L = 1$^-$ baryon supermultiplet and we shall discuss their result below. We should note here that the qq tensor force has zero expectation value for any S = 1/2 configuration, in particular for the state (1.10). The tensor force can contribute to this doublet splitting only in second-order, involving transitions to and from states in higher bands with odd $N \geq 3$.

The experimental situation for this Δ doublet is as follows. The Particle Data Group (8) have reported that the ΔS31 resonance mass lies within the range 1600-1650 MeV, and that the ΔD33 resonance mass lies within the range 1630-1740 MeV. This may make it appear that the data are consistent with degeneracy for these two states, or consistent with any mass difference δM_Δ = (ΔD33-ΔS31) up to as much as 140 MeV. However, Kelly has recently emphasized (9) that although different pion-nucleon scattering phase-shift analyses do differ quite widely in the absolute mass values they assign to these resonances, they do agree rather well in the value obtained for δM_Δ, the values found all being in the range +80 to +100 MeV. This appears a rather strong indication that the state ΔD33 is not degenerate with ΔS31 but lies ≃90 MeV above it.

2. The relativistic c.m. to Lab. transformation

Osborn (10), Brodsky and Primack (11) and others(11, 12) have emphasized the great care necessary concerning the relationship between c.m. variables and laboratory variables, when the particles have intrinsic spin. Conventionally, the calculation and specification of two-particle interactions are carried out in the rest frame for the two particles, since this is the most economical procedure. However, these interactions are then used for calculating the energy of an n-particle system in its rest frame, which is not the frame for which the two-particle potential was calculated. How should this correction be made ?

Consider first two interacting particles, with masses (m_1, m_2), momenta $(\underline{p}_1, \underline{p}_2)$ and spin vectors $(\underline{s}_1, \underline{s}_2)$, moving in the frame O. The velocity of their barycentric frame $\hat{O}$ is given by $\underline{v} = \underline{P}/M$, where $\underline{P} = \underline{p}_1 + \underline{p}_2$ and $M = (m_1 + m_2)$. The relationships between the dynamical operators $(\hat{\underline{r}}_i, \hat{\underline{p}}_i, \hat{\underline{s}}_i)$ in the frame $\hat{O}$ and those $(\underline{r}_i, \underline{p}_i, \underline{s}_i)$ in the frame O have been given explicitly by Osborn (10), up to and including terms of order $(v/c)^2$, where v denotes either $\bar{v}$ or $\hat{v}_i$, where $\hat{v}_i$ denotes the velocity of particle i in their barycentric frame. For our present discussion we need these relationships only up to order (v/c). The quantity of interest to us is the relative position operator $(\hat{\underline{r}}_1 - \hat{\underline{r}}_2)$, which we denote by $\hat{\underline{r}}_{12}$, and its transformation is given by (12)

$$\hat{\underline{r}}_{12} = (\underline{r}_1 - \underline{r}_2) - \frac{\underline{P}}{4M} \times \left(\frac{\underline{s}_1}{m_1} - \frac{\underline{s}_2}{m_2}\right) + O\left((v/c)^2\right), \qquad (2.1)$$

reducing to the form

$$\hat{\underline{r}}_{12} = \underline{r}_{12} - \frac{1}{8m^2} \underline{P} \times (\underline{s}_1 - \underline{s}_2) + O\left((v/c)^2\right), \qquad (2.2)$$

in the equal mass case. The additional terms arise because the two particles have intrinsic spins. They are due to Wigner rotation of these spins and so represent a relativistic effect; we note that they fall to zero in the limit $\{m_i\} \to \infty$, or as $c \to \infty$, which is the same thing.

Let us denote by $V(\hat{r}_{12})$ the leading central term of the potential between the two particles in the frame $\hat{O}$; it is a function of $\hat{r}_{12} = |\hat{\underline{r}}_{12}|$ only. Relativistic additions to it will not be changed by the transformation (2.1), since we consider the potential only to order (v/c), so we do not need to give them explicitly. In general, if the vector $\hat{\underline{r}}_{12}$ is changed by amount $\delta\hat{\underline{r}}_{12}$, the addition to the potential is, to first order, given by the expression

$$\delta V(\hat{r}_{12}) = \frac{1}{\hat{r}_{12}} \frac{dV}{d\hat{r}_{12}} \hat{\underline{r}}_{12} \cdot \delta\hat{\underline{r}}_{12}. \qquad (2.3)$$

Thus, in terms of the co-ordinates in the frame O, the addition to the potential V due to the effect of Wigner rotation is

$$\frac{-1}{8m^2 r_{12}} \frac{dV}{dr_{12}} (\underline{r}_{12} \cdot P \times (\underline{s}_1 - \underline{s}_2)) = -\frac{1}{4\sqrt{3}m^2} \frac{1}{r_{12}} \frac{dV}{dr_{12}} (\underline{S}^-_{12} \cdot \underline{\rho} \times \underline{p}_\lambda) \qquad (2.4)$$

This additional term has a purely kinematical origin, and must always be found to occur with the leading potential $V(r_{12})$ when this is calculated in the laboratory frame O rather than in the c.m. frame $\hat{O}$. In order $(v/c)^2$, there would occur further corrections to V, and also corrections of order (v/c) to the other Galilean-invariant potential terms which are themselves of order v/c relative to V, but we shall not discuss these further here. We note that the Galilean-noninvariant term is the same for V^{SO}_g and V^{SO}_c, as given in eqs. (1.9.1, 1.9.2), and agrees precisely with this term (2.4) of kinematic origin.

This discussion shows us that we may calculate the energy of a two-particle system in the laboratory frame O, if we so wish, and obtain the same value as results from the conventional calculation in the barycentric frame $\hat{O}$, at least to first order in v/c, provided that we add the Galilean-noninvariant term (2.4) to the two-particle Hamiltonian H_0. This term compensates precisely for the difference between the potential $V(r_{12})$ used in the laboratory frame O and the potential $V(\hat{r}_{12})$ used in the barycentric frame $\hat{O}$. If the energy of this two-particle system is calculated to be $\hat{M}$ in their barycentric frame $\hat{O}$, then we know that it must be $(\hat{M} + P^2/2\hat{M} + \ldots)$ in the laboratory frame O, so that the terms linear in $\underline{P}$ in the latter calculation must cancel precisely if no error has been made. A more physical case to consider is that of a meson, regarded as a quark-antiquark system. The net sum of all terms of order P in its energy as calculated in the frame O must also vanish, but this requirement goes far beyond the discussion of Close and Osborn (12), which is concerned only with the case of weak binding and which is valid only to order $(v_i/c)^2$.

Our conclusion is that these additional Galilean-noninvariant terms in the Hamiltonian H_0 calculated for the n-particle barycentric frame 0 can be used with confidence, and indeed must be included. Although Galilean-invariance is a proper requirement for two-particle interactions, the condition expressing this must take into account the relativistic effects of Wigner rotation, even in the non-relativistic regime where the Schrodinger equation is being used. We recall that the Galilean-invariant spin-orbit interactions obtained for scalar meson exchange and for vector meson exchange (with vector coupling) are themselves relativistic effects, from the standpoint of the Dirac theory of spin-$\frac{1}{2}$ particles.

3. Spin-orbit splitting in the N = 1, $L = 1^-$ supermultiplet states

We discuss briefly the predictions following from these spin-orbit potentials. We need do this only for the case of $SU(3)_f$ symmetry. The discrepancies which exist may be modified, but will not be removed, by more detailed calculations which take into account the mass difference between the s quark and the (u, d) quarks. The formulae needed are given in table VII of ref 14. The Galilean-invariant terms SO_L give contributions proportional to $(\delta - \gamma)$, and the Galilean-noninvariant terms SO_K give contributions proportional to $(\delta + 3\gamma)/4$, where δ arises from the potential V_g and γ from the potential V_c. The expressions for δ and γ depend on both the form of the shell-model orbitals used and the nature of the confining potential V_c used.

Isgur and Karl (14, 15) used a harmonic confining potential and shell-model orbitals which are eigenfunctions of this potential. Their expressions for δ and γ are

$$\text{(a)} \quad \delta = 4\alpha_s\omega^2/(3\surd(2\pi m\omega)) , \qquad \text{(b)} \; \gamma = \omega^2/2m, \tag{3.1}$$

where ω is the oscillator frequency associated with these orbitals. Their choice of parameters is

$$\omega = 520 \text{ MeV}, \qquad m = 420 \text{ MeV}, \qquad \alpha_s = 0.97 \tag{3.2}$$

which give δ = 300 MeV and γ = 215 MeV. As pointed out by Isgur and Karl (15) there is a substantial cancellation between the contributions of V_g and V_c to SO_L, since $(\delta - \gamma)$ = 85 MeV, but not for their contributions to SO_K, where $(\delta + 3\gamma)/4$ = 236 MeV. If SO_K is neglected, then the contribution of the spin-orbit forces would be relatively weak, which would be in accord with the Isgur-Karl philosophy of regarding the spin-spin and tensor forces as being the dominant spin-dependent interactions, but linked in a definite relationships. However, we shall include SO_K and consider three particularly clear cases :

(i) the ΔD33-ΔS31 mass difference arises only from SO_K, as remarked earlier. The calculated mass splitting is $(\delta + 3\gamma)/4$ = +236 MeV. This has the correct sign, but the magnitude is about three times too large.

(ii) The ND15-ΔS31 mass difference. With Fermi-Breit spin-spin forces, the ^{48} and ^{210} states are degenerate. These two particular states involve no mixing, and their mass separation is

$$\mathrm{ND15} - \Delta\mathrm{S31} = (+15 - 64 + 157.3)\ \mathrm{MeV} \simeq 108\ \mathrm{MeV},$$

the three terms being due to the tensor, SO_L and SO_K interactions, in turn. This is much too large, in comparison with the data, which suggests a value more like +20 MeV.

(iii) The ΛD03-ΛS01 mass difference. This is not directly affected by the tensor force. The spin-orbit forces give $\{3(\delta - \gamma)/2 - (\delta + 3\gamma)/4\} = -108\cdot5$ MeV. The SO_K term is large and dominant and has the wrong sign. This calculation is modified when the s-quark mass m_s is taken to be $m_s = m/0.6 = 700$ MeV, as found by Isgur and Karl (15); using their tables V and VI, we obtain a mass difference of -189 MeV, still with the wrong sign and larger than the $SU(3)_f$ value just given.

Gromes and Stamatescu (7) and recently Gromes (16) have made calculations for the N and Δ states of the N = 1, L = 1^- supermultiplet, using a two-body linear confining potential $V_c = \lambda r$. These calculations use harmonic orbitals whose frequency parameter ω is chosen by the variational principle, using a three-body Hamiltonian with two-body potentials $(V_g(ij) + \lambda r_{ij})$. Gromes and Stamatescu used the parameter values

$$m = 300\ \mathrm{MeV},\quad \lambda = 0.05\ \mathrm{GeV}^2,\quad b = 0.32\ \mathrm{GeV}\quad \text{and}\quad \alpha_s = 0.5$$

A reasonable spin-orbit splitting was found for (ΔD33 - ΔS31), of amount 72 MeV, and the authors reported that the tensor terms were negligible. Gromes subsequently carried out these calculations in more detail. His best choice of parameters to fit the data was

$$\omega = 186\ \mathrm{MeV},\quad m = 350\ \mathrm{MeV},\quad \alpha_s = 0.038,\quad = 0.028\ \mathrm{GeV}^2,$$

markedly different from those of Isgur and Karl. The values obtained for δ and γ were then 27.4 MeV and 33.0 MeV, respectively, an order of magnitude smaller than those used by Isgur and Karl. The mass differences of interest to us here were as follows :

$$\text{(i)}\ \Delta\mathrm{D33} - \Delta\mathrm{S31} = (\delta + 2.82\,\gamma)/4 = +30\ \mathrm{MeV},$$

$$\text{(ii)}\ \mathrm{ND15} - \Delta\mathrm{S31} = -\frac{1}{20}\delta + \frac{3}{4}(\delta - 0.93\,\gamma) + \frac{1}{6}(\delta + 2.282\,\gamma) = +16.1\ \mathrm{MeV},$$

the three terms being due to the tensor force, the SO_L and the SO_K interactions, in turn. These values are both compatible with the data. The strange baryons have not been discussed in this work.

There is still a clear difficulty concerning the spin-orbit interactions in the baryonic mass spectrum calculated for the first-excited supermultiplet, using the present three-quark model based on the one-gluon-exchange potential together with a confinement potential. The antisymmetric spin-orbit interaction is predicted to be much larger than the observed spectroscopy can allow, yet we see that it must exist, from rather general considerations. The major discrepancy is provided by the ΛD03 - ΛS01 mass separation. This is quite large (+115 ± 5 MeV) and can be attributed only to the existence of a strong spin-orbit interaction; yet the calculated separation, even larger in magnitude than the observed value, has the wrong sign. It is clear that we do not yet know the full dynamics underlying the spectrum of baryonic states(18).

References and notes

1. De Rujula, A., Georgi, H., and Glashow, S. : 1975, Phys. Rev. D12, 147.
2. Bethe, H.A., and Salpeter, E.E. : 1957, Quantum Mechanics of One- and Two-Electron Atoms, Academic Press, New York.
3. Reinders, L.J. : 1978, J. Phys. G4, 1241.
4. Eisenbud L., and Wigner, E.P. : 1941, Proc. Nat. Acad. Sci. 27, 281.
5. Dalitz, R.H., and Horgan, R.R. : 1973, Nucl. Phys. B66, 135.
6. Greenberg, O.W., and Resnikoff, M. : 1967, Phys. Rev. 163, 1844.
7. Gromes, D., and Stamatescu, I.O. : 1976, Nucl. Phys. B112, 213.
8. Particle Data Group : 1980, Rev. Mod. Phys. 52, S1.
9. Kelly, R. : 1979, private communication (September).
10. Osborn, H. : 1968, Phys. Rev. 176, 1523.
11. Brodsky, S.J., and Primack, J.R. : 1969, Ann. Phys. (N.Y.) 52, 315.
12. Close, F., and Osborn, H. : 1970, Phys. Rev. D2, 2127.
13. Krajak, R.A., and Foldy, L. : 1970, Phys. Rev. Letters 24, 545.
14. Isgur, N., and Karl, G. : 1977, Phys. Letters 72B, 109.
15. Isgur, N., and Karl, G. : 1978, Phys. Rev. 18, 4187.
16. Gromes, D. : 1977, Nucl. Phys. B130, 18.
17. Kim, H. : 1971, Phys. Letters 37B, 347.
18. Dr. A. Gal (July 1980), has drawn our attention to a discussion by H. Kim (17) of such Galilean-noninvariant components of the NN interaction and of their possible role in nuclear theory.

KAON PHYSICS. SOME COMMENTS ON ITS STATUS AND PERSPECTIVES

G. Violini
Istituto di Fisica - Università di Roma
Istituto Nazionale di Fisica Nucleare - Sezione di Roma

After this week of talks, all accompanied by lively discussions, it is a difficult task to try to summarize all that. Moreover, with the help of all speakers and our scientific secretaries I hope that very soon this matter could be read in the Proceedings.

Instead, I prefer to take this opportunity to make some more general remarks, and not only to refer in detail to the various talks which were presented in the workshop. These talks indeed provide a rather wide view of what is presently being done in kaon physics. Even if this will probably be somewhat superfluous, I shall try to give a brief account of at least some points of those talks.

First of all, I would like to say a few words on why this meeting was held. The proposal emerged during a recent visit of mine to Aarhus, when I had very interesting conversations with Dumbrajs and Oades and we thought that it could have been good to extend this type of contact to other people working on KN dispersion theory. Thus the meeting was originally planned to be essentially a theoretical one. We intended also to have some experimental part, but, as you have seen, despite the very short notice we have been very happy to receive a substantial contribution to our workshop by a quite large number of experimental groups.

Another feature which is worth some comments in the number of communications not referring to KN physics.

The reason for having them in the program is rather transparent. There are good possibilities of transferring methods from one case to another and, in general, but mainly in theoretical research, it is certainly of interest to learn what is being done in, say, NN, πN, or e^+e^-.

A point which soon became apparent during the workshop is that many groups, both theoretical and experimental, have presented several new results. It is remarkable that not only those working on kaon-nucleon two-body interactions, but also people interested in KN $\rightarrow$ Kπp and πN $\rightarrow$

E. Ferrari and G. Violini (eds.), Low and Intermediate Energy Kaon-Nucleon Physics, 419–424.

$\longrightarrow \pi\pi p$, or in the search for strange baryons or in kaon capture in emulsions, have found this an adequate place to discuss their results.

What unified these contributions was the kaon physics, and indeed I will later say more about the high versatility of this particle and the interest it may present in different fields of physics.

However, there has probably also been another component.At present high-energy physics interests are probably too strongly polarized toward a different set of problems, and not much room is left for this type of physics in the current general meetings. Therefore, the favour found by this workshop is not surprising.

However I would like to make a comment about this trend of high-energy physics. It is obvious that frontier research is more appealing than systematic research, but it is also true that some part of what is done in frontier fields is actually of the systematic type and that certainly it would be at least narrow-minded to think that systematic research should be abandoned as not important and out of fashion.

Now KN physics presents several reasons which suggest the importance of this type of research.

Probably the most immediate one, suggested by the talks which were presented on πN phase shift analyses, is the comparison between the situations for the πN and KN processes. It is evident that the phase shift analyses for the former process are much more sophisticated and reach much higher energies. Therefore this is a field of KN physics which will certainly have some development in the near future. This development will also be favoured by the large number of experimental data which have been appearing and which have not yet been incorporated in the current analyses. We heard here of two such analyses in progress and it is very likely that in future meetings new data like those which were presented here on KN polarization will no longer be compared with the old BGRT phases.

However, I think that it has to be said that there are more fundamental reasons for our interest in KN interactions than simply of sort of natural curiosity or the need to reach the same level as for πN analyses.

These reasons are related to what is likely to be a development of the experimental facilities, namely the possible construction of kaon factories. This should enable a much larger use of kaons, and already much theoretical work suggests their use as a probe for nuclear structure. This indeed should develop further the field which was touched upon in one of the talks we heard, that of kaon-nucleus physics. Certainly the present situation, where only for a few nuclei there is a large experimental information, is going to change and a more extensive use of analyticity in the theoretical analyses can be easily envisaged. The development of this work is also probably going to require (and to give) an improvement of our understanding of the elementary kaon-nucleon interaction.

The possible availability of much more intense kaon beams, as will be the case at the Vancouver factory, about which a workshop was held at TRIUMF last summer, or that discussed for BNL, certainly shed new light on many of the problems we discussed here as well as on several others, for which I refer you to the Proceedings of the Vancouver workshop.

I would like to mention a few of these possible fields of work, which are likely to awaken a new and large interest in our research area in a few years from now. Incidentally, I would like to add that this perspective is probably one of the strongest reasons for maintaining groups of people doing this kind of research both at the theoretical and the experimental level.

Obviously kaon factories where very intense kaon beams are available around 700 MeV/c would contribute to that improvement of KN phase shift analysis I was speaking about. This will possibly improve our understanding of the old problem of the continuation below threshold of the KN scattering amplitude as well as our knowledge about the existence of Y^* and possibly Z^*. This latter point is rather obvious and I would only like to recall that it might also be of interest in connection with QCD.

For what concerns the first point, it is clear that the procedure used in Martin's new K-matrix analysis, according to which the lack of uniqueness in the determination of the parameters can be eliminated by adding theoretical (or phenomenological) constraints, should very probably benefit from an improved knowledge of the scattering amplitude in the GeV region.

There will also be other important applications, besides the scattering of kaons on nuclei, in hypernuclear physics and in the study of exotic atoms. Both these fields were extensively reviewed and this meeting continued in some sense the active discussion at the Jablonna Conference of last fall, with the only difference that in Rome the workshop perhaps focused more on kaon-nucleon physics, whereas in Poland the emphasis was more on hypernuclei and exotic atoms.

However, a point which is being elucidated more and more clearly is that one must have a simultaneous understanding of both aspects of kaon physics, that provided by low-energy processes and that provided by the above-mentioned phenomena. This is especially important in connection with the well-known discrepancy between the results of X ray de-excitation of kaonic hydrogen and low-energy experiments on $\bar{K}N$ scattering. This problem is nowadays receiving quite a lot of attention, but whatever the final solution, there is no doubt that it must explain both aspects of the puzzle. Today's talk by Gensini does illustrate how strong is the correlation between different aspects of KN physics.

Anyway, returning to the X ray aspect, in this workshop we heard of a possible line of interpretation of it, namely that Y_0^* (1405) could be a CDD pole. This is certainly an interesting possibility, also in the light of Dalitz's talk on the nature of this resonance, but certainly it must be clarified by an actual fit to the data that this point of view can be used for a global fit and that the discrepancy between the required behaviour in that model of the real part of the amplitude and that fixed by existing Coulomb interference determinations of it may be solved. Further evidence that the problem may meet serious difficulty to be explained comes also from the remark that in K-matrix analyses the condition of $a_p \approx 0$ and the production ratio of strange baryons would require $a_o \approx a_1 \approx -$ 1fm which is quite different from the current values.

To these problems, which would all benefit from the possible new facilities about which I was speaking, I would like to add another one, which was not discussed here, just in order to give an example which illustrates why it is very likely that kaon physics may receive much attention in the forthcoming years.

This is the possibility of studying the rarest kaon decays; there is a problem for which obviously an increase factor 100 or 1000 in beam intensity is crucial but which is also of high relevance for the current theories of elementary particles.

To this experimental perspective one should add the possibility of improving our knowledge of low-energy $\bar{K}N$ interactions, for which little hope was left in Miller's talk. Besides the use of these new facilities, these perspectives offer in my opinion a strong justification for this type of meeting and indicate that it is not an accident that in the last year there were three such meetings.

The talk I was just mentioning presented a number of new experimental results. This was not rare in our workshop and you may have noticed that several 1980 analyses were quoted in the invited talks (those marked with the mysterious + in the program) or were the object of the other contributions. This was true for both experimental and theoretical works.

Among the results which were reached during the workshop I would like to mention one, which was prompted by a remark of A. Martin.

This results is the explanation of the disturbing discrepancy between different estimates of $D_p(0)$, the real part of the Kp scattering amplitude for $\omega=0$.

It has become clear that this substraction constant is not particularly significant when in a once-subtracted dispersion relation, one makes the assumption that Λ and Σ poles can be replaced by an effective pole, because then the constant which is well determined by the analysis of the experimental real part, the C of Queen's talk, would give rise to different values of $D_p(0)$, depending on where the effective pole is sitting. Thus it was possible to trace back the difference between the analyses of Baillon et al. and Di Claudio et al. on one hand and that of A. Martin on the other, to the fact that the former authors put the effective pole at $\bar{\omega}=(\omega_\Lambda + \omega_\Sigma)/2 \simeq 100$ MeV, whereas Martin puts it at the position $\omega_\Lambda \simeq 64$ MeV. This will require a reexamination of some conclusions of the most recent determinations of the chiral symmetry breaking term for KN scattering. To conclude my talk, I would like to mention briefly what are the perspectives of theoretical work in this field, referring in particular to my sector of activity, KN dispersion theory, on which we heard an accurate review of recent results by our chairman for today.

In this field I think that the next step should be to arrive at a simultaneous analysis of all the data. There already exist programs in this direction; thus we are going to extend very soon the analysis of K^o_L regeneration on hydrogen by including also $K^\pm$ n elastic scattering data, but there is no doubt that one must soon make a full analysis of elastic and charge-exchange scattering and of regeneration.

It is an easy prediction that the very lively discussions on the zeroes of the amplitudes will lead to further results.

In the field of dispersion theory there is however one sector which was not mentioned in much detail, except in some discussions, and which in my opinion deserves much attention. It is that of model function calculations.

Too often methods in dispersive calculations arrive at numerical results through very complicated procedures which make it very hard to attribute errors to the results. The technique of estimating the biases of the methods by studying model functions which incorporate the most relevant features of the true amplitudes has already proven to be very effective in a number of cases in KN scattering. For example, it has shed much light on the reliability of the discrepancy function analyses which has been such a popular tool to reduce the uncertainties related to the presence of the unphysical cut.

Probably this method should find wider application, also in view of the fact that from some discussion held here it is possible to predict that there is going to be some future work in two fields which are almost virgin at the moment. These fields are respectively that of non-forward dispersion relations and that of applications of dispersion relations to other channels.

To conclude, let me thank all the speakers and chairmen, as well as our scientific secretaries Atkin and Di Claudio; recording the discussions was indeed a very heavy job. I am sure I also interpret your sentiments in giving our warmest thanks to our very efficient secretaries, Mrs. Di Silvestro and Mrs. Fascetti, whose contribution to the success of this workshop was essential.

Finally, let me thank the Sezione di Roma of I.N.F.N., the University of Rome and the I.C.T.P. for their generous support of our meeting.

References

- Proceedings of this workshop.

- Proceedings of the International Conference on the Hypernuclear and Low-Energy Kaon Physics, Jablonna, Poland, 1979 (to be published in Nukleonika, Editor: Wycech, S.).

- Proceedings of the Vancouver Kaon factory workshop, Editor Craddock, M.K., TRI-79-1, December 1979.

DISCUSSION

- N. M. Queen :

As chairman of this concluding session, I would like to take this opportunity to say a few words on behalf of all the participants. It is clear that this meeting has been a very rewarding and stimulating experience for a large number of people, and I suspect that the material presented here will provide a valuable source of information to many physicists working in the field of kaon-nucleon physics during the next few years. I would like to propose a special vote of thanks to Professor E. Ferrari and Professor G. Violini, whose organizational efforts have made all this possible.

INDEX OF NAMES

INDEX OF SUBJECTS

GPSR Compliance

The European Union's (EU) General Product Safety Regulation (GPSR) is a set of rules that requires consumer products to be safe and our obligations to ensure this.

If you have any concerns about our products, you can contact us on ProductSafety@springernature.com

In case Publisher is established outside the EU, the EU authorized representative is:

Springer Nature Customer Service Center GmbH
Europaplatz 3
69115 Heidelberg, Germany

Batch number: 10761599

Printed by Printforce, the Netherlands